M o n o g r a p h s   i n

V o l u m e   7

The Theory of Multiple Zeta Values with Applications in Combinatorics

Monographs in Number Theory

ISSN 1793-8341

Published

Vol. 1   Analytic Number Theory: An Introductory Course
           by Paul T. Bateman & Harold G. Diamond

Vol. 2   Topics in Number Theory
           by Minking Eie

Vol. 3   Analytic Number Theory for Undergraduates
           by Heng Huat Chan

Vol. 4   Number Theory: An Elementary Introduction Through
           Diophantine Problems
           by Daniel Duverney

Vol. 5   Hecke's Theory of Modular Forms and Dirichlet Series
           (2nd Printing with Revisions)
           by Bruce C. Berndt & Marvin I. Knopp

Vol. 6   Development of Elliptic Functions According to Ramanujan
           by K. Venkatachaliengar, edited and revised by Shaun Cooper

Vol. 7   The Theory of Multiple Zeta Values with Applications in Combinatorics
           by Minking Eie

Monographs in

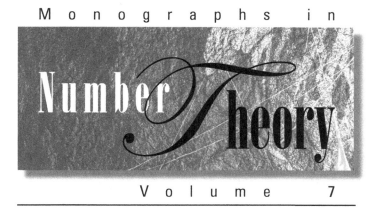

Volume 7

The Theory of Multiple Zeta Values with Applications in Combinatorics

Minking Eie

National Chung Cheng University, Taiwan

NEW JERSEY · LONDON · SINGAPORE · BEIJING · SHANGHAI · HONG KONG · TAIPEI · CHENNAI

Published by

World Scientific Publishing Co. Pte. Ltd.

5 Toh Tuck Link, Singapore 596224

USA office: 27 Warren Street, Suite 401-402, Hackensack, NJ 07601

UK office: 57 Shelton Street, Covent Garden, London WC2H 9HE

British Library Cataloguing-in-Publication Data
A catalogue record for this book is available from the British Library.

Monographs in Number Theory — Vol. 7
THE THEORY OF MULTIPLE ZETA VALUES WITH APPLICATIONS IN COMBINATORICS

Copyright © 2013 by World Scientific Publishing Co. Pte. Ltd.

ISBN 978-981-4472-63-0

Printed in Singapore

Preface

This is a record of my research on multiple zeta values since 2007. In a series of lecture notes used in classes or in seminars on number theory, I provided detailed proofs for both newly found and well-known results. These lecture notes were extremely useful and helpful to my graduate students in learning number theory and writing their thesis. Some particularly interesting sections were published as research papers. These lecture notes are now reworked into a book, which I hope will also be helpful for researchers. This book contains the following important topics concerning multiple zeta values and applications.

1. The duality theorem, the sum formula and the restricted sum formula of multiple zeta values.

2. Shuffle relations produced from shuffle products of multiple zeta values.

3. Double weighted sum formulas of multiple zeta values.

4. Applications of shuffle products in combinatorics.

5. Combinatorial identities of convolution type.

6. Generalizations of Pascal identity.

Around 2005, my former Ph.D student Kwang-Wu Chen and I found that Euler double sums of odd weight can be obtained from integral transforms of products of two Bernoulli polynomials, one with even index and the other with odd index. The evaluations of double Euler sums of odd weight are equivalent to express the corresponding products of Bernoulli polynomials into linear combinations of Bernoulli polynomials of odd indices. Therefore we published several papers concerning evaluations of Euler double sums as well as their analogues. Among other things, a Dirichlet character is added to the first summation of Euler double sums so that it can be evaluated such that either the character is even and the weight is odd, or the character is odd and the weight is even.

Multiple zeta values or r-fold Euler sums are natural generalizations of Euler double sums which arose from the knot theory with close relation to Feynman diagrams in quantum physics. In 2007, we began to develop effective ways to explicitly evaluate multiple zeta values of depth $3, 4$ or higher. It ended up with the introduction of multiple zeta values with parameters, something like to replace Riemann zeta functions by Hurwitz zeta functions. We noticed that multiple zeta values of the form $\zeta(\{1\}^m, n+2)$ were much easier to be evaluated. Also certain analogues of $\zeta(\{1\}^m, n+2)$ with parameters can also be evaluated easily. Additional differentiations are needed in order to evaluate multiple zeta values with the sum of depth and weight is odd.

Due to Kontsevich, multiple zeta values are expressed as iterated integrals over simplices of weight dimensions of strings of differential forms. Only two kinds of differential forms appear: $dt/(1 - t)$ or dt/t. Moreover it always begins with $dt/(1 - t)$ and ends up with dt/t. Once multiple zeta values are expressed as iterated integrals over simplices, the shuffle product of two multiple zeta values is equivalent to find all possible interlacings of two sets of variables. So two multiple zeta values of weight m and n will produce $\binom{m+n}{m}$ multiple zeta values of weight $m + n$ after their shuffle product. Based on such a simple fact, we are able to produce combinatorial identities from shuffle relations obtained from shuffle products of certain multiple zeta values.

Some particular multiple zeta values such as $\zeta(\{1\}^m, n + 2)$ or sums of multiple zeta values can be expressed as integrals in one or two variables so that their shuffle products can be carried out more efficiently. By counting numbers of multiple zeta values produced from shuffle products, we may obtain combinatorial identities with a single binomial coefficient on one side and sums of products of binomial coefficients on the other side.

The theory of multiple zeta values is not fully developed. Devoted wholly to this subject, this book aims to introduce a systematic theory of multiple zeta values as well as applications of shuffle products to combinatorics. Hopefully, this will lead to further developments of the theory and provide inspiration to both experts and amateurs.

Finally, I would like to thank my graduate students who helped to type the whole book, thank my Ph.D student Chung for compiling the files and thank my colleague Professor Chang for the final editing.

Minking Eie

October 29, 2012

Department of Mathematics

National Chung Cheng University

Contents

Preface   **v**

I   Basic Theory of Multiple Zeta Values   **1**

0   The Time Before Multiple Zeta Values   **5**

     0.1   The Evaluation of Euler Double Sums   6

     0.2   Vandermonde Convolution   9

     0.3   Zeta Functions Associated with Multiple Zeta Values   11

     0.4   Messages from Modular Forms   12

1   Introduction to the Theory of Multiple Zeta Values   **15**

     1.1   Introduction and Notations   16

     1.2   Drinfeld Integral Representations of Multiple Zeta Values .   19

     1.3   Double Weighted Sum Formulas   24

     1.4   The Expectations of Binomial Distributions   27

     1.5   Exercises   32

2   The Sum Formula   **35**

     2.1   Through the Integral Representations   35

     2.2   Another Proof of the Sum Formula   38

2.3   Evaluation of Multiple Zeta Values of Height One   45

2.4   Exercises   50

II   Shuffle Relations among Multiple Zeta Values   53

3   Some Shuffle Relations   57

3.1   Shuffle Relations of Multiple Zeta Values   57

3.2   An Application of Double Weighted Sums   62

3.3   Shuffle Relations of Two Sums of Multiple Zeta Values . . .   68

3.4   A Vector Version of the Restricted Sum Formula   75

3.5   Exercises   84

4   Euler Decomposition Theorem   85

4.1   A Shuffle Relation with Two Parameters   85

4.2   Integrals with Three Factors   95

4.3   Generalizations of Euler Decomposition Theorem   102

4.4   Applications of the Decomposition Theorem   110

4.5   Applications of Another Decomposition Theorem   114

4.6   Exercises   120

5   Multiple Zeta Values of Height Two   123

5.1   Sums of Multiple Zeta Values of Height Two   123

5.2   Weighted Sums of Multiple Zeta Values of Height Two . . .   132

5.3   The Shuffle Product Formula of a Sum and Others   134

5.4   Exercises   137

III   Applications of Shuffle Relations in Combinatorics   139

6   Generalizations of Pascal Identity   143

6.1   Applications of Shuffle Products in Combinatorics   144

6.2   Hypergeometric Distribution   157

6.3   The Generating Function of Three Variables   165

6.4   Exercises   172

7 Combinatorial Identities of Convolution Type   173

7.1 Some Particular Combinatorial Identities 173

7.2 A Generating Function for Products 177

7.3 A Combinatorial Identity of Convolution Type 184

7.4 Another Generating Function of Three Variables 188

7.5 Exercises 194

8 Vector Versions of Some Combinatorial Identities   197

8.1 The Shuffle Product of Two Sums 198

8.2 More Combinatorial Identities of Convolution Type 205

8.3 Vector Versions of Pascal Identity 214

8.4 Problems on Combinatorial Identity 221

Appendices   251

A Singular Modular Forms on the Exceptional Domain   253

A.1 Cayley Numbers and Integral Cayley Numbers 254

A.2 The Exceptional Domain 256

A.3 The Theory of Jacobi Forms 258

A.4 A Final Application 260

B Shuffle Product Formulas of Multiple Zeta Values   265

B.1 Introduction 266

B.2 The Shuffle Product Formula of Two Multiple Zeta Values . 267

B.3 Some Basic Shuffle Relations 270

B.4 Shuffle Relations of Two Sums of Multiple Zeta Values . . . 274

B.5 The Generating Function of Height One 277

C The Sum Formula and Their Generalizations   281

C.1 The Double Integral Representation of a Sum 282

C.2 Euler Sums with Two Branches 284

C.3 Another Application of Euler Sums with Two Branches . . 286

C.4 The Restricted Sum Formula 289

C.5 A Generating Function for Sums of Multiple Zeta Values . . 291

C.6 The Vector Version of the Restricted Sum Formula 292

C.7 Another Generating Function 293

Bibliography   **295**

Index   **299**

Part I

Basic Theory of Multiple Zeta Values

The theory of multiple zeta values began with the evaluation of Euler double sums in terms of special values at positive integers of Riemann zeta function, proposed by Goldbach in 1742 to Euler. Until 1992, multi-versions of Euler double sums appeared and are called multiple zeta values or r-fold Euler sums.

A multiple zeta value of depth r and weight w can be expressed as a sum of multiple zeta values of lower depth and the same weight when $r + w$ is odd. Some particular multiple zeta values and sums of multiple zeta values can be expressed in terms of single zeta values, the special values at positive integers of Riemann zeta function. All these assertions need further investigation through integral representation of multiple zeta values due to Kontsevich around 1996.

The shuffle product of two multiple zeta values enables us to express the product of two multiple zeta values of weight m and n as a sum of $\binom{m+n}{m}$ zeta values of weight $m + n$. How to carry out the shuffle product becomes an intricate problem. Fortunately, we are able to develop an alternative to overcome such difficulty.

CHAPTER 0

The Time Before Multiple Zeta Values

According to the book *Number Theory* by Andre Weil, Euler discovered his famous result

$$\zeta(2) = 1 + \frac{1}{2^2} + \frac{1}{3^2} + \cdots + \frac{1}{n^2} + \cdots = \frac{\pi^2}{6}$$

in 1735. He also calculated the values $\zeta(2n)$ up to

$$\zeta(12) = \frac{691\pi^{12}}{6825 \times 93555}.$$

For a pair of positive integers p and q with $q \geq 2$, the classical Euler double sum is defined as

$$S_{p,q} = \sum_{k=1}^{\infty} \frac{1}{k^q} \sum_{j=1}^{k} \frac{1}{j^p}.$$

It was a problem proposed in 1742 from Goldbach to Euler in an attempt to evaluate $S_{p,q}$ in terms of special values at positive integers of Riemann zeta function. In 1775, Euler proved the case $p = 1$ and gave a general formula for $S_{p,q}$ when the weight $p + q$ is odd without any proof. Nielsen was the first to fill in the gap by giving the correct version and a proof.

0.1   The Evaluation of Euler Double Sums

Each double Euler sum

$$S_{p,q} = \zeta(p+q) + \sum_{j=1}^{\infty}\sum_{k=1}^{\infty} \frac{1}{j^p(j+k)^q},$$

corresponds to a unique rational function

$$\frac{1}{X^p(X+T)^q}.$$

The partial fraction decomposition of such rational function then leads to an identity among Euler double sum of the same weight. For example when $p = 1$ and $q = n \geq 2$, the partial fraction decomposition is

$$\frac{1}{X(X+T)^n} = \frac{1}{T^n}\left(\frac{1}{X} - \frac{1}{X+T}\right) - \sum_{r=2}^{n} \frac{1}{T^{n+1-r}(X+T)^r}$$

and the corresponding identity is

$$S_{1,n} - \zeta(n+1) = S_{1,n} - \sum_{r=2}^{n} \{S_{n+1-r,r} - \zeta(n+1)\}$$

or

$$S_{1,n} = n\zeta(n+1) - \sum_{r=2}^{n-1} S_{n+1-r,r}.$$

Therefore, the evaluation

$$S_{1,n} = \frac{n+2}{2}\zeta(n+2) - \frac{1}{2}\sum_{r=2}^{n-1} \zeta(r)\zeta(n+1-r)$$

by Euler then follows from the reflection formula

$$S_{p,q} + S_{q,p} = \zeta(p)\zeta(q) + \zeta(p+q).$$

Bernoulli polynomial $B_n(x)$ $(n = 0, 1, 2, \ldots)$ defined by

$$\frac{te^{xt}}{e^t - 1} = \sum_{n=0}^{\infty} \frac{B_n(x)t^n}{n!}, \quad |t| < 2\pi,$$

have the Fourier expansions

$$B_{2m}(x) = \frac{(-1)^{m-1} 2 \cdot (2m)!}{(2\pi)^{2m}} \sum_{p=1}^{\infty} \frac{\cos(2p\pi x)}{p^{2m}}, \quad 0 \le x \le 1,$$

and

$$B_{2n+1}(x) = \frac{(-1)^{n+1} 2 \cdot (2n+1)!}{(2\pi)^{2n+1}} \sum_{k=1}^{\infty} \frac{\sin(2k\pi x)}{k^{2n+1}}, \quad 0 \le x \le 1.$$

Along with the basic identity

$$\sin 2k\pi x \cot \pi x = 1 + 2 \sum_{j=1}^{k} \cos(2j\pi x) - \cos(2k\pi x),$$

we obtain the relation

$$\int_0^1 B_{2m}(x) B_{2n+1}(x) \cot(\pi x) dx$$
$$= \frac{(-1)^{m+n} 4 (2m)!(2n+1)!}{(2\pi)^{2m+2n+1}} \left\{ S_{2m,2n+1} - \frac{1}{2}\zeta(2m+2n+1) \right\}.$$

Consequently, the evaluation of Euler sums of odd weight $w = 2m + 2n + 1$

$$S_{2m,2n+1} = \frac{1}{2}\zeta(w) + \sum_{k=0}^{m} \binom{w - 2k - 1}{2n} \zeta(2k)\zeta(w - 2k)$$
$$+ \sum_{k=0}^{n} \binom{w - 2k - 1}{2m - 1} \zeta(2k)\zeta(w - 2k)$$

follows from the identity

$$B_{2m}(x) B_{2n+1}(x) = 2m \sum_{k=0}^{n} \binom{2n+1}{2k} B_{2k} \cdot \frac{B_{w-2k}(x)}{w - 2k}$$
$$+ (2n+1) \sum_{k=0}^{m} \binom{2m}{2k} B_{2k} \cdot \frac{B_{w-2k}(x)}{w - 2k}.$$

There is a famous identity on Bernoulli numbers due to Euler

$$\sum_{k=1}^{n-1} \frac{(2n)!}{(2k)!(2n-2k)!} B_{2k} B_{2n-2k} = -(2n+1) B_{2n}, \quad n \ge 2,$$

which is equivalent to the identity

$$\sum_{k=1}^{n-1} \zeta(2k)\zeta(2n - 2k) = \frac{2n + 1}{2}\zeta(2n)$$

since

$$\zeta(2k) = \frac{(-1)^{k-1}(2\pi)^{2k} B_{2k}}{2(2k)!}.$$

Similar identities on Bernoulli polynomials are easily obtained through special values at negative integers of certain zeta functions. For example, we have

$$\sum_{p+q=2n+1} \frac{(2n-1)!}{p!q!} B_p(x)B_q(x) = -\frac{B_{2n+1}(2x)}{2n+1} + 2B_1(x)\frac{B_{2n}(2x)}{2n}$$

or

$$\sum_{k=1}^{n-1} \frac{(2n-1)!}{(2k+1)!(2n-2k)!} B_{2k+1}(x)B_{2n-2k}(x)$$

$$= -\frac{(2n-1)!}{(2n+1)!}B_{2n+1}(x) - \frac{1}{2}\frac{B_{2n+1}(2x)}{2n+1} + B_1(x)\frac{1}{2n}\{B_{2n}(2x) - B_{2n}\}$$

$$- B_1(x)\frac{1}{2n}\{B_{2n}(x) - B_{2n}\}.$$

Multiplying both sides by $\cot(\pi x)$ and then integrating with respect to x from 0 to 1/2, we get the identity

$$\left\{E_{1,2n} - \frac{1}{4}\zeta(2n+1)\right\} - \left\{S_{1,2n} - \frac{1}{2}\zeta(2n+1)\right\}$$

$$= \frac{n+1}{2}\zeta(2n+1) - \sum_{k=1}^{n-1}\left\{S_{2n-2k,2k+1} - \frac{1}{2}\zeta(2n+1)\right\}.$$

Here the extended Euler sums $E_{p,q}$ are defined by

$$E_{p,q} = \sum_{k=1}^{\infty} \frac{1}{k^q} \sum_{j=1}^{2k} \frac{1}{j^p}$$

and

$$T_{p,q} = \sum_{k=1}^{\infty} \frac{1}{k^q} \sum_{j=1}^{[k/2]} \frac{1}{j^p}.$$

For any positive rational number $r = a/b$,

$$E_{p,q}^{(r)} = \sum_{k=1}^{\infty} \frac{1}{k^q} \sum_{j=1}^{[kr]} \frac{1}{j^p}.$$

These come from integral transforms of a product of a pair of Bernoulli polynomials like $S_{p,q}$, i.e.

$$\int_0^1 B_{2m}\left(\{bx\}\right) B_{2n+1}\left(\{ax\}\right) \cot(\pi x) dx$$
$$= \frac{(-1)^{m+n} 4(2m)!(2n+1)!}{(2\pi)^{2m+2n+1}} \left\{ E_{2m,2n+1}^{(r)} - \frac{1}{2a^{2m}b^{2n+1}} \zeta(2m+2n+1) \right\}.$$

0.2 Vandermonde Convolution

Around 2008, while doing shuffle products of multiple zeta values, I discovered the following identity

$$\sum_{a+b=k} \binom{m+a}{m} (\mu+1)^b = \frac{(m+k+1)!}{m!k!} \int_0^1 x^m (\mu+1-\mu x)^k dx$$

and concluded that

$$\sum_{a+b=k} \binom{m+a}{m} \binom{b}{j} = \binom{m+k+1}{m+j+1}$$

by comparing the coefficient of μ^j ($0 \le j \le k$) of both sides of the identity. This formula is quite similar to the classical Vandermonde convolution

$$\sum_{0 \le k \le \ell} \binom{\ell-k}{m} \binom{q+k}{n} = \binom{\ell+q+1}{m+n+1}, \quad n \ge q \ge 0,$$

which is known to Chu Shih-Chieh as early as 1303. This formula got its name because Alexandre Vandermonde wrote a paper about it in the late 1700s.

On the other hand, there is an identity

$$\sum_{g=0}^{n} \binom{G}{g} \binom{B}{n-g} = \binom{G+B}{n}$$

from the hypergeometric distribution of theory of probability. I did not consider that the afore-mentioned result as new. However, I continued my research in shuffle products of multiple zeta values and obtained some general forms which are beyond the scope of Vandermonde convolution. Here we list a few among them.

(1) For all integers m, n, k, j with $m, n \geq 0$ and $0 \leq j \leq k$,

$$\sum_{a+b=k} \binom{m+a}{m}\binom{n+b}{n}\binom{b}{j} = \binom{n+j}{j}\binom{m+n+k+1}{k-j}.$$

(2)

$$\sum_{|\boldsymbol{\beta}|=r} \binom{\alpha_1+\beta_1}{\alpha_1}\binom{\alpha_2+\beta_2}{\alpha_2}\cdots\binom{\alpha_n+\beta_n}{\alpha_n} = \binom{|\boldsymbol{\alpha}|+r+n-1}{r}.$$

(3) Let $\boldsymbol{p} = (p_1, p_2, \ldots, p_n)$ be n-tuple of nonnegative integers. Then

$$\sum_{\boldsymbol{\alpha_1}+\boldsymbol{\alpha_2}+\cdots+\boldsymbol{\alpha_m}=\boldsymbol{p}} M(\boldsymbol{\alpha_1})M(\boldsymbol{\alpha_2})\cdots M(\boldsymbol{\alpha_m}) = M(\boldsymbol{p})\binom{|\boldsymbol{p}|+m-1}{|\boldsymbol{p}|}$$

with

$$M(\boldsymbol{g}) = \binom{g_1 + g_2 + \cdots + g_n}{g_1, g_2, \ldots, g_n} = \frac{(g_1 + g_2 + \cdots + g_n)!}{g_1! g_2! \ldots g_n!}$$

if $\boldsymbol{g} = (g_1, g_2, \ldots, g_n)$.

(4) Let $\boldsymbol{k}$ be n-tuple of nonnegative integers. Then

$$\sum_{\boldsymbol{\alpha}+\boldsymbol{\beta}=\boldsymbol{k}} \binom{m+\alpha_1+\alpha_2+\cdots+\alpha_n}{m, \alpha_1, \alpha_2, \ldots, \alpha_n}\binom{p+\beta_1+\beta_2+\cdots+\beta_n}{p, \beta_1, \beta_2, \ldots, \beta_n}$$

$$\times \binom{\beta_1}{j_1}\binom{\beta_2}{j_2}\cdots\binom{\beta_n}{j_n}$$

$$= \binom{p+j_1+\cdots+j_n}{p, j_1, \ldots, j_n}\binom{m+p+k_1+\cdots+k_n+1}{k_1-j_1, k_2-j_2, \ldots, k_n-j_n, m+p+|j|+1}$$

for all nonnegative integers m and p.

(5) Let $\boldsymbol{q}$ be n-tuple of nonnegative integers. Then

$$\sum_{\boldsymbol{\alpha}+\boldsymbol{\beta}=\boldsymbol{q}} \binom{|\boldsymbol{\alpha}|+m+p+1}{m}\binom{\alpha_1+\alpha_2+\cdots+\alpha_n}{\alpha_1,\alpha_2,\ldots,\alpha_n}$$

$$\times \binom{\beta_1+\beta_2+\cdots+\beta_n}{\beta_1,\beta_2,\ldots,\beta_n}\binom{|\boldsymbol{\beta}|+n+\ell+2}{n+1}$$

$$=\binom{q_1+q_2+\cdots+q_n}{q_1,q_2,\ldots,q_n}\sum_{a+b=|\boldsymbol{q}|}\binom{a+m+p+1}{m}\binom{b+n+\ell+2}{n+1}$$

for all nonnegative integers m,n,p,ℓ.

All these combinatorial identities will be proved later as applications of shuffle products of multiple zeta values to combinatorics.

0.3   Zeta Functions Associated with Multiple Zeta Values

Multiple zeta values are special values at positive integers of certain convergent series of several variables. For positive integers $\alpha_1,\alpha_2,\ldots,\alpha_r$ with $\alpha_r \geq 2$, a multiple zeta value of depth r and weight $|\boldsymbol{\alpha}| = \alpha_1+\alpha_2+\cdots+\alpha_r$ is defined as

$$\zeta(\alpha_1,\alpha_2,\ldots,\alpha_r) \equiv \sum_{1\leq n_1<n_2<\ldots<n_r} n_1^{-\alpha_1} n_2^{-\alpha_2}\cdots n_r^{-\alpha_r}$$

or in free dummy variable as

$$\sum_{k_1=1}^{\infty}\sum_{k_2=1}^{\infty}\cdots\sum_{k_r=1}^{\infty} k_1^{-\alpha_1}(k_1+k_2)^{-\alpha_2}\cdots(k_1+k_2+\cdots+k_r)^{-\alpha_r}.$$

Associated with the multiple zeta value $\zeta(\alpha_1,\alpha_2,\ldots,\alpha_r)$, we define the zeta function

$$\zeta(\alpha;s) = \sum_{k_1=1}^{\infty}\sum_{k_2=1}^{\infty}\cdots\sum_{k_n=1}^{\infty} k_1^{-\alpha_1 s}(k_1+k_2)^{-\alpha_2 s}\cdots(k_1+k_2+\cdots+k_r)^{-\alpha_r s}.$$

Obviously, the zeta function $\zeta(\alpha;s)$ is analytic for Re $s \geq 1$ and

$$\zeta(\alpha;s)\prod_{i=1}^{r}\Gamma(\alpha_i s)$$

$$=\int_0^{\infty}\int_0^{\infty}\cdots\int_0^{\infty}\frac{t_1^{\alpha_1 s-1}t_2^{\alpha_2 s-1}\cdots t_r^{\alpha_r s-1}dt_1 dt_2\ldots dt_r}{(e^{t_1+\cdots+t_r}-1)(e^{t_2+\cdots+t_r}-1)\cdots(e^{t_r}-1)}.$$

With the changes of variables:

$$(t_1, t_2, \ldots, t_r) = t(u_1, u_2, \ldots, u_r)$$

with $t \geq 0$ and

$$u_j \geq 0, \quad 1 \leq j \leq r, \quad u_1 + u_2 + \cdots + u_r = 1,$$

the integral representation of $\zeta(\alpha; s)$ is transformed into

$$\int_0^\infty \frac{t^{|\alpha|s+r-1}dt}{e^t - 1} \int_{E^r} \frac{u_1^{\alpha_1 s - 1} \cdots u_r^{\alpha_r s - 1} du_1 du_2 \ldots du_r}{(e^{t(u_2 + \cdots + u_r)} - 1) \cdots (e^{t u_r} - 1)}$$

where

$$E^r = \{(u_1, u_2, \ldots, u_r) | u_j \geq 0, \ u_1 + u_2 + \cdots + u_r = 1\}.$$

Consequently, the zeta function has meromorphic analytic continuation in the whole complex plane as a function of s and its special values at negative integers can be evaluated in terms of Bernoulli numbers.

0.4   Messages from Modular Forms

The classical Eisenstein series defined by

$$G_k(z) = \sum_{(m,n) \neq (0,0)} (mz + n)^{-k}, \quad k \geq 4, \quad k : \text{even},$$

provided basic examples of modular forms of one variable on the upper half plane. They have Fourier expansions as

$$G_k(z) = 2\zeta(k) + \frac{2(-2\pi i)^k}{(k-1)!} \sum_{n=1}^\infty \sum_{m=1}^\infty n^{k-1} e^{2\pi i m n z}.$$

Hence

$$G_k(i\infty) = \lim_{\lambda \to \infty} G(i\lambda) = 2\zeta(k),$$

and the zeta function associated with $G_k(z)$, up to an constant, is equal to

$$Z(s) = \sum_{m=1}^\infty \sum_{n=1}^\infty n^{k-1}(mn)^{-s} = \zeta(s)\zeta(s - k + 1).$$

Of course, such a zeta function has a functional equation

$$(2\pi)^{-(k-s)}\Gamma(k-s)Z(k-s) = i^k(2\pi)^{-s}\Gamma(s)Z(s)$$

which follows from the transform

$$G_k\left(-\frac{1}{z}\right) = z^k G_k(z).$$

Relations among Eisenstein series then lead to interesting identities of Bernoulli numbers. For example, the identity

$$\frac{1}{6}(2n+1)(2n-1)(2n-6)G_{2n}(z) = \sum_{k=2}^{n-2}(2k-1)(2n-2k-1)G_{2k}(z)G_{2n-2k}(z)$$

implies that for $n \geq 4$,

$$\sum_{k=2}^{n-2}\frac{(2n-4)!}{(2k-2)!(2n-2k-2)!}\frac{B_{2k}}{2k}\frac{B_{2n-2k}}{2n-2k} = \left(-\frac{B_{2n}}{2n}\right)\frac{(2n+1)(2n-6)}{6(2n-2)(2n-3)}.$$

Similar identities concerning Bernoulli polynomials can be produced directly from considerations of some particular zeta functions. For example, for a pair of positive integers p, q, we have

$$\zeta(ps; x)\zeta(qs; x) + \zeta(ps + qs; x)$$

$$= \sum_{n_1=0}^{\infty}\sum_{n_2=0}^{\infty}[(n_1 + n_2 + x)^p(n_2 + x)^q]^{-s}$$

$$+ \sum_{n_1=0}^{\infty}\sum_{n_2=0}^{\infty}[(n_1 + n_2 + x)^q(n_2 + x)^p]^{-s},$$

since

$$\zeta(ps; x)\zeta(qs; x)$$

$$= \sum_{n_1=0}^{\infty}\sum_{n_2=0}^{\infty}[(n_1 + x)^p(n_2 + x)^q]^{-s}$$

$$= \sum_{n_2 \leq n_1}[(n_1 + x)^p(n_2 + x)^q]^{-s} + \sum_{n_1 \leq n_2}[(n_2 + x)^q(n_1 + x)^p]^{-s}$$

$$- \sum_{n_1 = n_2}[(n_1 + x)^p(n_2 + x)^q]^{-s}.$$

Set $s = -1$ and $p = 2m, q = 2n + 1$, we get the following identity for $w = 2m + 2n + 1$,

$$B_{2m}(x)B_{2n+1}(x) = 2m \sum_{k=0}^{n} \binom{2n+1}{2k} B_{2k} \frac{B_{w-2k}(x)}{w - 2k}$$

$$+ (2n+1) \sum_{k=0}^{m} \binom{2m}{2k} B_{2k} \frac{B_{w-2k}(x)}{w - 2k}$$

which is equivalent to the evaluation of $S_{2m,2n+1}$:

$$S_{2m,2n+1} = \frac{1}{2}\zeta(w) + \sum_{k=0}^{m} \binom{w - 2k - 1}{2n} \zeta(2k)\zeta(w - 2k)$$

$$+ \sum_{k=0}^{n} \binom{w - 2k - 1}{2m - 1} \zeta(2k)\zeta(w - 2k).$$

On the other hand, a direct calculation showed that

$$\zeta((p, q); -1)$$

$$= \frac{1}{q + 1} \sum_{\beta=0}^{q+1} (-1)^{p+q} \binom{q + 1}{\beta} B_{\beta} \frac{B_{p+q+2-\beta}}{p + q + 2 - \beta} + \frac{q(-1)^{q}}{p + q} p!q! \frac{B_{p+q+2}}{(p + q + 2)!}.$$

It is interesting to note the similarity between expressions of $S_{p,q}$ and $\zeta((p, q); -1) + \zeta((q, p); -1)$ when $p + q$ is odd. So it is not unreasonable to infer that the special values at negative integers of $\zeta(\alpha; s)$ are related to the multiple zeta value $\zeta(\alpha)$.

CHAPTER 1

Introduction to the Theory of Multiple Zeta Values

Multiple zeta values are multi-version of the classical Euler double sums

$$S_{p,q} = \sum_{k=1}^{\infty} \frac{1}{k^q} \sum_{j=1}^{k} \frac{1}{j^p}$$

which are problems proposed by Goldbach in 1742 in an attempt to evaluate $S_{p,q}$ in terms of single zeta values, the special values at positive integers of the Riemann zeta function

$$\zeta(s) = \sum_{n=1}^{\infty} \frac{1}{n^s}.$$

With Drinfeld integral representation of multiple zeta values, we are able to express some particular multiple zeta values or sum of multiple zeta values as integrals over simplices of two dimension, so that shuffle products of multiple zeta values can be carry out more efficiently.

1.1　Introduction and Notations

For an r-tuple of positive integers $\alpha = (\alpha_1, \alpha_2, \ldots, \alpha_r)$ with $\alpha_r \geq 2$, the multiple zeta value or r-fold Euler sum $\zeta(\alpha)$ [6, 5, 21, 31] is defined as

$$\zeta(\alpha) = \sum_{1 \leq n_1 < n_2 < \cdots < n_r} n_1^{-\alpha_1} n_2^{-\alpha_2} \cdots n_r^{-\alpha_r},$$

or in free dummy variables as

$$\zeta(\alpha) = \sum_{\mathbf{k} \in \mathbb{N}^r} k_1^{-\alpha_1} (k_1 + k_2)^{-\alpha_2} \cdots (k_1 + k_2 + \cdots + k_r)^{-\alpha_r},$$

or in traditional nested form as

$$\sum_{n_r=1}^{\infty} \frac{1}{n_r^{\alpha_r}} \cdots \sum_{n_2=1}^{n_3-1} \frac{1}{n_2^{\alpha_2}} \sum_{n_1=1}^{n_2-1} \frac{1}{n_1^{\alpha_1}}.$$

The numbers r and $|\alpha| = \alpha_1 + \alpha_2 + \cdots + \alpha_r$ are the depth and the weight of $\zeta(\alpha)$, respectively.

The case $r = 2$ went back to 1742 as a problem proposed by Goldbach to Euler. For a pair of positive integers p, q with $q \geq 2$, the classical Euler double sum [2, 3, 7, 10] is defined as

$$S_{p,q} = \sum_{k=1}^{\infty} \frac{1}{k^q} \sum_{j=1}^{k} \frac{1}{j^p}.$$

The purpose of the problem is to evaluate $S_{p,q}$ in terms of the special values at positive integers of the Riemann zeta function (or single zeta values) defined by

$$\zeta(s) = \sum_{n=1}^{\infty} \frac{1}{n^s}, \quad \mathrm{Re}\, s > 1.$$

For our convenience, we let $\{1\}^k$ be the k repetitions of 1, or more general, let $\{a\}^k$ be the k repetitions of a. For instance, we have

$$\zeta(\{1\}^3, 5) = \zeta(1,1,1,5) \quad \text{and} \quad \zeta(\{2\}^4, 6) = \zeta(2,2,2,2,6).$$

Some important results concerning multiple zeta values are worth mentioned here.

1. **C. Markett [26] in 1994:** The evaluation of $\zeta(1,1,n)$.

   C. Markett and later J. M. Borwein and R. Girgensohn [3] obtained among other thing, that for $n \geq 3$

   $$\zeta(1,1,n) = \frac{n(n+1)}{6}\zeta(n+2) - \frac{n-1}{2}\zeta(2)\zeta(n)$$
   $$- \frac{n}{4}\sum_{k=0}^{n-4}\zeta(n-k-1)\zeta(k+3)$$
   $$+ \frac{1}{6}\sum_{k=0}^{n-4}\zeta(n-k-2)\sum_{j=0}^{k}\zeta(k-j+2)\zeta(j+2).$$

2. **Granville [20] in 1997:** The sum formula, i.e. for a pair of positive integers m, r with $m \geq r$,

   $$\sum_{|\boldsymbol{\alpha}|=m}\zeta(\alpha_1,\alpha_2,\ldots,\alpha_r+1) = \zeta(m+1).$$

3. **Drinfeld [9] in 1996:** Drinfeld duality theorem, i.e. for nonnegative integers m and n,

   $$\zeta(\{1\}^m, n+2) = \zeta(\{1\}^n, m+2).$$

4. **Ohno [28] in 2000:** Ohno's generalization of the duality theorem and sum formula. Let (a_1,b_1), (a_2,b_2), $\ldots$, (a_r,b_r) be r pairs of nonnegative integers. Suppose that

   $$\mathbf{k} = (\{1\}^{a_1}, b_1 + 2, \{1\}^{a_2}, b_2 + 2, \ldots, \{1\}^{a_r}, b_r + 2)$$

   and $\mathbf{k}'$ be its dual,

   $$\mathbf{k}' = (\{1\}^{b_r}, a_r + 2, \{1\}^{b_{r-1}}, a_{r-1} + 2, \ldots, \{1\}^{b_1}, a_1 + 2).$$

   Then for any nonnegative integer m, we have

   $$\sum_{|\mathbf{c}|=m}\zeta(\mathbf{k}+\mathbf{c}) = \sum_{|\mathbf{d}|=m}\zeta(\mathbf{k}'+\mathbf{d}).$$

   When $m = 0$, the assertion $\zeta(\mathbf{k}) = \zeta(\mathbf{k}')$ is Drinfeld duality theorem.

5. **Tsumura [22] in 2004:** A combinational relation asserted that $\zeta(\alpha)$ can be expressed in multiple zeta values of lower depth provided that the sum of its depth and weight is odd.

6. **Eie, Liaw and Ong [27] in 2008:** Evaluations of triple Euler sums. For a positive integer n, we have

$$\zeta(1,1,2n) = \zeta(2n+2) - \sum_{\alpha=2}^{2n-1} (-1)^{\alpha} \zeta(1,\alpha)\zeta(2n+1-\alpha)$$

and

$$3\zeta(1,1,2n+1) = -\zeta(2n+3) - 2\{\zeta(1,2n+2) + \zeta(2,2n+1)\}$$
$$+ \sum_{\alpha=2}^{2n} (-1)^{\alpha} \zeta(1,\alpha)\zeta(2n+2-\alpha).$$

7. **Eie, Liaw and Ong [12, 15] in 2009:** The restricted sum formula, for integers m, p, q with $p \geq 1$ and $m \geq q \geq 1$,

$$\sum_{|\boldsymbol{\alpha}|=m} \zeta(\{1\}^{p}, \alpha_1, \alpha_2, \ldots, \alpha_q+1) = \sum_{|\mathbf{c}|=p+q} \zeta(c_1, c_2, \ldots, c_{p+1}+m-q+1).$$

A vector version of the restricted sum formula was given in 2010 by Eie and Wei [16].

8. **Eie, Ong and Wei [14] in 2010:** Evaluations of quadruple Euler sums. For a positive integer n, we have

$$2\zeta(1,1,1,2n) = \zeta(2n+3) + \{\zeta(1,2n+2) + \zeta(2,2n+1) + \zeta(3,2n)\}$$
$$- \sum_{\alpha=2}^{2n-1} (-1)^{\ell} \zeta(1,1,\alpha)\zeta(2n+1-\alpha)$$

and

$$\zeta(1,1,1,2n+1) = -\zeta(2n+4) + \sum_{\alpha=2}^{2n} (-1)^{\ell} \zeta(1,1,\alpha)\zeta(2n+2-\alpha)$$
$$- \frac{1}{2} \sum_{\alpha=2}^{2n} (-1)^{\ell} \zeta(1,\alpha)\zeta(1,2n+2-\alpha).$$

1.2 Drinfeld Integral Representations of Multiple Zeta Values

Multiple zeta values can be represented by Drinfeld iterated integrals over the simplex defined by

$$0 < t_1 < t_2 < \cdots < t_{|\alpha|} < 1,$$

or symbolically by

$$\zeta(\alpha_1, \alpha_2, \ldots, \alpha_r) = \int_0^1 \Omega_1 \Omega_2 \cdots \Omega_{|\alpha|}$$

with

$$\Omega_j = \begin{cases} \frac{dt_j}{1-t_j}, & \text{if } j = 1, \alpha_1 + 1, \ldots, \alpha_1 + \alpha_2 + \cdots + \alpha_{r-1} + 1; \\ \frac{dt_j}{t_j}, & \text{otherwise.} \end{cases}$$

In particular, for nonnegative integers m and n, we have

$$\zeta(\{1\}^m, n+2) = \int_{0<t_1<\cdots<t_{m+n+2}<1} \prod_{j=1}^{m+1} \frac{dt_j}{1-t_j} \prod_{k=m+2}^{m+n+2} \frac{dt_k}{t_k}.$$

Note that the change of variables

$$u_1 = 1 - t_{m+n+2}, \ u_2 = 1 - t_{m+n+1}, \ \ldots, \ u_{m+n+2} = 1 - t_1,$$

yields the duality theorem

$$\zeta(\{1\}^m, n+2) = \zeta(\{1\}^n, m+2).$$

The general case $\zeta(\mathbf{k}) = \zeta(\mathbf{k}')$ can be obtained in the same way.

Some of multiple zeta values can be evaluated through duality theorem. For example, for any nonnegative integer m,

$$\zeta(\{1\}^m, 2) = \zeta(m+2)$$

or for positive integer n

$$\zeta(\{1, 2\}^n) = \zeta(\{3\}^n).$$

In 1992, Hoffman [24] proved that

$$\zeta(\{2\}^n) = \frac{\pi^{2n}}{(2n+1)!}$$

which was previously conjectured by C. Moen. Such kind of evaluations comes directly from the product expression of sine function as

$$\frac{\sin \pi x}{\pi x} = \prod_{n=1}^{\infty} \left(1 - \frac{x^2}{n^2}\right)$$

which is equal to

$$\sum_{n=0}^{\infty} (-1)^n \zeta(\{2\}^n) x^{2n}.$$

The following self-dual evaluation, previously conjectured by D. Zagier

$$\zeta(\{1,3\}^n) = 4^{-n}\zeta(\{4\}^n) = \frac{2\pi^{4n}}{(4n+2)!}$$

was proved by Borwein and others [4]. Some particular multiple zeta values as well as sums of multiple zeta values can be expressed in integrals in simple ways.

Proposition 1.2.1. [15] *For nonnegative integers m, n, we have*

$$\zeta(\{1\}^m, n+2) = \frac{1}{m!n!} \int_{0<t_1<t_2<1} \left(\log \frac{1}{1-t_1}\right)^m \left(\log \frac{1}{t_2}\right)^n \frac{dt_1}{1-t_1} \frac{dt_2}{t_2}$$

$$= \frac{1}{m!(n+1)!} \int_0^1 \left(\log \frac{1}{1-t}\right)^m \left(\log \frac{1}{t}\right)^{n+1} \frac{dt}{1-t}.$$

Proposition 1.2.2. [15] *For nonnegative integer p and positive integers m, n, r with $m \geq r$, we have*

$$\sum_{|\alpha|=m} \zeta(\{1\}^p, \alpha_1, \alpha_2, \ldots, \alpha_r + n)$$

$$= \frac{1}{p!(r-1)!(m-r)!(n-1)!} \int_{0<t_1<t_2<1} \left(\log \frac{1}{1-t_1}\right)^p \left(\log \frac{1-t_1}{1-t_2}\right)^{r-1}$$

$$\times \left(\log \frac{t_2}{t_1}\right)^{m-r} \left(\log \frac{1}{t_2}\right)^{n-1} \frac{dt_1 dt_2}{(1-t_1)t_2}.$$

Once multiple zeta values are expressed in Drinfeld iterated integrals, the shuffle product formula of two multiple zeta values takes the form

$$\int_0^1 \Omega_1\Omega_2\cdots\Omega_m \int_0^1 \Omega_{m+1}\Omega_{m+2}\cdots\Omega_{m+n} = \sum_\sigma \int_0^1 \Omega_{\sigma(1)}\Omega_{\sigma(2)}\cdots\Omega_{\sigma(m+n)},$$

where σ ranges over all $(m+n)!/(m!n!)$ permutations on the set $\{1, 2, \ldots, m+n\}$ which preserves of orders of $\Omega_1\Omega_2\cdots\Omega_m$ and $\Omega_{m+1}\Omega_{m+2}\cdots\Omega_{m+n}$, i.e., for $1 \le i < j \le m$ or $m+1 \le i < j \le m+n$, we have

$$\sigma^{-1}(i) < \sigma^{-1}(j).$$

In general, it is not easy to perform the shuffle precess for two arbitrary multiple zeta values. So we choose multiple zeta values of height one (multiple zeta values of the form $\zeta(\{1\}^m, n+2)$) for our candidates. Here we mention two basic shuffle relations to illustrate what we mean.

Theorem 1.2.3 (Euler's Decomposition Theorem). [3] *For a pair of positive integers p, q,*

$$\zeta(p+1)\zeta(q+1) = \sum_{|\alpha|=p+q+1} \binom{\alpha_2}{q}\zeta(\alpha_1, \alpha_2+1) + \sum_{|\alpha|=p+q+1} \binom{\alpha_2}{p}\zeta(\alpha_1, \alpha_2+1).$$

Proof. By Proposition 1.2.1, we have

$$\zeta(p+1)\zeta(q+1) = \frac{1}{p!q!}\int_0^1\int_0^1 \left(\log\frac{1}{t}\right)^p\left(\log\frac{1}{u}\right)^q \frac{dt}{1-t}\frac{du}{1-u}.$$

Subdivide the region of integration into two simplices as

$$D_1: \ 0 < t < u < 1 \quad \text{and} \quad D_2: \ 0 < u < t < 1.$$

On the simplex D_1, we replace the factor $\left(\log\frac{1}{t}\right)^p$ by its binomial expression

$$\sum_{a=0}^p \frac{p!}{a!(p-a)!}\left(\log\frac{u}{t}\right)^a\left(\log\frac{1}{u}\right)^{p-a},$$

so that the value of integration over D_1 is

$$\sum_{a=0}^p \binom{p+q-a}{q}\zeta(a+1, p+q-a+1)$$

or

$$\sum_{|\alpha|=p+q+1} \binom{\alpha_2}{q} \zeta(\alpha_1, \alpha_2 + 1).$$

On the other hand, the value of the integration over D_2 is

$$\sum_{b=0}^{q} \binom{p+q-b}{p} \zeta(b+1, p+q-b+1)$$

or

$$\sum_{|\alpha|=p+q+1} \binom{\alpha_2}{p} \zeta(\alpha_1, \alpha_2 + 1).$$

Of course, the value of the integral is equal to the sum of the values of integration over D_1 and D_2. Consequently, our assertion follows.   □

Remark 1.2.4. In Euler's decomposition theorem, if we let $p = \ell + 1$ and $q = (r - \ell) + 1$ with $0 \le \ell \le r$, we get

$$\zeta(\ell+2)\zeta(r-\ell+2) = \sum_{|\alpha|=r+3} \zeta(\alpha_1, \alpha_2 + 1) \left\{ \binom{\alpha_2}{\ell+1} + \binom{\alpha_2}{r-\ell+1} \right\}.$$

Then sum over all ℓ with $0 \le \ell \le r$, it yields

$$\sum_{\ell=0}^{r} \zeta(\ell+2)\zeta(r-\ell+2) = \sum_{|\alpha|=r+3} (2^{\alpha_2+1} - 2)\zeta(\alpha_1, \alpha_2 + 1) - 2\zeta(1, r+3).$$

With the evaluation of $\zeta(1, r + 3)$ given by

$$\zeta(1, r+3) = \frac{r+3}{2}\zeta(r+4) - \frac{1}{2}\sum_{\ell=0}^{r} \zeta(\ell+2)\zeta(r-\ell+2),$$

we get

$$\sum_{|\alpha|=r+3} 2^{\alpha_2}\zeta(\alpha_1, \alpha_2 + 1) = \frac{r+5}{2}\zeta(r+4).$$

This is the weighted Euler sum formula proved by Y. Ohno and W. Zudilin [29] in 2008.

Theorem 1.2.5. [18] *For a pair of nonnegative integers k and r, we have*

(1.2.1)

$$\sum_{p+q=k} \binom{p}{j} \sum_{|\alpha|=q+r+3} \zeta(\{1\}^p, \alpha_0, \ldots, \alpha_q, \alpha_{q+1}+1)\binom{\alpha_{q+1}}{\ell+1}$$

$$+ \sum_{p+q=k} \binom{p}{k-j} \sum_{|\alpha|=q+r+3} \zeta(\{1\}^p, \alpha_0, \ldots, \alpha_q, \alpha_{q+1}+1)\binom{\alpha_{q+1}}{r-\ell+1}$$

$$= \zeta(\{1\}^j, r-\ell+2)\zeta(\{1\}^{k-j}, \ell+2)$$

for all $0 \le j \le k$ and $0 \le \ell \le r$.

Proof. By Proposition 1.2.1, we have

$$\zeta(\{1\}^j, r-\ell+2)\zeta(\{1\}^{k-j}, \ell+2)$$

$$= \frac{1}{j!(k-j)!(r-\ell+1)!(\ell+1)!} \int_0^1 \int_0^1 \left(\log\frac{1}{1-t}\right)^j \left(\log\frac{1}{t}\right)^{r-\ell+1}$$

$$\times \left(\log\frac{1}{1-u}\right)^{k-j} \left(\log\frac{1}{u}\right)^{\ell+1} \frac{dtdu}{(1-t)(1-u)}.$$

Again, we decompose the region of integration, the square $[0,1]^2$, into two simplices as follows:

$$D_1 : 0 < t < u < 1 \quad \text{and} \quad D_2 : 0 < u < t < 1.$$

On the simplex D_1, we replace the factors

$$\left(\log\frac{1}{1-u}\right)^{k-j} \quad \text{and} \quad \left(\log\frac{1}{t}\right)^{r-\ell+1},$$

by their binomial expansions

$$\sum_{a+b=k-j} \frac{(k-j)!}{a!b!} \left(\log\frac{1}{1-t}\right)^a \left(\log\frac{1-t}{1-u}\right)^b$$

and

$$\sum_{m+n=r-\ell+1} \frac{(r-\ell+1)!}{m!n!} \left(\log\frac{u}{t}\right)^m \left(\log\frac{1}{u}\right)^n.$$

So that the value of integration over D_1 is

$$\sum_{a+b=k-j} \sum_{m+n=r-\ell+1} \binom{j+a}{j}\binom{n+\ell+1}{\ell+1}$$

$$\times \sum_{|\alpha|=b+m+1} \zeta(\{1\}^{j+a}, \alpha_0, \alpha_1, \ldots, \alpha_b, n+\ell+2),$$

or

$$\sum_{p+q=k} \binom{p}{j} \sum_{|\boldsymbol{\alpha}|=q+r+3} \zeta(\{1\}^p, \alpha_0, \alpha_1, \ldots, \alpha_q, \alpha_{q+1}+1)\binom{\alpha_{q+1}}{\ell+1}.$$

In a similar way, we may evaluate the value of the integration over D_2 and our assertion follows.   □

1.3   Double Weighted Sum Formulas

Most of double weighted sum formula can be obtained from the formula appeared in Theorem 1.2.5. For our convenience, we introduce a new double weighted sum. For a pair of nonnegative integers k, r and complex numbers μ, λ, we let

$$E(\mu, \lambda) = \sum_{p+q=k} \mu^p \sum_{|\boldsymbol{\alpha}|=q+r+3} \zeta(\{1\}^p, \alpha_0, \ldots, \alpha_q, \alpha_{q+1}+1)\lambda^{\alpha_{q+1}}.$$

When there is no ambiguity, we write $\zeta(\{1\}^p, \boldsymbol{\alpha})$ instead of $\zeta(\{1\}^p, \alpha_0, \ldots, \alpha_q, \alpha_{q+1}+1)$.

Here are some weighted sum formula easily obtained from the formula in Theorem 1.2.5.

(1) Just summing over all $0 \leq j \leq k$ and $0 \leq \ell \leq r$, we get

(1.3.1)
$$\sum_{p+q=k} 2^p \sum_{|\boldsymbol{\alpha}|=q+r+3} (2^{\alpha_{q+1}} - 1)\zeta(\{1\}^p, \boldsymbol{\alpha})$$
$$= (2^{k+1} - 1)\zeta(\{1\}^{k+1}, r+3)$$
$$+ \frac{1}{2}\sum_{j=0}^{k}\sum_{\ell=0}^{r} \zeta(\{1\}^j, r-\ell+2)\zeta(\{1\}^{k-j}, \ell+2).$$

On the other hand, when r is even, summing over all $0 \leq j \leq k$, and $0 \leq \ell \leq r$ with ℓ even, we get

(1.3.2)
$$\sum_{p+q=k} 2^p \sum_{|\boldsymbol{\alpha}|=q+r+3} 2^{\alpha_{q+1}}\zeta(\{1\}^p, \boldsymbol{\alpha})$$
$$= \sum_{j=0}^{k}\sum_{\substack{\ell=0 \\ \ell:\text{even}}}^{r} \zeta(\{1\}^j, r-\ell+2)\zeta(\{1\}^{k-j}, \ell+2).$$

Therefore, when r is even, the difference of (1.3.1) and (1.3.2) gives

$$\sum_{p+q=k} 2^p \sum_{|\alpha|=q+r+3} \zeta(\{1\}^p, \alpha)$$

(1.3.3)
$$= -\left(2^{k+1} - 1\right) \zeta(\{1\}^{k+1}, r+3)$$

$$+ \frac{1}{2} \sum_{j=0}^{k} \sum_{\ell=0}^{r} (-1)^\ell \zeta(\{1\}^j, r-\ell+2) \zeta(\{1\}^{k-j}, \ell+2).$$

(2) Multiplying both sides of the identity in Theorem 1.2.5 by $(-1)^{j+\ell}$ and then summing over all $0 \le j \le k$ and $0 \le \ell \le r$, we get

$$\{1 + (-1)^{k+r}\} \sum_{|\alpha|=k+r+3} \zeta(\alpha_0, \alpha_1, \ldots, \alpha_k, \alpha_{k+1}+1)$$

$$+ \{(-1)^k + (-1)^r\} \zeta(\{1\}^{k+1}, r+3)$$

$$= \sum_{j=0}^{k} \sum_{\ell=0}^{r} (-1)^{j+\ell} \zeta(\{1\}^j, r-\ell+2) \zeta(\{1\}^{k-j}, \ell+2),$$

using the facts that

$$\sum_{j=0}^{k} \binom{p}{j}(-1)^j = \begin{cases} 1, & \text{if } p = 0; \\ 0, & \text{otherwise,} \end{cases}$$

$$\sum_{j=0}^{k} \binom{p}{k-j}(-1)^j = \begin{cases} (-1)^k, & \text{if } p = 0; \\ 0, & \text{otherwise,} \end{cases}$$

and

$$\sum_{\ell=0}^{r} \binom{\alpha}{\ell+1}(-1)^\ell = \begin{cases} 1, & \text{if } \alpha \le r+1; \\ 1 + (-1)^r, & \text{if } \alpha = r+2, \end{cases}$$

$$\sum_{\ell=0}^{r} \binom{\alpha}{r-\ell+1}(-1)^\ell = \begin{cases} (-1)^r, & \text{if } \alpha \le r+1; \\ (-1)^r + 1, & \text{if } \alpha = r+2. \end{cases}$$

Consequently, when $k+r$ is even, the sum formula

$$\sum_{|\alpha|=k+r+3} \zeta(\alpha_0, \ldots, \alpha_k, \alpha_{k+1}+1) = \zeta(k+r+4)$$

is equivalent to the evaluation that

$$\zeta(\{1\}^{k+1}, r+3)$$
$$= (-1)^{k+1}\zeta(k+r+4)$$
$$+ \frac{(-1)^k}{2}\sum_{j=0}^{k}\sum_{\ell=0}^{r}(-1)^{j+\ell}\zeta(\{1\}^j, r-\ell+2)\zeta(\{1\}^{k-j}, \ell+2).$$

(3) Let μ, λ be complex numbers with $\mu\lambda \neq 0$. Multiplying both sides of the identity of Theorem 1.2.5 by $\mu^j\lambda^{\ell+1}$ and then summing over all $0 \leq j \leq k$ and $0 \leq \ell \leq r$, we get

(1.3.4)
$$\{E(\mu+1, \lambda+1) - E(\mu+1, 1)\}$$
$$+ \mu^k\lambda^{r+2}\left\{E\left(\frac{\mu+1}{\mu}, \frac{\lambda+1}{\lambda}\right) - E\left(\frac{\mu+1}{\mu}, 1\right)\right\}$$
$$= \left\{\frac{1}{\mu}[(\mu+1)^{k+1}-1]\lambda^{r+2} + [(\mu+1)^{k+1} - \mu^{k+1}]\right\}\zeta(\{1\}^{k+1}, r+3)$$
$$+ \sum_{j=0}^{k}\sum_{\ell=0}^{r}\mu^j\lambda^{\ell+1}\zeta(\{1\}^j, r-\ell+2)\zeta(\{1\}^{k-j}, \ell+2),$$

using the facts that

$$\sum_{j=0}^{k}\binom{p}{j}\mu^j = \sum_{j=0}^{p}\binom{p}{j}\mu^j = (\mu+1)^p,$$

$$\sum_{j=0}^{k}\binom{p}{k-j}\mu^j = \sum_{j=0}^{k}\binom{p}{j}\mu^{k-j} = \mu^k\left(1+\frac{1}{\mu}\right)^p,$$

$$\sum_{\ell=0}^{r}\binom{\alpha}{\ell+1}\lambda^{\ell+1} = \begin{cases} (\lambda+1)^\alpha - 1, & \text{if } \alpha \leq r+1; \\ (\lambda+1)^{r+2} - 1 - \lambda^{r+2}, & \text{if } \alpha = r+2, \end{cases}$$

and

$$\sum_{\ell=0}^{r}\binom{\alpha}{r-\ell+1}\lambda^{\ell+1} = \begin{cases} \lambda^{r+2}\left\{\left(\frac{\lambda+1}{\lambda}\right)^\alpha - 1\right\}, & \text{if } \alpha \leq r+1; \\ \lambda^{r+2}\left\{\left(\frac{\lambda+1}{\lambda}\right)^{r+2} - 1\right\} - 1, & \text{if } \alpha = r+2. \end{cases}$$

A lot of weighted sum formulas can be obtained from (1.3.4) by specifying the values of μ and λ. For example, let $\mu = -1$ and $\lambda = 1$. We get

$$\sum_{|\alpha|=k+r+3} 2^{\alpha_{k+1}}\zeta(\alpha_0,\ldots,\alpha_k,\alpha_{k+1}+1)$$

$$= \zeta(k+r+4) + \zeta(\{1\}^{k+1},r+3)$$

$$+ \frac{1}{2}\sum_{j=0}^{k}\sum_{\ell=0}^{r}(-1)^j\zeta(\{1\}^j,r-\ell+2)\zeta(\{1\}^{k-j},\ell+2)$$

if k is even.

On the other hand, if we let $\mu=1$ and $\lambda=-1$, we get

$$\sum_{p+q=k} 2^p \sum_{|\alpha|=q+r+3} \zeta(\{1\}^p,\alpha)$$

$$= -\left(2^{k+1}-1\right)\zeta(\{1\}^{k+1},r+3)$$

$$+ \frac{1}{2}\sum_{j=0}^{k}\sum_{\ell=0}^{r}(-1)^\ell\zeta(\{1\}^j,r-\ell+2)\zeta(\{1\}^{k-j},\ell+2)$$

if r is even. This is the same identity as (1.3.3).

1.4 The Expectations of Binomial Distributions

According to the theory of probability, the expectation of n trials of a binomial distribution with the probability of success a $(0 < a < 1)$ is na, i.e.,

$$\sum_{j=1}^{n} \binom{n}{j} j a^j (1-a)^{n-j} = na.$$

The following Lemmas are immediate consequences of this fact.

Lemma 1.4.1. *For nonnegative integers k,p with $0 \le p \le k$, we have*

$$\sum_{j=0}^{k} \binom{p}{j} j a^j (1-a)^{k-j} = pa(1-a)^{k-p}$$

and

$$\sum_{j=0}^{k} \binom{p}{k-j} j a^j (1-a)^{k-j} = ka^{k-p} - p(1-a)a^{k-p}.$$

Lemma 1.4.2. *For positive integers α with $\alpha \leq r + 2$, we have*

$$\sum_{\ell=0}^{r} \binom{\alpha}{\ell+1}(\ell+1)a^{\ell+1}(1-a)^{r-\ell+1}$$

$$= \begin{cases} \alpha a(1-a)^{r+2-\alpha}, & \text{if } \alpha \leq r+1; \\ (r+2)a - (r+2)a^{r+2}, & \text{if } \alpha = r+2, \end{cases}$$

and

$$\sum_{\ell=0}^{r} \binom{\alpha}{r-\ell+1}(\ell+1)a^{\ell+1}(1-a)^{r-\ell+1}$$

$$= (r+2)\left(a^{r+2-\alpha} - a^{r+2}\right) - \alpha(1-a)a^{r+2-\alpha}.$$

The following theorems are based on Lemma 1.4.1, Lemma 1.4.2 and the formula in Theorem 1.2.5.

Theorem 1.4.3. *For a pair of nonnegative integers k, r and a real number a, we have*

$$\sum_{p+q=k} \left\{ pa(1-a)^q + (-1)^r ka^q + (-1)^{r+1}p(1-a)a^q \right\}$$

$$\times \sum_{|\alpha|=q+r+3} \zeta(\{1\}^p, \alpha_0, \dots, \alpha_q, \alpha_{q+1}+1)$$

$$= -\sum_{p+q=k} \left\{ (-1)^r pa(1-a)^q + ka^q - p(1-a)a^q \right\} \zeta(\{1\}^{k+1}, r+3)$$

$$+ \sum_{j=1}^{k}\sum_{\ell=0}^{r}(-1)^\ell ja^j(1-a)^{k-j}\zeta(\{1\}^j, r-\ell+2)\zeta(\{1\}^{k-j}, \ell+2).$$

Theorem 1.4.4. *For a pair of nonnegative integers k, r and a real number a, we have*

$$\sum_{|\alpha|=k+r+3} \zeta(\alpha_0, \dots, \alpha_k, \alpha_{k+1}+1)$$

$$\times \left\{ \alpha_{k+1}a(1-a)^{r+2-\alpha_{k+1}} + (-1)^{k+1}\alpha_{k+1}(1-a)a^{r+2-\alpha_{k+1}} \right\}$$

$$+ (-1)^k(r+2)\sum_{|\alpha|=k+r+3} \zeta(\alpha_0, \dots, \alpha_k, \alpha_{k+1}+1)\left\{ a^{r+2-\alpha_{k+1}} - a^{r+2} \right\}$$

$$= (r+2)a^{r+2}\zeta(\{1\}^{k+1}, r+3)$$

$$+ \sum_{j=0}^{k}\sum_{\ell=0}^{r}(-1)^j(\ell+1)a^{\ell+1}(1-a)^{r-\ell+1}\zeta(\{1\}^j, r-\ell+2)\zeta(\{1\}^{k-j}, \ell+2).$$

Corollary 1.4.5. *Suppose that k and r are nonnegative integers and r is even. Then*

$$\sum_{p+q=k} \frac{k}{q+1} \sum_{|\alpha|=q+r+3} \zeta(\{1\}^p, \alpha_0, \ldots, \alpha_q, \alpha_{q+1}+1)$$

$$= -\left(\sum_{q=0}^{k} \frac{k}{q+1}\right) \zeta(\{1\}^{k+1}, r+3)$$

$$+ \sum_{j=1}^{k}\sum_{\ell=0}^{r}(-1)^{\ell}\, \frac{j \cdot j!(k-j)!}{k \cdot (k+1)!}\zeta(\{1\}^j, r-\ell+2)\zeta(\{1\}^{k-j}, \ell+2).$$

Proof. Integrating with respect to a form 0 to 1 with the identity of Theorem 1.4.3, we get our assertion. □

Corollary 1.4.6. *Suppose that k and r are nonnegative integers with k even. Then*

$$\sum_{|\alpha|=k+r+3} \zeta(\alpha_0, \ldots, \alpha_k, \alpha_{k+1}+1)\frac{\alpha_{k+1}}{r+3-\alpha_{k+1}}$$

$$= \zeta(\{1\}^{k+1}, r+3) + \sum_{j=1}^{k}\sum_{\ell=0}^{r}(-1)^j \frac{(\ell+1)\cdot(\ell+1)!(r-\ell+1)!}{(r+2)\cdot(r+2)!}$$

$$\times \zeta(\{1\}^j, r-\ell+2)\zeta(\{1\}^{k-j}, \ell+2).$$

Proof. Integrating with respect to a form 0 to 1 and the identity of Theorem 1.4.4, we get our required assertion. □

Again, we consider the hypergeometric distribution from theory of probability. The probability of getting g good elements and b bad elements from a population of size N containing G good and B bad elements is

$$\binom{G}{g}\binom{B}{b}\bigg/\binom{G+B}{g+b}.$$

This implies that

$$\sum_{g=0}^{n}\binom{G}{g}\binom{B}{n-g} = \binom{G+B}{n}.$$

Also the mathematical expectation of hypergeometric distribution is

$$n \cdot \frac{G}{G+B},$$

or equivalently

$$\sum_{g=0}^{n} \binom{G}{g}\binom{B}{n-g}g = \frac{nG}{G+B}\binom{G+B}{n}.$$

The following lemmas are immediate consequences of the above assertions

Lemma 1.4.7. *For nonnegative integers k, p with $0 \leq p \leq k$, we have*

(1.4.1) $$\sum_{j=0}^{k} \binom{p}{j}\binom{k}{k-j} = \sum_{j=0}^{k} \binom{p}{k-j}\binom{k}{k-j} = \binom{k+p}{k},$$

(1.4.2) $$\sum_{j=0}^{k} \binom{p}{j}\binom{k}{k-j}(k-j) = \frac{k^2}{k+p}\binom{k+p}{k}$$

and

(1.4.3) $$\sum_{j=0}^{k} \binom{p}{k-j}\binom{k}{k-j}(k-j) = \frac{kp}{k+p}\binom{k+p}{k}.$$

Lemma 1.4.8. *Let r, α be nonnegative integers with $0 \leq \alpha \leq r+2$. Then*

(1.4.4)
$$\sum_{\ell=0}^{r} \binom{\alpha}{\ell+1}\binom{r+2}{r-\ell+1} = \sum_{\ell=0}^{r} \binom{\alpha}{r-\ell+1}\binom{r+2}{r-\ell+1}$$
$$= \begin{cases} \binom{\alpha+r+2}{r+2} - 1, & \text{if } \alpha \leq r+1; \\ \binom{2r+4}{r+2} - 2, & \text{if } \alpha = r+2, \end{cases}$$

and

$$\sum_{\ell=0}^{r} \binom{\alpha}{\ell+1}\binom{r+2}{r-\ell+1}(\ell+1) = \begin{cases} \frac{(r+2)\alpha}{\alpha+r+2}\binom{\alpha+r+2}{r+2}, & \text{if } \alpha \leq r+1; \\ \frac{r+2}{2}\binom{2r+4}{r+2} - (r+2), & \text{if } \alpha = r+2. \end{cases}$$

Based on Lemmas 1.4.7 and 1.4.8, we obtain the following double weighted sum formulas.

Theorem 1.4.9. *Suppose that k and r are integers with $k, r \geq 0$. Then*

$$\sum_{p+q=k} \binom{k+p}{k} \sum_{|\boldsymbol{\alpha}|=q+r+3} \zeta(\{1\}^p, \alpha_0, \ldots, \alpha_q, \alpha_{q+1}+1)$$

$$\times \left[\binom{\alpha_{q+1}+r+2}{r+2} - 1\right]$$

$$= \binom{2k+1}{k} \zeta(\{1\}^{k+1}, r+3)$$

$$+ \frac{1}{2} \sum_{j=0}^{k} \sum_{\ell=0}^{r} \binom{k}{j}\binom{r+2}{\ell+1} \zeta(\{1\}^j, r-\ell+2)\zeta(\{1\}^{k-j}, \ell+2).$$

Proof. Multiplying both sides of (1.2.1) by $\binom{k}{j}\binom{r+2}{\ell+1}$ and then summing over $0 \leq j \leq k$ and $0 \leq \ell \leq r$, we get our identity in light of (1.4.1) and (1.4.4). □

Theorem 1.4.10. *Suppose that k and r are nonnegative integers with r even. Then*

$$\sum_{p+q=k} \binom{k+p}{k} \sum_{|\boldsymbol{\alpha}|=q+r+3} \zeta(\{1\}^p, \alpha_0, \ldots, \alpha_q, \alpha_{q+1}+1)$$

$$= -\binom{2k+1}{k} \zeta(\{1\}^{k+1}, r+3)$$

$$+ \frac{1}{2} \sum_{j=0}^{k} \sum_{\ell=0}^{r} \binom{k}{j}(-1)^\ell \zeta(\{1\}^j, r-\ell+2)\zeta(\{1\}^{k-j}, \ell+2).$$

Theorem 1.4.11. *Suppose that k and r are nonnegative integers with k even. Then*

$$\sum_{|\boldsymbol{\alpha}|=k+r+3} \zeta(\alpha_0, \ldots, \alpha_k, \alpha_{k+1}+1)\binom{\alpha_{k+1}+r+2}{r+2}$$

$$= \zeta(k+r+4) + \zeta(\{1\}^{k+1}, r+3)$$

$$+ \frac{1}{2} \sum_{j=0}^{k} \sum_{\ell=0}^{r} (-1)^j \binom{r+2}{\ell+1} \zeta(\{1\}^j, r-\ell+2)\zeta(\{1\}^{k-j}, \ell+2).$$

Combining Theorems 1.4.10 and 1.4.11, we obtain the following.

Theorem 1.4.12. *Suppose that k and r are nonnegative integers with r even. Then*

$$\sum_{p+q=k} \binom{k+p}{k} \sum_{|\alpha|=q+r+3} \zeta(\{1\}^p, \alpha_0, \ldots, \alpha_q, \alpha_{q+1}+1)\binom{\alpha_{q+1}+r+2}{r+2}$$

$$= \frac{1}{2}\sum_{j=0}^{k}\sum_{\ell=0}^{r} \binom{k}{j}\left\{\binom{r+2}{\ell+1} + (-1)^\ell\right\}\zeta(\{1\}^j, r-\ell+2)\zeta(\{1\}^{k-j}, \ell+2).$$

In the above theorems, there are restrictions on k or r so that some interesting cases are left out. Here we retrieve some of them.

Theorem 1.4.13. *Suppose that k and r are nonnegative integers with r odd. Then*

$$\sum_{p+q=k} \frac{k^2-kp}{k+p}\binom{k+p}{k} \sum_{|\alpha|=q+r+3} \zeta(\{1\}^p, \alpha_0, \ldots, \alpha_q, \alpha_{q+1}+1)$$

$$= \sum_{p+q=k} \frac{k^2-kp}{k+p}\binom{k+p}{k}\zeta(\{1\}^{k+1}, r+3)$$

$$+ \sum_{j=0}^{k}\sum_{\ell=0}^{r} \binom{k}{j}(k-j)(-1)^\ell\zeta(\{1\}^j, r-\ell+2)\zeta(\{1\}^{k-j}, \ell+2).$$

Proof. Multiplying both sides of (1.2.1) by $\binom{k}{j}j(-1)^\ell$ and then summing over $0 \le j \le k$ and $0 \le \ell \le r$ with the help of (1.4.2) and (1.4.3), we get our identity.   □

1.5   Exercises

1. Begin with

$$\zeta(p+1)\zeta(q+1) = \frac{1}{p!q!}\int_0^1\int_0^1 \left(\log\frac{1}{1-t}\right)^p \left(\log\frac{1}{u}\right)^q \frac{dt}{t}\frac{du}{(1-u)}$$

to prove the identity

$$\sum_{a+b=p}\sum_{|\alpha|=b+q+1} \zeta(\{1\}^a, \alpha_0, \ldots, \alpha_b+1)\alpha_b + \zeta(\{1\}^{p-1}, 2, q+1)$$

$$= \zeta(p+1)\zeta(q+1).$$

2. Prove that for positive integers p and q,

$$\sum_{a+b=p}\sum_{|\alpha|=b+q+1}\zeta(\{1\}^a,\alpha_0,\alpha_1,\ldots,\alpha_b+1)\alpha_b$$

$$=\sum_{c+d=q}\sum_{|\beta|=d+p+1}\zeta(\{1\}^c,\beta_0,\beta_1,\ldots,\beta_d+1)\beta_d.$$

3. For any positive integer c, prove that

$$\sum_{j=0}^{k}\binom{p}{j}(-1)^j\frac{1}{j+c+1}=\frac{p!c!}{(p+c+1)!},$$

$$\sum_{j=0}^{k}\binom{p}{k-j}(-1)^j\frac{1}{j+c+1}=\frac{(-1)^{k-p}(k-p+c)!p!}{(k+c+1)!}$$

and the identity

$$\sum_{p+q=k}\left\{\frac{p!c!}{(p+c+1)!}+(-1)^q\frac{(k-p+c)!p!}{(k+c+1)!}\right\}$$

$$\times\sum_{|\alpha|=q+r+3}\zeta(\{1\}^p,\alpha_0,\ldots,\alpha_q,\alpha_{q+1}+1)$$

$$=\left\{\frac{(k+2)!(c-1)!}{(k+c+2)!}+\frac{(-1)^{k+1}}{k+c+1}-\frac{1}{c}\right\}\zeta(\{1\}^{k+1},r+3)$$

$$+\sum_{j=0}^{k}\sum_{\ell=0}^{r}\frac{(-1)^{j+\ell}}{j+c+1}\zeta(\{1\}^j,r-\ell+2)\zeta(\{1\}^{k-j},\ell+2)$$

when r is even.

4. For any positive real number d, prove that

$$\sum_{j=0}^{k}\binom{p}{j}\frac{(-1)^j}{jd+1}=\frac{p!\,\Gamma(1/d)}{d\,\Gamma(p+1+1/d)},$$

$$\sum_{j=0}^{k}\binom{p}{k-j}\frac{(-1)^j}{jd+1}=\frac{(-1)^{k-p}p!\,\Gamma(k-p+1/d)}{d\,\Gamma(k+1+1/d)},$$

and the identity

$$\sum_{p+q=k} \left\{ \frac{p!\,\Gamma\,(1/d)}{d\,\Gamma\,(p+1+1/d)} + \frac{(-1)^{q+r}p!\,\Gamma\,(k-p+1/d)}{d\,\Gamma\,(k+1+1/d)} \right\}$$

$$\times \sum_{|\alpha|=q+r+3} \zeta(\{1\}^p, \alpha_0, \ldots, \alpha_q, \alpha_{q+1}+1)$$

$$= \left\{ \sum_{j=1}^{k+1} \binom{k+1}{j} \frac{(-1)^j}{d(j-1)+1} - \frac{(k+1)!\,\Gamma\,(1/d)}{d\,\Gamma\,(k+2+1/d)} + \frac{(-1)^{k+1}}{d(k+1)+1} \right\}$$

$$\times \zeta(\{1\}^{k+1}, r+3)$$

$$+ \sum_{j=0}^{k} \sum_{\ell=0}^{r} \frac{(-1)^{j+\ell}}{jd+1} \zeta(\{1\}^j, r-\ell+2)\zeta(\{1\}^{k-j}, \ell+2).$$

5. Prove the counterpart of Theorem 1.4.11 when k is odd.

6. Prove the counterpart of Theorem 1.4.12 when r is odd.

CHAPTER 2

The Sum Formula

The sum formula asserted that the sum of multiple zeta values of the same depth r and the weight w is equal to $\zeta(w)$.

Through shuffle products of multiple zeta values of height one, we are able to express a sum as a linear combination of products of multiple zeta values of height one so that the sum formula is closely related to the evaluations of multiple zeta values of height one.

2.1 Through the Integral Representations

The sum formula

$$\sum_{|\boldsymbol{\alpha}|=m} \zeta(\alpha_1, \alpha_2, \ldots, \alpha_{r-1}, \alpha_r + 1) = \zeta(m+1)$$

originally conjectured by C. Moen and M. Schmidt independently around 1994, was proved for $r = 2$ by Euler long time ago and for $r = 3$ by Hoffman and Moen in 1996. A. Granville proved the general cases in 1997 and he mentioned that it was proved independently by Zagier in one of his unpublished papers. Zagier had also made this remark: Although this proof is not very long, it seems too complicated compared with elegance of the statement.

Here we adopted the proof from Eie and Wei's paper: A short proof for the sum formula and its generalization [15].

Proof of the restricted sum formula. First we express the sum of multiple zeta values as a double integral

$$\sum_{|\boldsymbol{\alpha}|=m} \zeta(\{1\}^p, \alpha_1, \alpha_2, \ldots, \alpha_q + 1)$$

$$= \frac{1}{p!(q-1)!(m-q)!}$$

$$\times \int_{0<t_1<t_2<1} \left(\log\frac{1}{1-t_1}\right)^p \left(\log\frac{1-t_1}{1-t_2}\right)^{q-1} \left(\log\frac{t_2}{t_1}\right)^{m-q} \frac{dt_1 dt_2}{(1-t_1)t_2}$$

by Proposition 1.2.2. Now make a change of variables in the integral with new variables

$$x_1 = \log\frac{1}{1-t_1} \quad \text{and} \quad x_2 = \log\frac{t_2}{t_1}.$$

Then we have

$$dx_1 dx_2 = \frac{dt_1}{(1-t_1)} \frac{dt_2}{t_2},$$

the factor $\log\frac{1-t_1}{1-t_2}$ is transformed into $\log\frac{1}{e^{x_1}+e^{x_2}-e^{x_1+x_2}}$ and the simplex $0 < t_1 < t_2 < 1$ is transformed into the domain

$$D : x_1 > 0, \ x_2 > 0 \quad \text{and} \quad e^{x_1} + e^{x_2} > e^{x_1+x_2}.$$

Consequently, the integral is equal to

$$\frac{1}{p!(q-1)!(m-q)!} \int_D x_1{}^p x_2{}^{m-q} \left(\log\frac{1}{e^{x_1}+e^{x_2}-e^{x_1+x_2}}\right)^{q-1} dx_1 dx_2.$$

Next, we make another change of variables so that D is transformed back into the simplex defined by $0 < u_1 < u_2 < 1$:

$$x_1 = \log\frac{1-u_1}{1-u_2} \quad \text{and} \quad x_2 = \log\frac{1}{u_2}.$$

Then a direct calculation leads to

$$\log\frac{1}{e^{x_1}+e^{x_2}-e^{x_1+x_2}} = \log\frac{u_2}{u_1}, \ dx_1 dx_2 = \frac{du_1 du_2}{(1-u_1)u_2}$$

and the integral over D is transformed into

$$\frac{1}{p!(q-1)!(m-q)!} \int_{0<u_1<u_2<1} \left(\log\frac{1-u_1}{1-u_2}\right)^p \left(\log\frac{u_2}{u_1}\right)^{q-1}$$
$$\times \left(\log\frac{1}{u_2}\right)^{m-q} \frac{du_1 du_2}{(1-u_1)u_2}.$$

In terms of multiple zeta values, it is

$$\sum_{|c|=p+q} \zeta(c_1,\ldots,c_p,c_{p+1}+m-q+1).$$

$\square$

Here we mention an equivalence of the sum formula. As the sum

$$\sum_{|\alpha|=m+n+1} \zeta(\alpha_0,\alpha_1,\ldots,\alpha_m+1)$$

has the integral representation

$$\frac{1}{m!n!} \int_{0<t_1<t_2<1} \left(\log\frac{1-t_1}{1-t_2}\right)^m \left(\log\frac{t_2}{t_1}\right)^n \frac{dt_1 dt_2}{(1-t_1)t_2},$$

so that we have for sufficient small $|x|$ and $|y|$,

$$\sum_{m=0}^{\infty}\sum_{n=0}^{\infty} \sum_{|\alpha|=m+n+1} \zeta(\alpha_0,\alpha_1,\ldots,\alpha_m+1)x^m y^n$$
$$= \int_{0<t_1<t_2<1} \left(\frac{1-t_1}{1-t_2}\right)^x \left(\frac{t_2}{t_1}\right)^y \frac{dt_1 dt_2}{(1-t_1)t_2}.$$

Under two consecutive transforms as given before, the above integral is transformed into

$$\int_{0<u_1<u_2<1} \left(\frac{u_2}{u_1}\right)^x \left(\frac{1}{u_2}\right)^y \frac{du_1 du_2}{(1-u_1)u_2},$$

which is equal to

$$\sum_{k=1}^{\infty} \frac{1}{(k-x)(k-y)}.$$

The power series expansion

$$\frac{1}{(k-x)(k-y)} = \sum_{m=0}^{\infty}\sum_{n=0}^{\infty} k^{-(m+n+2)}x^m y^n,$$

then leads to the identity

$$\sum_{m=0}^{\infty}\sum_{n=0}^{\infty}\sum_{|\boldsymbol{\alpha}|=m+n+1}\zeta(\alpha_0,\alpha_1,\ldots,\alpha_m+1)x^my^n = \sum_{m=0}^{\infty}\sum_{n=0}^{\infty}\zeta(m+n+2)x^my^n.$$

2.2   Another Proof of the Sum Formula

In the following, we shall provide a proof of the sum formula in an elementary way. Especially, we discard the usage of Drinfeld integral representations of multiple zeta values. First we need the following proposition to express the sum as r particular multiple zeta values of special form.

Proposition 2.2.1. *Let $a_1, a_2, \ldots, a_r$ be r different complex numbers. Then for any nonnegative integer m, we have*

$$\sum_{|\boldsymbol{\alpha}|=m} a_1^{\alpha_1} a_2^{\alpha_2} \cdots a_r^{\alpha_r} = \sum_{j=1}^{r} \frac{a_j^{m+r-1}}{\prod_{p\neq j}(a_j - a_p)}.$$

Proof. We prove the assertion by induction on r. For $r = 2$, we have

$$\sum_{|\boldsymbol{\alpha}|=m} a_1^{\alpha_1} a_2^{\alpha_2} = a_1^m + a_1^{m-1}a_2 + \cdots + a_2^m$$

$$= \frac{a_1^{m+1} - a_2^{m+1}}{a_1 - a_2}$$

$$= \frac{a_1^{m+1}}{a_1 - a_2} + \frac{a_2^{m+1}}{a_2 - a_1}.$$

Therefore, the assertion is valid for $r = 2$.

Suppose that it is valid for $r = k \geq 2$. Then for $r = k+1$,

$$\sum_{|\boldsymbol{\alpha}|=m} a_1^{\alpha_1} a_2^{\alpha_2} \cdots a_k^{\alpha_k} a_{k+1}^{\alpha_{k+1}} = \sum_{j=0}^{m} a_{k+1}^j \sum_{|\boldsymbol{\alpha}|=m-j} a_1^{\alpha_1} a_2^{\alpha_2} \cdots a_k^{\alpha_k}$$

$$= \sum_{j=0}^{m} a_{k+1}^j \sum_{\ell=1}^{k} \frac{a_\ell^{m+k-j-1}}{\prod_{\substack{p\neq\ell \\ 1\leq p\leq k}}(a_\ell - a_p)}.$$

According to the case $r = 2$,

$$\sum_{j=0}^{m} a_{k+1}^j a_\ell^{m+k-j-1} = \frac{a_\ell^{m+k} - a_{k+1}^{m+k}}{a_\ell - a_{k+1}}$$

and the sum is rewritten as

$$\sum_{\ell=1}^{k} \frac{a_\ell^{m+k}}{\displaystyle\prod_{\substack{p\neq\ell\\1\leq p\leq k+1}} (a_\ell - a_p)} + \sum_{\ell=1}^{k} \frac{a_{k+1}^{m+k}}{(a_{k+1} - a_\ell) \displaystyle\prod_{\substack{p\neq\ell\\1\leq p\leq k}} (a_\ell - a_p)}.$$

It remains to prove the second term is equal to

$$\frac{a_{k+1}^{m+k}}{\displaystyle\prod_{1\leq p\leq k} (a_{k+1} - a_p)}$$

or more precisely

$$\sum_{\ell=1}^{k} \left((a_{k+1} - a_\ell) \prod_{\substack{p\neq\ell\\1\leq p\leq k}} (a_\ell - a_p) \right)^{-1} = \left(\prod_{1\leq p\leq k} (a_{k+1} - a_p) \right)^{-1}.$$

Let

$$f(x) = \sum_{\ell=1}^{k} \prod_{\substack{p\neq\ell\\1\leq p\leq k}} \frac{x - a_p}{a_\ell - a_p}.$$

As a polynomial function, the degree of $f(x)$ is less than or equal to $k-1$, but

$$f(a_1) = f(a_2) = \cdots = f(a_k) = 1.$$

So we conclude that $f(x) = 1$ and hence $f(a_{k+1}) = 1$. This is precisely what we want and hence $r = k + 1$ is valid. □

Euler sums with two branches first appeared in the proof of the sum formula. For a pair of integers $p, q \geq 0$ and positive integer $n \geq 2$, the Euler sum with two branches is defined as

$$G_n(p, q) = \sum_{k_0=1}^{\infty} \frac{1}{k_0^n} \sum_{\mathbf{k}} \frac{1}{k_1 k_2 \cdots k_q} \sum_{\boldsymbol{\ell}} \frac{1}{\ell_1 \ell_2 \cdots \ell_p},$$

where $\mathbf{k} = (k_1, k_2, \ldots, k_q)$ ranges over all positive integers with

$$1 \leq k_q \leq k_{q-1} \leq \cdots \leq k_1 \leq k_0$$

and $\boldsymbol{\ell} = (\ell_1, \ell_2, \ldots, \ell_p)$ ranges over all positive integers with

$$1 \leq \ell_p < \ell_{p-1} < \cdots < \ell_1 < k_0.$$

When $q = 0$, it is just the multiple zeta value $\zeta(\{1\}^p, n)$. Also when $p = 0$, $G_n(0, q)$ has the decomposition

$$\sum_{r=1}^{q+1} \sum_{|\alpha|=q+1} \zeta(\alpha_1, \alpha_2, \ldots, \alpha_r + n - 1).$$

Some important properties of $G_n(p, q)$ are needed in our proof of the sum formula, so we list them here.

1. Drinfeld integral representation of $G_n(p, q)$: $G_n(p, q)$ has another expression in free dummy variables as

$$\sum_{k \in \mathbb{N}^{p+1}} \sum_{j \in \mathbb{N}^q} \frac{1}{\sigma_1 \sigma_2 \cdots \sigma_p \sigma_{p+1}^{n-1} \tau_1 \cdots \tau_q(\sigma_{p+1} + \tau_q)}$$

   with $\sigma_j = k_1 + \cdots + k_j$ and $\tau_\ell = j_1 + \cdots + j_\ell$. Then Drinfeld iterated integral representation of $G_n(p, q)$ is given by

$$\int_D \prod_{j=1}^{p+1} \frac{dt_j}{1 - t_j} \prod_{k=p+2}^{n+p-1} \frac{dt_k}{t_k} \prod_{\ell=1}^{q} \frac{du_\ell}{1 - u_\ell} \frac{dt_{p+n}}{t_{p+n}},$$

   where D is a region of $\mathbb{R}^{p+q+n}$ defined by

$$0 < t_1 < t_2 < \cdots < t_{p+n} < 1, \quad 0 < u_1 < u_2 < \cdots < u_q < t_{p+n}.$$

   Fixing t_{p+1} and t_{p+n} as t_1 and t_2 and then integrating with respect to the remaining variables, we obtain another integral representation of $G_n(p, q)$ as

$$\frac{1}{p!q!(n-2)!} \int_{0 < t_1 < t_2 < 1} \left(\log \frac{1}{1 - t_1} \right)^p \left(\log \frac{1}{1 - t_2} \right)^q$$
$$\times \left(\log \frac{t_2}{t_1} \right)^{n-2} \frac{dt_1 dt_2}{(1 - t_1)t_2}.$$

2. Evaluation of $G_n(p, q)$: Consider the double integral

$$\frac{1}{p!q!n!} \int_0^1 \int_0^1 \left(\log \frac{1}{1 - t} \right)^p \left(\log \frac{1}{1 - u} \right)^q \left(\log \frac{u}{t} \right)^n \frac{dt\, du}{t\, u}$$

   with p, q, n are positive integers. Such an integral is separable and its value is

$$\sum_{r=0}^{n} (-1)^r \zeta(\{1\}^{p-1}, n - r + 2) \zeta(\{1\}^{q-1}, r + 2).$$

Divide the region of integration $[0,1] \times [0,1]$ into two simplices $D_1 : 0 < t < u < 1$ and $D_2 : 0 < u < t < 1$. On the simplex D_1, we rewrite the integral as

$$\frac{1}{(p-1)!q!(n+1)!} \int_{D_1} \left(\log \frac{1}{1-t} \right)^{p-1} \left(\log \frac{1}{1-u} \right)^{q}$$
$$\times \left(\log \frac{u}{t} \right)^{n+1} \frac{dt}{1-t} \frac{du}{u}$$

so that it is equal to $G_{n+3}(p-1,q)$.

On the other hand, the value of the integration over D_2 is $(-1)^n G_{n+3}(q-1,p)$. Consequently, we conclude that

$$G_{n+3}(p-1,q) + (-1)^n G_{n+3}(q-1,p)$$
$$= \sum_{r=0}^{n} (-1)^r \zeta(\{1\}^{p-1}, n-r+2)\zeta(\{1\}^{q-1}, r+2).$$

For practicality, we rewrite the above result as follows.

Proposition 2.2.2. *For all positive integers m, n and p with $m \geq 2$ and $1 \leq p \leq m-1$, we have*

$$G_{2n}(p, m-p) - G_{2n}(m-1-p, p+1)$$
$$= \sum_{\alpha=2}^{2n-1} (-1)^{\alpha+1} \zeta(\{1\}^p, 2n+1-\alpha)\zeta(\{1\}^{m-p-1}, \alpha)$$

and

$$G_{2n+1}(p, m-p) + G_{2n+1}(m-1-p, p+1)$$
$$= \sum_{\alpha=2}^{2n} (-1)^{\alpha} \zeta(\{1\}^p, 2n+2-\alpha)\zeta(\{1\}^{m-p-1}, \alpha).$$

3. Decomposition of $G_n(p,q)$: $G_n(p,q)$ has the following decomposition

$$G_n(p,q) = \sum_{r=p+1}^{p+q+1} \binom{r-1}{p} \sum_{|\alpha|=p+q+1} \zeta(\alpha_1, \ldots, \alpha_r + n - 1).$$

Here is a proof. First we express $G_n(p,q)$ as a double integral

$$\frac{1}{p!q!(n-2)!}\int_{0<t_1<t_2<1}\left(\log\frac{1}{1-t_1}\right)^p\left(\log\frac{1}{1-t_2}\right)^q$$
$$\times\left(\log\frac{t_2}{t_1}\right)^{n-2}\frac{dt_1dt_2}{(1-t_1)t_2}.$$

Next, replace the factor $[\log 1/(1-t_2)]^q$ by its binomial expression as

$$\sum_{b=0}^{q}\binom{q}{b}\left(\log\frac{1}{1-t_1}\right)^{q-b}\left(\log\frac{1-t_1}{1-t_2}\right)^b.$$

The resulting integral represents the sum of multiple zeta values

$$\sum_{b=0}^{q}\binom{p+q-b}{p}\sum_{|\alpha|=b+n-1}\zeta(\{1\}^{p+q-b},\alpha_0,\alpha_1,\ldots,\alpha_b+1).$$

Now make a change of variables

$$u_1=1-t_2\quad\text{and}\quad u_2=1-t_1,$$

so that the integral is transformed into

$$\sum_{b=0}^{q}\frac{1}{p!b!(q-b)!(n-2)!}\int_{0<u_1<u_2<1}\left(\log\frac{1-u_1}{1-u_2}\right)^{n-2}\left(\log\frac{u_2}{u_1}\right)^b$$
$$\times\left(\log\frac{1}{u_2}\right)^{p+q-b}\frac{du_1du_2}{(1-u_1)u_2}.$$

This is the integral expression of the sum of multiple zeta values

$$\sum_{b=0}^{q}\binom{p+q-b}{p}\sum_{|\alpha|=b+n-1}\zeta(\alpha_0,\alpha_1,\ldots,\alpha_{n-2}+p+q-b+1).$$

Finally, we apply Ohno's generalization of the duality theorem and sum formula with $\mathbf{k}=(\{1\}^{n-2},p+q-b+2)$ and $\mathbf{k}'=(\{1\}^{p+q-b},n)$ to transform the above sum into

$$\sum_{r=p+1}^{p+q+1}\binom{r-1}{p}\sum_{|\alpha|=p+q+1}\zeta(\alpha_1,\alpha_2,\ldots,\alpha_r+n-1).$$

4. The decomposition of $G_n(p,q)$ along with pairwise evaluations, as shown in 2, can be used to evaluate multiple zeta values of the form $\zeta(\{1\}^m, n+2)$. Here we evaluate $\zeta(1,1,2n)$ and $\zeta(1,1,2n+1)$ as examples. First we have the decompositions

$$G_{2n}(0,2) = \zeta(1,1,2n) + \{\zeta(2,2n) + \zeta(1,2n+1)\} + \zeta(2n+2),$$
$$G_{2n}(1,1) = 2\zeta(1,1,2n) + \{\zeta(2,2n) + \zeta(1,2n+1)\}.$$

The difference then gives

$$G_{2n}(1,1) - G_{2n}(0,2) = \zeta(1,1,2n) - \zeta(2n+2).$$

But on the other hand, we have

$$G_{2n}(1,1) - G_{2n}(0,2) = \sum_{\alpha=2}^{2n-1} (-1)^{\alpha+1}\zeta(1,\alpha)\zeta(2n+1-\alpha).$$

It follows that

$$\zeta(1,1,2n) = \zeta(2n+2) - \sum_{\alpha=2}^{2n-1} (-1)^{\alpha}\zeta(1,\alpha)\zeta(2n+1-\alpha).$$

In the same manner, we obtain

$$3\zeta(1,1,2n+1) = -\{\zeta(2n+3) + 2\zeta(1,2n+2) + 2\zeta(2,2n+1)\}$$
$$+ \sum_{\alpha=2}^{2n}(-1)^{\alpha}\zeta(1,\alpha)\zeta(2n+2-\alpha).$$

Now we go back to the proof of the sum formula.

Proof of the sum formula. Let

$$a_1 = \frac{1}{k_1}, \; a_2 = \frac{1}{k_1+k_2}, \; \ldots, \; a_r = \frac{1}{k_1+k_2+\cdots+k_r}.$$

Then

$$\sum_{|\alpha|=m} \zeta(\alpha_1,\alpha_2,\ldots,\alpha_r+1) = \sum_{k\in\mathbb{N}^r} a_1\cdots a_{r-1}a_r^2 \sum_{|\beta|=m-r} a_1^{\beta_1}a_2^{\beta_2}\cdots a_r^{\beta_r}.$$

By Proposition 2.2.1, the inner sum is equal to

$$\sum_{j=1}^{r} \frac{a_j^{m-1}}{\prod_{\ell\neq j}(a_j - a_\ell)}.$$

When $j = r$, the corresponding series is just

$$\sum_{\mathbf{k}\in\mathbb{N}^r} \frac{(-1)^{r-1}}{k_r(k_r + k_{r-1})\cdots(k_r + k_{r-1} + \cdots + k_2)(k_1 + k_2 + \cdots + k_r)^{m-r+2}}$$

which is $(-1)^{r-1}\zeta(\{1\}^{r-1}, m - r + 2)$ or $(-1)^{r-1}G_{m-r+2}(r - 1, 0)$.

On the other hand, when $j = 1$, the corresponding series is

$$\sum_{\mathbf{k}\in\mathbb{N}^r} \frac{1}{k_1^{m-r+1}k_2(k_2 + k_3)\cdots(k_2 + k_3 + \cdots + k_r)(k_1 + k_2 + \cdots + k_r)}$$

which is $G_{m-r+2}(0, r - 1)$. In general, when $1 < j < r$, the corresponding series is

$$\sum_{\mathbf{k}\in\mathbb{N}^r} \frac{(-1)^{j-1}}{B(\mathbf{k})(k_1 + k_2 + \cdots + k_j)^{m-r+1}C(\mathbf{k})(k_1 + k_2 + \cdots + k_r)}$$

with

$$B(\mathbf{k}) = \frac{(-1)^{j-1}\prod_{\ell<j}(a_j - a_\ell)}{a_1\cdots a_{j-1}a_j} = k_j(k_j + k_{j-1})\cdots(k_j + k_{j-1} + \cdots + k_2)$$

and

$$C(\mathbf{k}) = \frac{\prod_{\ell>j}(a_j - a_\ell)}{a_j a_{j+1}\cdots a_r} = k_{j+1}(k_{j+1} + k_{j+2})\cdots(k_{j+1} + \cdots + k_r),$$

so it is equal to $(-1)^{j+1}G_{m-r+2}(j - 1, r - j)$. Consequently, we conclude that

$$\sum_{|\boldsymbol{\alpha}|=m} \zeta(\alpha_1, \alpha_2, \ldots, \alpha_r + 1) = \sum_{j=1}^{r}(-1)^{j-1}G_{m-r+2}(j - 1, r - j).$$

Now substituting $G_{m-r+2}(j - 1, r - j)$ by its decomposition

$$\sum_{k=j}^{r}\binom{k-1}{j-1}\sum_{|\boldsymbol{\alpha}|=r}\zeta(\alpha_1, \alpha_2, \ldots, \alpha_k + m - r + 1)$$

and exchanging the order of summations, we get that

$$\sum_{|\boldsymbol{\alpha}|=m} \zeta(\alpha_1, \alpha_2, \ldots, \alpha_r + 1)$$

$$= \sum_{k=1}^{r}\sum_{j=1}^{k}(-1)^{j-1}\binom{k-1}{j-1}\sum_{|\boldsymbol{\alpha}|=r}\zeta(\alpha_1, \alpha_2, \ldots, \alpha_k + m - r + 1).$$

Note that the second alternating sum

$$\sum_{j=1}^{k}(-1)^{j-1}\binom{k-1}{j-1}$$

is zero unless $k = 1$. It follows that the sum is

$$\sum_{|\alpha|=r} \zeta(\alpha + m - r + 1) \quad \text{or} \quad \zeta(m+1).$$

$\square$

2.3 Evaluation of Multiple Zeta Values of Height One

The height of a multiple zeta values $\zeta(\alpha_1, \alpha_2, \ldots, \alpha_r)$ is defined as the number of elements is the set $\{\alpha_j \mid \alpha_j \geq 2, \ 1 \leq j \leq r\}$. Therefore, multiple zeta values of height one appear to be the form $\zeta(\{1\}^m, n+2)$. Such kind of multiple zeta values is of special interest owing to their simplicity in their integral representations. Also they can be expressed as integrals with just one variable so that a generating function is obtained in a elementary way.

Lemma 2.3.1. *For a pair of nonnegative integers m and n, we have*

$$\zeta(\{1\}^m, n+2) = \frac{1}{m!(n+1)!} \int_0^1 \left(\log\frac{1}{1-t}\right)^m \left(\log\frac{1}{t}\right)^{n+1} \frac{dt}{1-t}.$$

Proof. As a first step, we express $\zeta(\{1\}^m, n+2)$ as a Drinfeld iterated integral as

$$I = \int_{0<t_1<\cdots<t_{m+n+2}<1} \prod_{j=1}^{m+1} \frac{dt_j}{1-t_j} \prod_{k=m+2}^{m+n+2} \frac{dt_k}{t_k}.$$

Fix t_{m+1} as t. For any permutation σ of the set $\{1, 2, \ldots, m\}$ the integral

$$\int_{0<t_{\sigma(1)}<\cdots<t_{\sigma(m)}<t<\cdots<t_{m+n+2}<1} \prod_{j=1}^{m+1} \frac{dt_j}{1-t_j} \prod_{k=m+2}^{m+n+2} \frac{dt_k}{t_k}$$

has the same value $\zeta(\{1\}^m, n+2)$. Note that

$$[0,t]^m = \bigcup_{\sigma} \{(t_1, t_2, \ldots, t_m) \mid 0 < t_{\sigma(1)} < \cdots < t_{\sigma(m)} < t\}$$

and hence

$$I = \frac{1}{m!} \int_{0<t<t_{m+2}<\cdots<t_{m+n+2}<1} \left\{ \prod_{j=1}^{m} \int_{0}^{t} \frac{dt_j}{1-t_j} \right\} \frac{dt}{1-t} \prod_{k=m+2}^{m+n+2} \frac{dt_k}{t_k}$$

$$= \frac{1}{m!} \int_{0<t<t_{m+2}<\cdots<t_{m+n+2}<1} \left(\log \frac{1}{1-t} \right)^{m} \frac{dt}{1-t} \prod_{k=m+2}^{m+n+2} \frac{dt_k}{t_k}.$$

In the similar way, we integrate with respect to t_k $(k = m+2, \ldots, m+n+2)$ from t to 1 then divided by $(n+1)!$ and get our assertion. $\qquad\square$

Theorem 2.3.2. *The generating function of $\zeta(\{1\}^{m}, n+2)$ is given by*

$$\sum_{m=0}^{\infty} \sum_{n=0}^{\infty} \zeta(\{1\}^{m}, n+2) x^{m+1} y^{n+1}$$

$$= 1 - \frac{\Gamma(1-x)\Gamma(1-y)}{\Gamma(1-x-y)}$$

$$= 1 - \exp \left\{ \sum_{k=2}^{\infty} \left(x^k + y^k - (x+y)^k \right) \frac{\zeta(k)}{k} \right\}.$$

Proof. For the time being, we assume that $x < 0$. By the Lemma 2.3.1, we have

$$\sum_{m=0}^{\infty} \sum_{n=0}^{\infty} \zeta(\{1\}^{m}, n+2) x^m y^{n+1}$$

$$= \sum_{m=0}^{\infty} \sum_{n=0}^{\infty} \frac{x^m y^{n+1}}{m!(n+1)!} \int_{0}^{1} \left(\log \frac{1}{1-t} \right)^{m} \left(\log \frac{1}{t} \right)^{n+1} \frac{dt}{1-t}$$

$$= \int_{0}^{1} (1-t)^{-x-1} (t^{-y} - 1) dt.$$

Under the condition $x < 0$, the above integral can be separated into two convergent improper integrals and the value is

$$\frac{1}{x} + \frac{\Gamma(-x)\Gamma(1-y)}{\Gamma(1-x-y)}.$$

Multiplying both sides by x and employing the functional equation

$$(-x)\Gamma(-x) = \Gamma(1-x),$$

we conclude that the generating function is

$$1 - \frac{\Gamma(1-x)\Gamma(1-y)}{\Gamma(1-x-y)}.$$

The condition $x < 0$ can be removed since both sides are analytic functions of x and y when $|x|$ and $|y|$ are sufficiently small.

Finally, we use the infinite product of gamma function

$$\frac{1}{\Gamma(1+s)} = e^{\gamma s} \prod_{n=1}^{\infty} (1 + \frac{s}{n}) e^{-s/n},$$

where γ is Euler's constant defined by

$$\gamma = \lim_{n \to \infty} \left(\sum_{k=1}^{n} \frac{1}{k} - \log n \right),$$

to change the quotient of gamma functions

$$\frac{\Gamma(1-x)\Gamma(1-y)}{\Gamma(1-x-y)}$$

into

$$\exp \left\{ \sum_{k=2}^{\infty} \left(x^k + y^k - (x+y)^k \right) \frac{\zeta(k)}{k} \right\}. \qquad \square$$

As the generating function is invariant under the interexchange of x and y, this implies the Drinfeld dual theorem:

$$\zeta(\{1\}^m, n+2) = \zeta(\{1\}^n, m+2).$$

Besides, $\zeta(\{1\}^m, n+2)$ can be evaluated in terms of single zeta values. However, it is difficult to do so directly.

Next, we shall express $\zeta(\{1\}^m, n+2)$ in terms of multiple zeta values of the same form, but with lower depth. This proves that $\zeta(\{1\}^m, n+2)$ can be expressed in terms of single zeta values recursively.

Theorem 2.3.3. [11] *For a pair of nonnegative integers k and r with $k+r$ even, then*

$$\zeta(\{1\}^{k+1}, r+3) = (-1)^{k+1}\zeta(k+r+4) + \frac{(-1)^k}{2}$$

$$\times \sum_{j=0}^{k} \sum_{\ell=0}^{r} (-1)^{j+\ell}\zeta(\{1\}^j, r-\ell+2)\zeta(\{1\}^{k-j}, \ell+2).$$

Proof. We begin with the shuffle product formula in (1.2.5). Multiply both sides of (1.2.5) by $(-1)^{j+\ell}$ and then sum over all j, ℓ with $0 \leq j \leq k$, $0 \leq \ell \leq r$. A direct calculation yields

$$\sum_{j=0}^{k} \binom{p}{j} (-1)^j = \begin{cases} 1, & \text{if } p = 0; \\ 0, & \text{if } p \geq 1, \end{cases}$$

and

$$\sum_{j=0}^{k} \binom{p}{k-j} (-1)^j = \begin{cases} (-1)^k, & \text{if } p = 0; \\ 0, & \text{if } p \geq 1. \end{cases}$$

On the other hand, we have for $1 \leq \alpha \leq r+2$

$$\sum_{\ell=0}^{r} \binom{\alpha}{\ell+1} (-1)^\ell = \begin{cases} 1, & \text{if } \alpha \leq r+1; \\ 1 + (-1)^r, & \text{if } \alpha = r+2, \end{cases}$$

and

$$\sum_{\ell=0}^{r} \binom{\alpha}{r-\ell+1} (-1)^\ell = \begin{cases} (-1)^r, & \text{if } \alpha \leq r+1; \\ 1 + (-1)^r, & \text{if } \alpha = r+2. \end{cases}$$

This leads to the relation

$$\{1 + (-1)^{k+r}\} \sum_{|\boldsymbol{\alpha}|=k+r+3} \zeta(\alpha_0, \ldots, \alpha_k, \alpha_{k+1} + 1)$$

$$+ \{(-1)^k + (-1)^r\} \zeta(\{1\}^{k+1}, r+3)$$

$$= \sum_{j=0}^{k} \sum_{\ell=0}^{r} (-1)^{j+\ell} \zeta(\{1\}^j, r-\ell+2) \zeta(\{1\}^{k-j}, \ell+2).$$

Our assertion then follows from the sum formula

$$\sum_{|\boldsymbol{\alpha}|=k+r+3} \zeta(\alpha_0, \ldots, \alpha_k, \alpha_{k+1} + 1) = \zeta(k+r+4). \qquad \square$$

Of course, the cases when $k+r$ is odd cannot be excluded here. In this cases, we multiply both sides of the identity (1.2.1) by $(-1)^{j+\ell}(j+1)$ and then sum over all j, ℓ with $0 \leq j \leq k$ and $0 \leq \ell \leq r$. For this, we need the following lemma.

Lemma 2.3.4. *For nonnegative integers k, p with $0 \leq p \leq k$, we have*

$$\sum_{j=0}^{k} \binom{p}{j}(-1)^j(j+1) = \begin{cases} 1, & \text{if } p = 0; \\ -1, & \text{if } p = 1; \\ 0, & \text{if } p \geq 2, \end{cases}$$

and

$$\sum_{j=0}^{k} \binom{p}{k-j}(-1)^j(j+1) = \begin{cases} (-1)^k(k+1), & \text{if } p = 0; \\ (-1)^k, & \text{if } p = 1; \\ 0, & \text{if } p \geq 2. \end{cases}$$

Proof. Observe the first sum is equal to

$$\frac{d}{dx} x(1-x)^p \bigg|_{x=1},$$

and the second sum can be separated into

$$(-1)^k(k+2)\sum_{j=0}^{k} \binom{p}{j}(-1)^j - (-1)^k \sum_{j=0}^{k} \binom{p}{j}(-1)^j(j+1).$$

Our assertion then follows from direct computations. $\qquad\square$

Theorem 2.3.5. *Suppose that k and r are a pair of nonnegative integers with $k + r$ odd. Then*

$$(k+2)\zeta(\{1\}^{k+1}, r+3)$$
$$= (-1)^k k\zeta(k+r+4) + (-1)^k 2 \sum_{|c|=k+2} \zeta(c_1, c_2 + r + 2)$$
$$+ (-1)^k \sum_{j=0}^{k}\sum_{\ell=0}^{r}(-1)^{j+\ell}(j+1)\zeta(\{1\}^j, r-\ell+2)\zeta(\{1\}^{k-j}, \ell+2).$$

Proof. Again we begin with the shuffle product formula as before. Multiply both sides by $(-1)^{j+\ell}(j+1)$ and then sum over all j and ℓ with $0 \leq j \leq k$

and $0 \le \ell \le r$. By Lemma 2.3.4, we get the shuffle relation

$$\{1 + (-1)^{k+r}(k+1)\} \sum_{|\boldsymbol{\alpha}|=k+r+3} \zeta(\alpha_0, \ldots, \alpha_k, \alpha_{k+1}+1)$$

$$+ \{-1 + (-1)^{k+r}\} \sum_{|\boldsymbol{\beta}|=k+r+2} \zeta(1, \beta_0, \ldots, \beta_{k-1}, \beta_k + 1)$$

$$+ (-1)^k(k+2)\zeta(\{1\}^{k+1}, r+3)$$

$$= \sum_{j=0}^{k} \sum_{\ell=0}^{r} (-1)^{j+\ell}(j+1)\zeta(\{1\}^j, r-\ell+2)\zeta(\{1\}^{k-j}, \ell+2).$$

Our assertion then follows from the sum formula and restricted sum formula

$$\sum_{|\boldsymbol{\beta}|=k+r+2} \zeta(1, \beta_0, \ldots, \beta_{k-1}, \beta_k + 1) = \sum_{|\boldsymbol{c}|=k+2} \zeta(c_1, c_2 + r + 2). \qquad \square$$

2.4   Exercises

1. Prove that for positive integers m, n, r with $m \ge r$

$$\sum_{|\boldsymbol{\alpha}|=m} \zeta(\alpha_1, \ldots, \alpha_r + n) = \sum_{|\boldsymbol{\beta}|=m} \zeta(\{1\}^{n-1}, \beta_1, \ldots, \beta_{m-r+1}+1)$$

$$= \sum_{|\boldsymbol{c}|=m+n-r} \zeta(c_1, c_2, \ldots, c_n + r).$$

2. Give a similar proof as above for the restricted sum formula. In particular, first prove the identity

$$\sum_{|\boldsymbol{\alpha}|=m} \zeta(\{1\}^p, \alpha_1, \alpha_2, \ldots, \alpha_q + 1)$$

$$= \sum_{j=1}^{q} \binom{p+j-1}{p} (-1)^{j-1} G_{m-q+2}(p+j-1, q-j).$$

3. Evaluate the multiple zeta values $\zeta(1,1,1,2n)$ and $\zeta(1,1,1,2n+1)$ for any positive integer n.

4. Prove that

$$\frac{\Gamma(1-x)\Gamma(1-y)}{\Gamma(1-x-y)} = \exp\left\{\sum_{k=2}^{\infty} (x^k + y^k - (x+y)^k)\frac{\zeta(k)}{k}\right\}.$$

$\Bigg($Hint: For $|x| < 1$, we have the power series expansion

$$\log(1 + x) = x - \frac{x^2}{2} + \frac{x^3}{3} - \frac{x^4}{4} + \cdots + \frac{(-1)^{n+1}x^n}{n} + \cdots .\Bigg)$$

5. Prove the following weighted sum formula

$$(-1)^k(r + 2) \sum_{|\alpha|=k+r+3} \zeta(\alpha_0, \alpha_1 \ldots, \alpha_{k+1} + 1)(2^{\alpha_{k+1}} - 1)$$

$$+ \left[(-1)^{k+1} - 1\right] \sum_{|\alpha|=k+r+3} \zeta(\alpha_0, \ldots, \alpha_{k+1} + 1)\alpha_{k+1}(2^{\alpha_{k+1}} - 1)$$

$$= \sum_{j=0}^{k}\sum_{\ell=0}^{r}(-1)^j(\ell + 1)\zeta(\{1\}^j, r - \ell + 2)\zeta(\{1\}^{k-j}, \ell + 2).$$

Part II

Shuffle Relations among Multiple Zeta Values

We are going to begin with some particular integrals over a product of two simplices of dimension two. The integrands of all the integrals are separable and the values of integrals can be expressed as linear combinations of multiple zeta values. Now we decompose the product of simplices into 6 simplices of dimension 4 obtained from all possible interlacings of two sets of two variables. Evaluate the corresponding integrals over each simplex in terms of multiple zeta values, this leads to shuffle relations with linear combinations of products of multiple zeta values of height one on one side and 6 sums of multiple zeta values on the other side.

Even the sum of multiple zeta values of the same weight and the same depth comes from shuffle products described as above. Among other things, we produce a lot of weighted sum formulas never seen before. Also some classical shuffle relations such as Euler decomposition theorem can be re-proved in a simple way.

CHAPTER 3

Some Shuffle Relations

Through shuffle product process of multiple zeta values, we are able to express the product of two multiple zeta values of weight w_1 and w_2 as a sum of $\binom{w_1+w_2}{w_1}$ multiple zeta values of weight $w_1 + w_2$.

Not only a sum of multiple zeta values of the same weight and the same depth can be equal to a single zeta value, but also several weighted sums produced from shuffle product process are multiples of singular zeta values.

3.1 Shuffle Relations of Multiple Zeta Values

Shuffle relations of multiple zeta values such as

(3.1.1)
$$\{E(\mu + 1, \lambda + 1) - E(\mu + 1, 1)\}$$
$$+ \mu^k \lambda^{r+2} \left\{ E\left(\frac{\mu + 1}{\mu}, \frac{\lambda + 1}{\lambda}\right) - E\left(\frac{\mu + 1}{\mu}, 1\right) \right\}$$
$$= \left\{ \frac{1}{\mu}[(\mu + 1)^{k+1} - 1]\lambda^{r+2} + [(\mu + 1)^{k+1} - \mu^{k+1}] \right\} \zeta(\{1\}^{k+1}, r + 3)$$
$$+ \sum_{j=0}^{k} \sum_{\ell=0}^{r} \mu^j \lambda^{\ell+1} \zeta(\{1\}^j, r - \ell + 2)\zeta(\{1\}^{k-j}, \ell + 2)$$

comes more naturally from a usual shuffle process to be demonstrated below. Let R_1 be the simplex $0 < t_1 < t_2 < 1$ and R_2 be the simplex $0 < u_1 < u_2 < 1$. For a pair of nonnegative integers k and r, we consider the integral on $R_1 \times R_2$ defined by

$$\frac{1}{k!r!} \iint_{R_1 \times R_2} \left(\mu \log \frac{1}{1-t_1} + \log \frac{1}{1-u_1} \right)^k \left(\log \frac{1}{t_2} + \lambda \log \frac{1}{u_2} \right)^r$$
$$\times \frac{dt_1 dt_2}{(1-t_1)t_2} \frac{du_1 du_2}{(1-u_1)u_2}.$$

After expanding two factors in the integrand, we see immediately that the integral is separable and the value is

$$(3.1.2) \qquad \sum_{j=0}^{k} \sum_{\ell=0}^{r} \mu^j \lambda^\ell \zeta(\{1\}^j, r-\ell+2) \zeta(\{1\}^{k-j}, \ell+2)$$

by Proposition 1.2.1.

As the replacement of shuffle process, we decompose $R_1 \times R_2$ into 6 simplices D_j ($j = 1, 2, 3, 4, 5, 6$) obtained from all possible interlacings of variables t_1, t_2 and u_1, u_2 as follows:

$$D_1 : 0 < t_1 < t_2 < u_1 < u_2 < 1,$$
$$D_2 : 0 < u_1 < u_2 < t_1 < t_2 < 1,$$
$$D_3 : 0 < t_1 < u_1 < t_2 < u_2 < 1,$$
$$D_4 : 0 < t_1 < u_1 < u_2 < t_2 < 1,$$
$$D_5 : 0 < u_1 < t_1 < u_2 < t_2 < 1 \text{ and}$$
$$D_6 : 0 < u_1 < t_1 < t_2 < u_2 < 1.$$

Note that $R_1 \times R_2$ is the union $\bigcup_{j=1}^{6} D_j$ and boundaries of D_j ($j = 1, 2, 3, 4, 5, 6$) so that the value of the integral is equal to the sum of the integrations over D_j ($j = 1, 2, 3, 4, 5, 6$). Now we proceed to evaluate the integrations over D_j ($j = 1, 2, 3, 4, 5, 6$) one by one.

On the simplex $D_1 : 0 < t_1 < t_2 < u_1 < u_2 < 1$, we need to substitute the factors of the integrand given by their multinomial expansions

$$\left(\mu \log \frac{1}{1 - t_1} + \log \frac{1}{1 - u_1} \right)^k$$

$$= \left((\mu + 1) \log \frac{1}{1 - t_1} + \log \frac{1 - t_1}{1 - t_2} + \log \frac{1 - t_2}{1 - u_1} \right)^k$$

$$= \sum_{a+b+c=k} (\mu + 1)^a \frac{k!}{a!b!c!} \left(\log \frac{1}{1 - t_1} \right)^a \left(\log \frac{1 - t_1}{1 - t_2} \right)^b \left(\log \frac{1 - t_2}{1 - u_1} \right)^c$$

and

$$\left(\log \frac{1}{t_2} + \lambda \log \frac{1}{u_2} \right)^r$$

$$= \left(\log \frac{u_1}{t_2} + \log \frac{u_2}{u_1} + (\lambda + 1) \log \frac{1}{u_2} \right)^r$$

$$= \sum_{\ell=0}^{r} \binom{r}{\ell} \left(\log \frac{u_1}{t_2} \right)^\ell \left(\log \frac{u_2}{u_1} + (\lambda + 1) \log \frac{1}{u_2} \right)^{r-\ell}$$

$$= \sum_{\ell=0}^{r} \binom{r}{\ell} \left(\log \frac{u_1}{t_2} \right)^\ell \sum_{d=0}^{r-\ell} \binom{r-\ell}{d} (\lambda + 1)^d \left(\log \frac{1}{u_2} \right)^d \left(\log \frac{u_2}{u_1} \right)^{r-\ell-d}.$$

In terms of multiple zeta values, the value of integration over D_1 is

$$\sum_{a+b+c=k} (\mu+1)^a \sum_{\ell=0}^{r} \sum_{|\alpha|=c+\ell+1} \zeta(\{1\}^{a+b}, \alpha_0+1, \alpha_1, \ldots, \alpha_c, r-\ell+2) \sum_{d=0}^{r-\ell} (\lambda+1)^d.$$

For fixed a, b, c and ℓ, we have

$$\sum_{|\alpha|=c+\ell+1} \zeta(\{1\}^{a+b}, \alpha_0 + 1, \alpha_1, \ldots, \alpha_c, r - \ell + 2)$$

$$= \sum_{|\alpha|=c+\ell+2} \zeta(\{1\}^{a+b}, \alpha_0, \alpha_1, \ldots, \alpha_c, r - \ell + 2)$$

$$- \sum_{|\alpha|=c+\ell+1} \zeta(\{1\}^{a+b+1}, \alpha_0, \alpha_1, \ldots, \alpha_{c-1}, r - \ell + 2).$$

Therefore, the value of integration over D_1 is rewritten as

$$\sum_{p+q=k} (\mu + 1)^p \sum_{\ell=0}^{r} \frac{1}{\lambda} \left[(\lambda + 1)^{r-\ell+1} - 1 \right]$$

$$\times \sum_{|\alpha|=q+\ell+2} \zeta(\{1\}^p, \alpha_0, \alpha_1, \ldots, \alpha_q, r - \ell + 2).$$

With $r - \ell + 1 = \alpha_{q+1}$ as a new dummy variable and new notations introduced before, the value is rewritten as

(3.1.3)
$$\frac{1}{\lambda} \left\{ E(\mu + 1, \lambda + 1) - E(\mu + 1, 1) \right\}$$
$$- \frac{1}{\lambda \mu} \left\{ (\mu + 1)^{k+1} - 1 \right\} \left\{ (\lambda + 1)^{r+2} - 1 \right\} \zeta(\{1\}^{k+1}, r + 3).$$

On the simplex $D_2 : 0 < u_1 < u_2 < t_1 < t_2 < 1$, this means to exchange the roles of t_1, t_2 and u_1, u_2. This is equivalent to replace the factor

$$\left(\mu \log \frac{1}{1 - t_1} + \log \frac{1}{1 - u_1} \right)^k \left(\log \frac{1}{t_2} + \lambda \log \frac{1}{u_2} \right)^r$$

by the factor

$$\mu^k \lambda^r \left(\frac{1}{\mu} \log \frac{1}{1 - u_1} + \log \frac{1}{1 - t_1} \right)^k \left(\log \frac{1}{u_2} + \frac{1}{\lambda} \log \frac{1}{t_2} \right)^r$$

and then carry out the shuffle process, the result is

$$\sum_{p+q=k} (\mu + 1)^p \mu^q \sum_{\ell=0}^{r} \lambda^\ell \left[(\lambda + 1)^{r-\ell+1} - \lambda^{r-\ell+1} \right]$$
$$\times \sum_{|\boldsymbol{\alpha}|=q+\ell+2} \zeta(\{1\}^p, \alpha_0, \alpha_1, \ldots, \alpha_q, r - \ell + 2)$$

or

(3.1.4)
$$\mu^k \lambda^{r+1} \left\{ E\left(\frac{\mu + 1}{\mu}, \frac{\lambda + 1}{\lambda} \right) - E\left(\frac{\mu + 1}{\mu}, 1 \right) \right\}$$
$$- \left\{ (\mu + 1)^{k+1} - \mu^{k+1} \right\} \frac{1}{\lambda} \left\{ (\lambda + 1)^{r+2} - \lambda^{r+2} \right\} \zeta(\{1\}^{k+1}, r + 3).$$

The integrations over D_3, D_4, D_5 and D_6 only produce multiples of $\zeta(\{1\}^{k+1}, r + 3)$. These values are

$$\frac{1}{\lambda \mu} \left\{ (\mu + 1)^{k+1} - 1 \right\} \left\{ (\lambda + 1)^{r+1} - 1 \right\} \zeta(\{1\}^{k+1}, r + 3),$$

$$\frac{1}{\mu} \left\{ (\mu + 1)^{k+1} - 1 \right\} \left\{ (\lambda + 1)^{r+1} - \lambda^{r+1} \right\} \zeta(\{1\}^{k+1}, r + 3),$$

(3.1.5)
$$\left\{ (\mu + 1)^{k+1} - \mu^{k+1} \right\} \left\{ (\lambda + 1)^{r+1} - \lambda^{r+1} \right\} \zeta(\{1\}^{k+1}, r + 3)$$

and

$$\frac{1}{\lambda} \left\{ (\mu + 1)^{k+1} - \mu^{k+1} \right\} \left\{ (\lambda + 1)^{r+1} - 1 \right\} \zeta(\{1\}^{k+1}, r + 3).$$

Consequently, our assertion follows from the value (3.1.2) is equal to the sum of (3.1.3), (3.1.4) and (3.1.5).

Remark 3.1.1. Multiplying both sides of (3.1.1) by μ and then differentiating with respect to μ, we obtain the following identity

$$\{E(\mu+1,\lambda+1) - E(\mu+1,1)\}$$

$$+ (k+1)\mu^k\lambda^{r+2}\left\{E\left(\frac{\mu+1}{\mu}, \frac{\lambda+1}{\lambda}\right) - E\left(\frac{\mu+1}{\mu}, 1\right)\right\}$$

$$+ \mu\left\{\frac{\partial}{\partial\mu}E(\mu+1,\lambda+1) - \frac{\partial}{\partial\mu}E(\mu+1,1)\right\}$$

$$- \mu^{k-1}\lambda^{r+2}\left\{\frac{\partial}{\partial\mu}E\left(\frac{\mu+1}{\mu}, \frac{\lambda+1}{\lambda}\right) - \frac{\partial}{\partial\mu}E\left(\frac{\mu+1}{\mu}, 1\right)\right\}$$

$$= \left\{(k+1)(\mu+1)^k\lambda^{r+2} + (\mu+1)^k[(k+2)\mu+1] - (k+2)\mu^{k+1}\right\}$$

$$\times \zeta(\{1\}^{k+1}, r+3)$$

$$+ \sum_{j=0}^{k}\sum_{\ell=0}^{r}(j+1)\mu^j\lambda^{\ell+1}\zeta(\{1\}^j, r-\ell+2)\zeta(\{1\}^{k-j}, \ell+2).$$

Some of double weighted sum formula can be obtained from above simply by specifying the values of μ and λ. For instance, letting $\mu = -1$ and $\lambda = 1$, we get that

$$k\sum_{|\boldsymbol{\alpha}|=k+r+3}(2^{\alpha_{k+1}}-1)\zeta(\alpha_0,\ldots,\alpha_k,\alpha_{k+1}+1)$$

$$+ 2\sum_{|\boldsymbol{\alpha}|=k+r+2}(2^{\alpha_k}-1)\zeta(1,\alpha_0,\ldots,\alpha_{k-1},\alpha_k+1)$$

$$= (k+2)\zeta(\{1\}^{k+1}, r+3)$$

$$+ \sum_{j=0}^{k}\sum_{\ell=0}^{r}(-1)^{j+1}(j+1)\zeta(\{1\}^j, r-\ell+2)\zeta(\{1\}^{k-j}, \ell+2)$$

when k is odd. Such kind of weighted sum formula is just the counterpart of the weighted sum formula

$$\sum_{|\boldsymbol{\alpha}|=k+r+3}(2^{\alpha_{k+1}}-1)\zeta(\alpha_0,\ldots,\alpha_k,\alpha_{k+1}+1)$$

$$= \zeta(\{1\}^{k+1}, r+3) + \frac{1}{2}\sum_{j=0}^{k}\sum_{\ell=0}^{r}(-1)^j\zeta(\{1\}^j, r-\ell+2)\zeta(\{1\}^{k-j}, \ell+2)$$

when k is even.

3.2   An Application of Double Weighted Sums

In the following considerations, we let

(3.2.1)
$$E(\lambda) = E(2, \lambda) = \sum_{p+q=k} 2^p \sum_{|\alpha|=q+r+3} \zeta(\{1\}^p, \alpha_0, \dots, \alpha_q, \alpha_{q+1}+1)\lambda^{\alpha_{q+1}}.$$

Setting $\mu = 1$ in (3.1.1), we get immediately that

(3.2.2)
$$E(\lambda+1) + \lambda^{r+2}E\left(\frac{\lambda+1}{\lambda}\right)$$
$$= (1 + \lambda^{r+2})\left\{E(1) + (2^{k+1}-1)\zeta(\{1\}^{k+1}, r+3)\right\}$$
$$+ \sum_{j=0}^{k}\sum_{\ell=0}^{r} \lambda^{\ell+1}\zeta(\{1\}^j, r-\ell+2)\zeta(\{1\}^{k-j}, \ell+2).$$

In an attempt to investigate the shuffle product formula produced from the consideration of the integral

(3.2.3)
$$\frac{1}{k!r!}\iint_{R_1 \times R_2}\left(\log\frac{1}{1-t_1} + \log\frac{1}{1-u_1}\right)^k$$
$$\times \left(\mu\log\frac{t_2}{t_1} + \lambda\log\frac{u_1}{t_2} + \nu\log\frac{u_2}{u_1}\right)^r \frac{dt_1 dt_2}{(1-t_1)t_2}\frac{du_1 du_2}{(1-u_1)u_2},$$

we have to evaluate the value of integration over the simplex $D_1 : 0 < t_1 < t_2 < u_1 < u_2 < 1$. In terms of multiple zeta values, the value is given by

(3.2.4)
$$\sum_{a+b+c=k} 2^a \sum_{m+n+p=r} \mu^m \lambda^n \nu^p$$
$$\times \sum_{\substack{|\alpha|=b+m+1 \\ |\beta|=c+n+1}} \zeta(\{1\}^a, \alpha_0, \dots, \alpha_{b-1}, \alpha_b + \beta_0, \beta_1, \dots, \beta_c, p+2)$$

which will be denoted by $P(\mu, \lambda, \nu)$. Such kind of sums of multiple zeta values are so complicate that it is almost hopeless to express them as weighted sums considered before. But, on the contrary $P(\mu, \lambda, \nu)$ can be expressed as a linear combination of $E(\nu/\mu)$ and $E(\nu/\lambda)$ in a simple way. Indeed, we have the following.

Theorem 3.2.1. *For complex numbers μ, λ, ν with $\mu \neq \lambda$ and $\mu\lambda\nu \neq 0$, we have*

$$P(\mu, \lambda, \nu) = \frac{1}{\nu(\lambda - \mu)} \left\{ \lambda^{r+2} E\left(\frac{\nu}{\lambda}\right) - \mu^{r+2} E\left(\frac{\nu}{\mu}\right) \right\}.$$

We need the following elementary identity which is a special case of the assertion in Proposition 2.2.1.

Lemma 3.2.2. *For different complex numbers μ, λ, ν and integer $k \geq 0$, we have*

$$\sum_{a+b+c=k} \mu^a \lambda^b \nu^c = \frac{\mu^{k+2}}{(\mu - \lambda)(\mu - \nu)} + \frac{\lambda^{k+2}}{(\lambda - \mu)(\lambda - \nu)} + \frac{\nu^{k+2}}{(\nu - \mu)(\nu - \lambda)}.$$

Besides, we need a special case.

Proposition 3.2.3. *For complex number $\lambda \neq 0$ and $\lambda \neq -1$, we have*

$$P(1, \lambda + 1, \lambda + 1) = \frac{1}{\lambda(\lambda + 1)} \left\{ (\lambda + 1)^{r+2} E(1) - E(\lambda + 1) \right\}.$$

Proof. For a pair of integers $k, r \geq 0$, we consider the integral

$$\frac{1}{k! r!} \iint_{R_1 \times R_2} \left(\log \frac{1}{1 - t_1} + \log \frac{1}{1 - u_1} \right)^k$$

$$\times \left(\log \frac{u_2}{t_1} + \lambda \log \frac{u_2}{t_2} \right)^r \frac{dt_1 dt_2}{(1 - t_1) t_2} \frac{du_1 du_2}{(1 - u_1) u_2}.$$

As

$$\left(\log \frac{u_2}{t_1} + \lambda \log \frac{u_2}{t_2} \right)^r$$

$$= \left(\log \frac{1}{t_1} + \lambda \log \frac{1}{t_2} - (\lambda + 1) \log \frac{1}{u_2} \right)^r$$

$$= \left(\log \frac{t_2}{t_1} + (\lambda + 1) \log \frac{1}{t_2} - (\lambda + 1) \log \frac{1}{u_2} \right)^r$$

$$= \sum_{\ell=0}^{r} (-1)^\ell \binom{r}{\ell} (\lambda + 1)^\ell \left(\log \frac{t_2}{t_1} + (\lambda + 1) \log \frac{1}{t_2} \right)^{r-\ell} \left(\log \frac{1}{u_2} \right)^\ell,$$

the value of the integral is

(3.2.5)
$$\frac{1}{\lambda}\sum_{j=0}^{k}\sum_{\ell=0}^{r}(-1)^{\ell}(\lambda+1)^{\ell}\left\{(\lambda+1)^{r-\ell+1}-1\right\}\zeta(\{1\}^{j},r-\ell+2)\zeta(\{1\}^{k-j},\ell+2).$$

As before, we decompose the region of integration $R_1 \times R_2$ into 6 simplices, denoted by D_j ($j = 1, 2, 3, 4, 5, 6$), and then perform integration over each simplex as a replacement of the shuffle process. By definition, the value of integration over D_1 is $P(1, \lambda+1, \lambda+1)$ since

$$\log\frac{u_2}{t_1}+\lambda\log\frac{u_2}{t_2}=\log\frac{t_2}{t_1}+(\lambda+1)\log\frac{u_1}{t_2}+(\lambda+1)\log\frac{u_2}{u_1}.$$

The value of integration over D_2 in terms of multiple zeta values is

$$(-1)^{r}\sum_{a+b+c=k}2^{a}\sum_{\ell=0}^{r}(\lambda+1)^{\ell}\lambda^{r-\ell}$$
$$\times\sum_{|\boldsymbol{\alpha}|=c+\ell+1}\zeta(\{1\}^{a+b},\alpha_0+1,\alpha_1,\ldots,\alpha_c,r-\ell+2)$$

or

$$(-1)^{r}\sum_{p+q=k}2^{p}\sum_{\ell=0}^{r}(\lambda+1)^{\ell}\lambda^{r-\ell}\sum_{|\boldsymbol{\alpha}|=q+\ell+2}\zeta(\{1\}^{p},\alpha_0,\ldots,\alpha_q,r-\ell+2).$$

In new notation, it is

$$(-1)^{r}\frac{(\lambda+1)^{r+1}}{\lambda}E\left(\frac{\lambda}{\lambda+1}\right)-(-1)^{r}\frac{\lambda^{r+1}}{\lambda+1}(2^{k+1}-1)\zeta(\{1\}^{k+1},r+3).$$

Integrations over D_3, D_4, D_5 and D_6 are performed one by one. Let I_j ($j = 3, 4, 5, 6$) be the values of integrations over D_j ($j = 3, 4, 5, 6$). Here are the final evaluations:

$$I_3=\frac{1}{\lambda(\lambda+1)}E(\lambda+1)-\frac{1}{\lambda}E(1),$$

$$I_4=\frac{1}{\lambda(\lambda+1)}E(-\lambda)+\frac{1}{\lambda+1}E(1),$$

$$I_5=\frac{(-1)^{r}\lambda^{r+1}+1}{\lambda+1}(2^{k+1}-1)\zeta(\{1\}^{k+1},r+3)$$

and

$$I_6=\frac{(\lambda+1)^{r+1}-1}{\lambda}(2^{k+1}-1)\zeta(\{1\}^{k+1},r+3).$$

Note that

(3.2.6)

$$
\frac{1}{\lambda(\lambda+1)}E(-\lambda) + \frac{(\lambda+1)^{r+1}}{\lambda}E\left(\frac{\lambda}{\lambda+1}\right)
$$

$$
= \frac{1}{\lambda(\lambda+1)}\left\{E(-\lambda) + (-\lambda-1)^{r+2}E\left(\frac{-\lambda}{-\lambda-1}\right)\right\}
$$

$$
= \frac{1}{\lambda(\lambda+1)}\left\{1 + (-\lambda-1)^{r+2}\right\}\left\{E(1) + (2^{k+1}-1)\zeta(\{1\}^{k+1},r+3)\right\}
$$

$$
+ \frac{1}{\lambda(\lambda+1)}\sum_{j=0}^{k}\sum_{\ell=0}^{r}(-\lambda-1)^{\ell+1}\zeta(\{1\}^{j},r-\ell+2)\zeta(\{1\}^{k-j},\ell+2).
$$

On the other hand, we separate (3.2.5) into two sums as

$$
\frac{(\lambda+1)^{r+1}}{\lambda}\sum_{j=0}^{k}\sum_{\ell=0}^{r}(-1)^{\ell}\zeta(\{1\}^{j},r-\ell+2)\zeta(\{1\}^{k-j},\ell+2)
$$

$$
+ \frac{1}{\lambda(\lambda+1)}\sum_{j=0}^{k}\sum_{\ell=0}^{r}(-\lambda-1)^{\ell+1}\zeta(\{1\}^{j},r-\ell+2)\zeta(\{1\}^{k-j},\ell+2).
$$

The first sum is equal to

$$
\frac{(\lambda+1)^{r+1}}{\lambda}[1 + (-1)^{r}]\left\{E(1) + (2^{k+1}-1)\zeta(\{1\}^{k+1},r+3)\right\}
$$

while the second sum will be canceled by a term in (3.2.6). After a careful cancelation, we get our assertion. $\square$

Proof of Theorem 3.2.1. Instead of considering the integral in (3.2.3), we begin with the integral

(3.2.7)

$$
\frac{1}{k!r!}\iint_{R_1\times R_2}\left(\log\frac{1}{1-t_1} + \log\frac{1}{1-u_1}\right)^{k}
$$

$$
\times\left(\log\frac{t_2}{t_1} + \lambda\log\frac{1}{t_2} + \nu\log\frac{u_2}{u_1}\right)^{r}\frac{dt_1 dt_2}{(1-t_1)t_2}\frac{du_1 du_2}{(1-u_1)u_2}
$$

where R_1 and R_2 are simplices given respectively by

$$
R_1:\ 0 < t_1 < t_2 < 1 \quad\text{and}\quad R_2:\ 0 < u_1 < u_2 < 1.
$$

As

$$\log \frac{t_2}{t_1} + \lambda \log \frac{1}{t_2} + \nu \log \frac{u_2}{u_1} = \log \frac{t_2}{t_1} + \lambda \log \frac{u_1}{t_2} + (\lambda + \nu) \log \frac{u_2}{u_1} + \lambda \log \frac{1}{u_2},$$

the value of integration over D_1 is

$$\sum_{a+b+c=k} 2^a \sum_{m+n+p=r} \frac{\lambda^n}{\nu} \left[(\lambda + \nu)^{p+1} - \lambda^{p+1} \right]$$

$$\times \sum_{\substack{|\boldsymbol{\alpha}|=b+m+1 \\ |\boldsymbol{\beta}|=c+n+1}} \zeta(\{1\}^a, \alpha_0, \ldots, \alpha_{b-1}, \alpha_b + \beta_0, \beta_1, \ldots, \beta_c, p+2)$$

or

$$\frac{\lambda+\nu}{\nu} P(1, \lambda, \lambda + \nu) - \frac{\lambda}{\nu} P(1, \lambda, \lambda)$$

according to the notation of (3.2.4).

The reason why we choose (3.2.7) instead of (3.2.3) is that it is much easier to evaluate the integration over $D_2 : 0 < u_1 < u_2 < t_1 < t_2 < 1$. On the simplex D_2, the value of integration is

$$\frac{1}{\lambda - 1} \sum_{a+b+c=k} 2^a \sum_{\ell=0}^{r} \nu^\ell (\lambda^{r-\ell+1} - 1)$$

$$\times \sum_{|\boldsymbol{\alpha}|=b+\ell+1} \zeta(\{1\}^a, \alpha_0, \ldots, \alpha_b + 1, \{1\}^c, r - \ell + 2).$$

Fixing $a = p$ and summing over b, c, the above sum is

$$\frac{1}{\lambda - 1} \sum_{p+q=k} 2^p \sum_{\ell=0}^{r} (\nu^\ell \lambda^{r-\ell+1} - \nu^\ell) \sum_{|\boldsymbol{\alpha}|=q+\ell+2} \zeta(\{1\}^p, \alpha_0, \ldots, \alpha_q, r - \ell + 2).$$

In notation of (3.2.1), it is

$$\frac{\nu^{r+1}}{\lambda - 1} E\left(\frac{\lambda}{\nu}\right) - \frac{\nu^{r+1}}{\lambda - 1} E\left(\frac{1}{\nu}\right)$$

$$- \left(\frac{\lambda^{r+2}}{\nu(\lambda - 1)} - \frac{1}{\nu(\lambda - 1)} \right) (2^{k+1} - 1) \zeta(\{1\}^{k+1}, r + 3).$$

The values of integrations over D_3, D_4, D_5 and D_6 are linear combinations of three $E(\lambda)$ for various λ. First we perform the integration over D_3. In terms of multiple zeta values, the value of integration over D_3 is

$$\sum_{p+q=k} 2^p \sum_{\ell=0}^{r} \sum_{|\boldsymbol{\alpha}|=q+\ell+1} \zeta(\{1\}^p, \alpha_0, \ldots, \alpha_q, r-\ell+3) \sum_{a+b+c=r-\ell} (\nu+1)^a (\lambda+\nu)^b \lambda^c.$$

By Lemma 3.2.2 the final summation is equal to

$$\frac{(\nu + 1)^{r-\ell+2}}{(\nu + 1 - \lambda)(1 - \lambda)} + \frac{(\lambda + \nu)^{r-\ell+2}}{(\lambda - 1)\nu} + \frac{\lambda^{r-\ell+2}}{\nu(\nu + 1 - \lambda)},$$

so that the value can be expressed as

$$\frac{1}{(\nu + 1 - \lambda)(1 - \lambda)}E(\nu + 1) + \frac{1}{(\lambda - 1)\nu}E(\lambda + \nu) + \frac{1}{\nu(\nu + 1 - \lambda)}E(\lambda).$$

In the similar way, the values of integrations over D_4, D_5 and D_6 are obtained one by one. Here are the values in terms of the function $E(\lambda)$:

$$\frac{1}{(\nu + 1 - \lambda)\nu}E(\nu + 1) + \frac{1}{\nu(\lambda - 1)}E(1) + \frac{1}{(\lambda - 1)(\lambda - 1 - \nu)}E(\lambda),$$

$$\frac{\nu^{r+2}}{(\nu - \lambda)\nu}E\left(\frac{\nu + 1}{\nu}\right) + \frac{\nu^{r+2}}{\nu(\lambda - 1)}E\left(\frac{1}{\nu}\right) + \frac{\nu^{r+2}}{(\lambda - 1)(\lambda - 1 - \nu)}E\left(\frac{\lambda}{\nu}\right)$$

and

$$\frac{\nu^{r+2}}{(\nu + 1 - \lambda)(1 - \lambda)}E\left(\frac{\nu + 1}{\nu}\right) + \frac{\nu^{r+2}}{(\lambda - 1)\nu}E\left(\frac{\lambda + \nu}{\nu}\right)$$

$$+ \frac{\nu^{r+2}}{(\nu + 1 - \lambda)\nu}E\left(\frac{\lambda}{\nu}\right).$$

On the other hand, the value of the integral (3.2.7) is

$$\frac{1}{\lambda - 1}\sum_{j=0}^{k}\sum_{\ell=0}^{r}\nu^\ell(\lambda^{r-\ell+1} - 1)\zeta(\{1\}^j, r - \ell + 2)\zeta(\{1\}^{k-j}, \ell + 2)$$

$$= \frac{\nu^{r+1}}{\lambda - 1}\sum_{j=0}^{k}\sum_{\ell=0}^{r}\frac{\lambda^{r-\ell+1} - 1}{\nu^{r-\ell+1}}\zeta(\{1\}^j, r - \ell + 2)\zeta(\{1\}^{k-j}, \ell + 2)$$

which can be expressed as linear combinations of

$$E\left(\frac{\lambda + \nu}{\nu}\right), \quad E\left(\frac{\lambda + \nu}{\lambda}\right), \quad E(\nu + 1) \quad \text{and} \quad E\left(\frac{\nu + 1}{\nu}\right)$$

according to (3.2.2).

Also in light of Proposition 1.2.1, we have

$$P(1, \lambda, \lambda) = \frac{1}{(\lambda - 1)\lambda}\left\{\lambda^{r+2}E(1) - E(\lambda)\right\}.$$

After a cancelation, we conclude that

$$P(1, \lambda, \lambda + \nu) = \frac{1}{(\lambda - 1)(\lambda + \nu)} \left\{ \lambda^{r+2} E\left(\frac{\lambda + \nu}{\lambda}\right) - E(\lambda + \nu) \right\},$$

and hence it follows that

$$P(\mu, \lambda, \nu) = \frac{1}{(\lambda - \mu)\nu} \left\{ \lambda^{r+2} E\left(\frac{\nu}{\lambda}\right) - \mu^{r+2} E\left(\frac{\nu}{\mu}\right) \right\}. \qquad \square$$

3.3   Shuffle Relations of Two Sums of Multiple Zeta Values

How to produce the shuffle relation of the sum of multiple zeta values? Recall that the particular sum

$$\sum_{|\alpha|=p+q+1} \zeta(\alpha_0, \alpha_1, \ldots, \alpha_p + 1)$$

can be expressed as the double integral

$$\frac{1}{p!q!} \int_{0<t_1<t_2<1} \left(\log \frac{1 - t_1}{1 - t_2}\right)^p \left(\log \frac{t_2}{t_1}\right)^q \frac{dt_1 dt_2}{(1 - t_1)t_2}.$$

Therefore for nonnegative integers p, q, m and n, we have

$$\zeta(p + q + 2)\zeta(m + n + 2)$$

$$= \sum_{|\alpha|=p+q+1} \zeta(\alpha_0, \alpha_1, \ldots, \alpha_p + 1) \sum_{|\beta|=m+n+1} \zeta(\beta_0, \beta_1, \ldots, \beta_m + 1)$$

$$= \frac{1}{p!q!m!n!} \iint_{R_1 \times R_2} \left(\log \frac{1 - t_1}{1 - t_2}\right)^p \left(\log \frac{t_2}{t_1}\right)^q \left(\log \frac{1 - u_1}{1 - u_2}\right)^m$$

$$\times \left(\log \frac{u_2}{u_1}\right)^n \frac{dt_1 dt_2}{(1 - t_1)t_2} \frac{du_1 du_2}{(1 - u_1)u_2}$$

when R_1 and R_2 are two dimensional simplices defined by $0 < t_1 < t_2 < 1$ and $0 < u_1 < u_2 < 1$, respectively. This leads to our consideration of a combination of products of two sums of multiple zeta values.

For a pair of nonnegative integers k, r with k even, we consider the integral

$$\frac{1}{k!r!} \iint_{R_1 \times R_2} \left(\log \frac{1-u_1}{1-u_2} - \log \frac{1-t_1}{1-t_2} \right)^k \left(\log \frac{t_2}{t_1} + \log \frac{u_2}{u_1} \right)^r$$
$$\times \frac{dt_1 dt_2}{(1-t_1)t_2} \frac{du_1 du_2}{(1-u_1)u_2}.$$

Such an integral is separable and its value is equal to

$$\sum_{j=0}^{k} \sum_{\ell=0}^{r} (-1)^j \zeta(j+r-\ell+2)\zeta(k-j+\ell+2).$$

Proceeding with the shuffle process as before. As k is even, the integrations over D_1: $0 < t_1 < t_2 < u_1 < u_2 < 1$ and D_2: $0 < u_1 < u_2 < t_1 < t_2 < 1$ are the same and the common value is

$$\sum_{j=0}^{k} \sum_{\ell=0}^{r} (-1)^j \sum_{\substack{|\boldsymbol{\alpha}|=j+r-\ell+1 \\ |\boldsymbol{\beta}|=k-j+\ell+1}} \zeta(\alpha_0, \ldots, \alpha_{j-1}, \alpha_j + 1, \beta_0, \ldots, \beta_{k-j-1}, \beta_{k-j} + 1).$$

By Ohno's generalization of the duality theorem and sum formula, it is equal to

$$\sum_{j=0}^{k} \sum_{\ell=0}^{r} (-1)^j \zeta(k-j+\ell+2, j+(r-\ell)+2).$$

In light of the reflection formula

$$\zeta(p,q) + \zeta(q,p) = \zeta(p)\zeta(q) - \zeta(p+q), \ p,q \geq 2,$$

it is equal to

$$\frac{1}{2} \sum_{j=0}^{k} \sum_{\ell=0}^{r} (-1)^j \zeta(j+r-\ell+2)\zeta(k-j+\ell+2) - \frac{r+1}{2}\zeta(k+r+4).$$

On the simplex $D_3 : 0 < t_1 < u_1 < t_2 < u_2 < 1$, we substitute the factors

$$\left(\log \frac{1-u_1}{1-u_2} - \log \frac{1-t_1}{1-t_2} \right)^k \quad \text{and} \quad \left(\log \frac{t_2}{t_1} + \log \frac{u_2}{u_1} \right)^r$$

by

$$\sum_{j=0}^{k} \binom{k}{j} (-1)^j \left(\log \frac{1-t_1}{1-u_1} \right)^j \left(\log \frac{1-t_2}{1-u_2} \right)^{k-j}$$

and

$$\sum_{a+b+c=r} \frac{r!}{a!b!c!} \left(\log \frac{u_1}{t_1} \right)^a \left(2 \log \frac{t_2}{u_1} \right)^b \left(\log \frac{u_2}{t_2} \right)^c.$$

So that in terms of multiple zeta values, the value of integration over D_3 is

$$\sum_{j=0}^{k} (-1)^j \sum_{a+b+c=r} 2^b \sum_{\substack{|\alpha|=j+a+1 \\ |\beta|=k-j+c+1}} \zeta(\alpha_0, \alpha_1, \ldots, \alpha_j, \beta_0+b+1, \beta_1, \ldots, \beta_{k-j}+1).$$

On the other hand, in a similar consideration, the value of integration over D_4: $0 < t_1 < u_1 < u_2 < t_2 < 1$ is

$$\sum_{j=0}^{k} \sum_{a+b+c=r} 2^b \sum_{\substack{|\alpha|=j+a+1 \\ |\beta|=k-j+c+1}} \zeta(\alpha_0, \alpha_1, \ldots, \alpha_j, \beta_0 + b + 1, \beta_1, \ldots, \beta_{k-j} + 1).$$

The value of integration over D_5 (resp. D_6) is the same as the value of integration over D_3 (resp. D_4). Therefore, the total value of integrations over D_3, D_4, D_5, D_6 is

$$2 \sum_{j=0}^{k} \left\{ 1 + (-1)^j \right\} \sum_{a+b+c=r} 2^b$$
$$\times \sum_{\substack{|\alpha|=j+a+1 \\ |\beta|=k-j+c+1}} \zeta(\alpha_0, \alpha_1, \ldots, \alpha_j, \beta_0 + b + 1, \beta_1, \ldots, \beta_{k-j} + 1).$$

Let α_{j+1} be a new variable in place of $\beta_0 + b + 1$. Then for $\alpha_{j+1} \geq 2$, the range of b is given by

$$0 \leq b \leq \alpha_{j+1} - 2$$

and hence

$$\sum_{b=0}^{\alpha_{j+1}-2} 2^b = 2^{\alpha_{j+1}-1} - 1.$$

Therefore the sum is equal to

$$4 \sum_{\substack{1 \leq j \leq k+1 \\ j:\text{odd}}} \sum_{|\boldsymbol{\alpha}|=k+r+3} (2^{\alpha_j-1} - 1)\zeta(\alpha_0, \alpha_1, \ldots, \alpha_k, \alpha_{k+1} + 1).$$

Consequently, we obtain the following shuffle relation from shuffle products of sums of multiple zeta values.

Theorem 3.3.1. *For a pair of nonnegative integers k, r with k even, we have*

$$\sum_{\substack{1 \leq j \leq k+1 \\ j:\text{odd}}} \sum_{|\boldsymbol{\alpha}|=k+r+3} \left(2^{\alpha_j-1} - 1\right) \zeta(\alpha_0, \alpha_1, \ldots, \alpha_{k+1}+1) = \frac{1}{4}(r+1)\zeta(k+r+4).$$

In light of the sum formula, we rewrite the above theorem as the following.

Theorem 3.3.2. *For a pair of positive integers n, k with $n \geq k$ and k is even, we have*

$$\sum_{|\boldsymbol{\alpha}|=n} \sum_{\substack{1 \leq j \leq k \\ j:\text{even}}} 2^{\alpha_j}\zeta(\alpha_1, \ldots, \alpha_{k-1}, \alpha_k + 1) = \frac{n+k}{2}\zeta(n+1).$$

Remark 3.3.3. The above theorem is a generalization of weighted Euler sum formula by Ohno and Zudilin [29] which asserted that

$$\sum_{|\boldsymbol{\alpha}|=n} 2^{\alpha_2}\zeta(\alpha_1, \alpha_2 + 1) = \frac{n+2}{2}\zeta(n+1).$$

The case of odd depth, i.e, when k is odd, is much more complicated. We begin with a family of integrals with a parameter μ given by

$$\frac{1}{k!r!} \iint_{R_1 \times R_2} \left(\mu \log \frac{1-t_1}{1-t_2} + \log \frac{1-u_1}{1-u_2}\right)^k \left(\log \frac{t_2}{t_1} + \log \frac{u_2}{u_1}\right)^r$$
$$\times \frac{dt_1 dt_2}{(1-t_1)t_2} \frac{du_1 du_2}{(1-u_1)u_2}.$$

After the shuffle process as carried out before, we obtain the following relation

(3.3.1)
$$\sum_{j=0}^{k} \mu^j \sum_{\ell=0}^{r} \left\{ \sum_{\substack{|\boldsymbol{\alpha}|=j+1+r-\ell \\ |\boldsymbol{\beta}|=k-j+1+\ell}} \zeta(\alpha_0,\ldots,\alpha_{j-1},\alpha_j+1,\beta_0,\ldots,\beta_{k-j-1},\beta_{k-j}+1) \right.$$

$$\left. + \sum_{\substack{|\boldsymbol{\alpha}|=k-j+1+\ell \\ |\boldsymbol{\beta}|=j+1+r-\ell}} \zeta(\alpha_0,\ldots,\alpha_{k-j-1},\alpha_{k-j}+1,\beta_0,\ldots,\beta_{j-1},\beta_j+1) \right\}$$

$$+ \sum_{a+b+c=k} \left\{ (\mu^a + \mu^c + \mu^{a+c} + 1)(\mu+1)^b \right\}$$

$$\times \sum_{m+n+\ell=r} 2^n \sum_{\substack{|\boldsymbol{\alpha}|=a+m+1 \\ |\boldsymbol{\beta}|=b+n+1 \\ |\boldsymbol{\gamma}|=c+\ell+1}} \zeta(\alpha_0,\ldots,\alpha_a,\beta_0,\ldots,\beta_b+\gamma_0,\gamma_1,\ldots,\gamma_c+1)$$

$$= \sum_{j=0}^{k} \sum_{\ell=0}^{r} \mu^j \zeta(j+r-\ell+2)\zeta(k-j+\ell+2).$$

Multiply both sides on the above relation by μ, differentiate with respect to μ and then set $\mu = -1$. By Ohno's generalization of duality theory and sum formula, the first sum is equal to

$$\sum_{j=0}^{k} \sum_{\ell=0}^{r} (j+1)(-1)^j$$

$$\times \left\{ \zeta(k-j+\ell+2, j+r-\ell+2) + \zeta(j+r-\ell+2, k-j+\ell+2) \right\}.$$

By the reflection formula, it is equal to

$$\sum_{j=0}^{k} \sum_{\ell=0}^{r} (-1)^j (j+1) \left\{ \zeta(j+r-\ell+2)\zeta(k-j+\ell+2) - \zeta(k+r+4) \right\}.$$

On the other hand, the value

$$\frac{d}{d\mu} \mu(\mu^a + \mu^c + \mu^{a+c} + 1)(\mu+1)^b \bigg|_{\mu=-1}$$

is zero unless $b=0$ or $b=1$. When $b=0$, its value is

$$(a+1)(-1)^a + (c+1)(-1)^c + (k+1)(-1)^k + 1$$

or $(2a - k)(-1)^a - k$ if k is odd. The corresponding sum of multiple zeta value is

$$\sum_{j=0}^{k} \{(2j - k)(-1)^j - k\} \sum_{|\alpha|=k+r+3} (2^{\alpha_{j+1}-1} - 1)\zeta(\alpha_0, \ldots, \alpha_k, \alpha_{k+1} + 1).$$

When $b = 1$, the value is

$$(-1)^{a+1} + (-1)^{c+1} + (-1)^{a+c+1} + (-1),$$

or simply

$$(-2)\{(-1)^a + 1\}.$$

The corresponding sum of multiple zeta value is

$$-4 \sum_{\substack{j=0 \\ j:\text{even}}}^{k-1} \sum_{|\alpha|=k+r+3} 2^{\alpha_{j+1}-1}(2^{\alpha_{j+2}-1} - 1)\zeta(\alpha_0, \ldots, \alpha_k, \alpha_{k+1} + 1).$$

Finally, when k is odd, we have

$$\sum_{j=0}^{k}(j + 1)(-1)^j = -\frac{k + 1}{2}.$$

Consequently, we obtain the counterpart of our weighted sum formula.

Theorem 3.3.4. *For a pair of nonnegative integers k, r with k odd, we have*

$$k \sum_{\substack{j=0 \\ j:\text{even}}}^{k} \sum_{|\alpha|=k+r+3} (2^{\alpha_{j+1}} - 2)\zeta(\alpha_0, \ldots, \alpha_k, \alpha_{k+1} + 1)$$

$$+ \sum_{j=0}^{k} j(-1)^{j+1} \sum_{|\alpha|=k+r+3} (2^{\alpha_{j+1}} - 2)\zeta(\alpha_0, \ldots, \alpha_k, \alpha_{k+1} + 1)$$

$$+ \sum_{\substack{j=0 \\ j:\text{even}}}^{k-1} \sum_{|\alpha|=k+r+3} 2^{\alpha_{j+1}}(2^{\alpha_{j+2}} - 2)\zeta(\alpha_0, \ldots, \alpha_k, \alpha_{k+1} + 1)$$

$$= \frac{(k + 1)(r + 1)}{2}\zeta(k + r + 4).$$

In light of the sum formula

$$\sum_{|\boldsymbol{\alpha}|=k+r+3} \zeta(\alpha_0,\ldots,\alpha_k,\alpha_{k+1}+1) = \zeta(k+r+4),$$

we conclude the following theorem.

Theorem 3.3.5. *For a pair of positive integers n,k with k odd and $n > k \geq 3$, we have*

$$(k-2)\sum_{\substack{j=2 \\ j:\text{even}}}^{k} \sum_{|\boldsymbol{\alpha}|=n} 2^{\alpha_j}\zeta(\alpha_1,\ldots,\alpha_{k-1},\alpha_k+1)$$

$$+\sum_{j=3}^{k}(j-2)(-1)^{j+1}\sum_{|\boldsymbol{\alpha}|=n} 2^{\alpha_j}\zeta(\alpha_1,\ldots,\alpha_{k-1},\alpha_k+1)$$

$$+\sum_{\substack{j=2 \\ j:\text{even}}}^{k} \sum_{|\boldsymbol{\alpha}|=n} 2^{\alpha_j}(2^{\alpha_{j+1}}-2)\zeta(\alpha_1,\ldots,\alpha_{k-1},\alpha_k+1)$$

$$= \frac{(k-1)(n+k-2)}{2}\zeta(n+1).$$

Another weighted sum formula comes from the shuffle relation with the parameter (3.3.1) directly. Letting $\mu = 1$ in (3.3.1) and then applying Ohno's generalization of duality theorem and sum formula as well as the reflection formula, we obtain

$$\sum_{a+b+c=k} 2^b \sum_{m+n+\ell=r} 2^n$$

$$\times \sum_{\substack{|\boldsymbol{\alpha}|=a+m+1 \\ |\boldsymbol{\beta}|=b+n+1 \\ |\boldsymbol{\gamma}|=c+\ell+1}} \zeta(\alpha_0,\ldots,\alpha_a,\beta_0,\ldots,\beta_b+\gamma_0,\gamma_1,\ldots,\gamma_c+1)$$

$$= \frac{(k+1)(r+1)}{4}\zeta(k+r+4),$$

which can be rewritten as

$$\sum_{j+\ell\leq k} \sum_{|\boldsymbol{\alpha}|=k+r+3} 2^{|\boldsymbol{\alpha}|_{j\ell}}(1-2^{1-\alpha_{j+\ell+1}})\zeta(\alpha_0,\ldots,\alpha_k,\alpha_{k+1}+1)$$

$$= \frac{(k+1)(r+1)}{2}\zeta(k+r+4)$$

with $|\boldsymbol{\alpha}|_{j\ell} = \alpha_{j+1} + \alpha_{j+2} + \cdots + \alpha_{j+\ell+1}$.

From the above weighted sum formula, we obtain the following theorem.

Theorem 3.3.6. *For a pair of positive integers n, k with $n > k \geq 2$, we have*

$$\sum_{\substack{1 \leq j\ell \\ j+\ell \leq k}} \sum_{|\boldsymbol{\alpha}|=n} 2^{|\boldsymbol{\alpha}|_{j\ell}}(1 - 2^{1-\alpha_{j+\ell}})\zeta(\alpha_1, \ldots, \alpha_{k-1}, \alpha_k + 1)$$

$$= \frac{(k-1)(n-k)}{2}\zeta(n+1)$$

with $|\boldsymbol{\alpha}|_{j\ell} = \alpha_{j+1} + \alpha_{j+2} + \cdots + \alpha_{j+\ell}$.

3.4   A Vector Version of the Restricted Sum Formula

In 2008, the sum formula was generalized by the author and others [12, 15] to a more general form known as the restricted sum formula

$$\sum_{|\boldsymbol{\alpha}|=m} \zeta(\{1\}^p, \alpha_1, \alpha_2, \ldots, \alpha_q + 1) = \sum_{|\mathbf{c}|=p+q} \zeta(c_1, c_2, \ldots, c_p, c_{p+1} + m - q + 1).$$

There is a vector version of the above formula. Let

$$\mathbf{p} = (p_1, p_2, \ldots, p_n), \quad \mathbf{q} = (q_1, q_2, \ldots, q_n) \quad \text{and} \quad \mathbf{r} = (r_1, r_2, \ldots, r_n)$$

be n-tuples of nonnegative integers. Suppose that

$$\boldsymbol{\alpha_j} = \left(\alpha_{j0}, \alpha_{j1}, \ldots, \alpha_{j,q_j} + 1\right), \quad j = 1, 2, \ldots, n$$

is a $1 \times (q_j + 1)$ row vector with $\alpha_{j\ell}$ positive integers, $0 \leq \ell \leq q_j$. Also suppose that

$$\mathbf{k} = (\{1\}^{p_1}, r_1 + 2, \{1\}^{p_2}, r_2 + 2, \ldots, \{1\}^{p_n}, r_n + 2)$$

and

$$\mathbf{k}' = (\{1\}^{r_n}, p_n + 2, \{1\}^{r_{n-1}}, p_{n-1} + 2, \ldots, \{1\}^{r_1}, p_1 + 2)$$

are a dual pair.

Theorem 3.4.1. [16] *Notations as shown above. Then for any nonnegative integer m, we have*

$$\sum_{|\mathbf{q}|=m} \sum_{\substack{|\alpha_j|=q_j+r_j+1 \\ 1 \le j \le n}} \zeta(\{1\}^{p_1}, \alpha_1, \{1\}^{p_2}, \alpha_2, \ldots, \{1\}^{p_n}, \alpha_n)$$

$$(3.4.1) \qquad = \sum_{|\mathbf{c}|=m} \zeta(\mathbf{k}+\mathbf{c})$$

$$= \sum_{|\mathbf{d}|=m} \zeta(\mathbf{k}'+\mathbf{d}).$$

Remark 3.4.2. The second part of the identity

$$\sum_{|\mathbf{c}|=m} \zeta(\mathbf{k}+\mathbf{c}) = \sum_{|\mathbf{d}|=m} \zeta(\mathbf{k}'+\mathbf{d})$$

is just Ohno's generalization of the duality theorem and sum formulae. Let $S(\mathbf{p}, \mathbf{r}; m)$ be the sum of multiple zeta values in the first term of (3.4.1). Once we express $S(\mathbf{p}, \mathbf{r}; m)$ as the integral

$$\sum_{|\mathbf{q}|=m} \int_{E_{2n}} \prod_{j=1}^{n} \frac{1}{p_j! q_j! r_j!}$$

$$\times \left(\log \frac{1-t_{2j-2}}{1-t_{2j-1}}\right)^{p_j} \left(\log \frac{1-t_{2j-1}}{1-t_{2j}}\right)^{q_j} \left(\log \frac{t_{2j}}{t_{2j-1}}\right)^{r_j} \frac{dt_{2j-1}dt_{2j}}{(1-t_{2j-1})t_{2j}},$$

where E_{2n} is the $2n$-dimensional simplex defined by

$$0 < t_1 < t_2 < \cdots < t_{2n-1} < t_{2n} < 1$$

and $t_0 = 0$ by convention, we see immediately that

$$S(\mathbf{p}, \mathbf{r}; m) = \sum_{|\mathbf{d}|=m} \zeta(\mathbf{k}'+\mathbf{d})$$

by the change of variables

$$u_1 = 1 - t_{2n}, \quad u_2 = 1 - t_{2n-1}, \quad \ldots, \quad u_{2n-1} = 1 - t_2, \quad u_{2n} = 1 - t_1.$$

Remark 3.4.3. Our theorem is equivalent to prove that $S(\mathbf{p}, \mathbf{r}; m)$ also has the integral representation

$$\sum_{|\mathbf{q}|=m} \int_{E_{2n}} \prod_{j=1}^{n} \frac{1}{p_j! q_j! r_j!}$$

$$\times \left(\log \frac{1-u_{2j-1}}{1-u_{2j}}\right)^{p_j} \left(\log \frac{u_{2j}}{u_{2j-1}}\right)^{q_j} \left(\log \frac{u_{2j+1}}{u_{2j}}\right)^{r_j} \frac{du_{2j-1}du_{2j}}{(1-u_{2j-1})u_{2j}},$$

where $u_{2n+1} = 1$ by convention.

The following lemma is crucial, without which we are unable to prove Theorem 3.4.1.

Lemma 3.4.4. *For a positive integer n and real variables $x_1, x_2, \ldots, x_{2n}$, define a sequence of functions $f_n(\mathbf{x}_{2n}) = f_n(x_1, x_2, \ldots, x_{2n})$ recursively by*

$$f_1(x_1, x_2) = e^{x_1} + e^{x_2} - e^{x_1+x_2},$$

$$f_{k+1}(\mathbf{x}_{2k+2}) = e^{x_1+x_3+\cdots+x_{2k+1}} + e^{x_{2k+2}} f_k(\mathbf{x}_{2k}) - e^{x_1+x_3+\cdots+x_{2k+1}+x_{2k+2}},$$

for $k = 1, 2, \ldots, n-1$. Then we have

$$\begin{aligned}
f_n(\mathbf{x}_{2n}) &= f_n(\sigma(\mathbf{x}_{2n})) \\
&:= f_n(x_{\sigma(1)}, x_{\sigma(2)}, \ldots, x_{\sigma(2n)}) \\
&= f_n(x_{2n}, x_{2n-1}, \ldots, x_1)
\end{aligned}$$

with the permutation

$$\sigma = \begin{pmatrix} 1 & 2 & \cdots & 2n \\ 2n & 2n-1 & \cdots & 1 \end{pmatrix}.$$

Proof. We shall prove by induction on $n \geq 2$ that

$$\begin{aligned}
f_{n+1}(\mathbf{x}_{2n+2}) &= e^{x_1+x_3+\cdots+x_{2n+1}} + e^{x_2+x_4+\cdots+x_{2n+2}} \\
&\quad - e^{x_1+x_{2n+2}} \left\{ e^{x_2+x_4+\cdots+x_{2n}} + e^{x_3+x_5+\cdots+x_{2n+1}} \right\} \\
&\quad + e^{x_1+x_{2n+2}} f_{n-1}(x_3, x_4, \ldots, x_{2n}),
\end{aligned}$$

and hence our assertion follows. As

$$\begin{aligned}
f_2(\mathbf{x}_4) &= e^{x_1+x_3} + e^{x_4} \left\{ e^{x_1} + e^{x_2} - e^{x_1+x_2} \right\} - e^{x_1+x_3+x_4} \\
&= e^{x_1+x_3} + e^{x_2+x_4} + e^{x_1+x_4} \left\{ 1 - e^{x_2} - e^{x_3} \right\},
\end{aligned}$$

so we have

$$\begin{aligned}
f_3(\mathbf{x}_6) &= e^{x_1+x_3+x_5} + e^{x_6} f_2(\mathbf{x}_4) - e^{x_1+x_3+x_5+x_6} \\
&= e^{x_1+x_3+x_5} - e^{x_1+x_3+x_5+x_6} \\
&\quad + e^{x_6} \left\{ e^{x_1+x_3} + e^{x_2+x_4} + e^{x_1+x_4} - e^{x_1+x_2+x_4} - e^{x_1+x_3+x_4} \right\} \\
&= e^{x_1+x_3+x_5} + e^{x_2+x_4+x_6} - e^{x_1+x_6} \left\{ e^{x_2+x_4} + e^{x_3+x_5} \right\} \\
&\quad + e^{x_1+x_6} f_1(x_3, x_4).
\end{aligned}$$

Suppose that it is true for $n = k - 1$ with $k \geq 2$. Then for $n = k$, we have

$$
\begin{aligned}
&f_{k+1}(\mathbf{x}_{2k+2}) \\
&= e^{x_1+x_3+\cdots+x_{2k+1}} + e^{x_{2k+2}} f_k(\mathbf{x}_{2k}) - e^{x_1+x_3+\cdots+x_{2k+1}+x_{2k+2}} \\
&= e^{x_1+x_3+\cdots+x_{2k+1}} + e^{x_{2k+2}} \left\{ e^{x_1+x_3+\cdots+x_{2k-1}} + e^{x_2+x_4+\cdots+x_{2k}} \right\} \\
&\quad - e^{x_1+x_{2k}+x_{2k+2}} \left\{ e^{x_2+x_4+\cdots+x_{2k-2}} + e^{x_3+x_5+\cdots+x_{2k-1}} \right\} \\
&\quad + e^{x_1+x_{2k}+x_{2k+2}} f_{k-2}(x_3, x_4, \ldots, x_{2k-2}) - e^{x_1+x_3+\cdots+x_{2k+1}+x_{2k+2}} \\
&= e^{x_1+x_3+\cdots+x_{2k+1}} + e^{x_2+x_4+\cdots+x_{2k+2}} \\
&\quad - e^{x_1+x_{2k+2}} \left\{ e^{x_2+x_4+\cdots+x_{2k}} + e^{x_3+x_5+\cdots+x_{2k+1}} \right\} \\
&\quad + e^{x_1+x_{2k+2}} \left\{ e^{x_3+x_5+\cdots+x_{2k-1}} \right. \\
&\qquad \left. + e^{x_{2k}} f_{k-2}(x_3, x_4, \ldots, x_{2k-2}) - e^{x_3+x_5+\cdots+x_{2k-1}+x_{2k}} \right\} \\
&= e^{x_1+x_3+\cdots+x_{2k+1}} + e^{x_2+x_4+\cdots+x_{2k+2}} \\
&\quad - e^{x_1+x_{2k+2}} \left\{ e^{x_2+x_4+\cdots+x_{2k}} + e^{x_3+x_5+\cdots+x_{2k+1}} \right\} \\
&\quad + e^{x_1+x_{2k+2}} f_{k-1}(x_3, x_4, \ldots, x_{2k}).
\end{aligned}
$$

$\square$

Proof of Theorem 3.4.1. First we express the sum of multiple zeta values $S(\mathbf{p}, \mathbf{r}; m)$ as the integral

$$
\sum_{|\mathbf{q}|=m} \int_{E_{2n}} \prod_{j=1}^{n} \frac{1}{p_j! q_j! r_j!} \left(\log \frac{1-t_{2j-2}}{1-t_{2j-1}} \right)^{p_j} \left(\log \frac{1-t_{2j-1}}{1-t_{2j}} \right)^{q_j}
$$
$$
\times \left(\log \frac{t_{2j}}{t_{2j-1}} \right)^{r_j} \frac{dt_{2j-1}\, dt_{2j}}{(1-t_{2j-1})t_{2j}}.
$$

With the change of variables

$$
x_{2j-1} = \log \frac{1-t_{2j-2}}{1-t_{2j-1}}, \quad x_{2j} = \log \frac{t_{2j}}{t_{2j-1}}, \quad j = 1, 2, \ldots, n,
$$

the integral is transformed into

$$
\sum_{|\mathbf{q}|=m} \int_{\substack{x_j>0,\, f_n(\mathbf{x}_{2n})>0 \\ j=1,2,\ldots,n}} \prod_{j=1}^{n} \frac{1}{p_j! q_j! r_j!} x_{2j-1}^{p_j} x_{2j}^{r_j}
$$
$$
\times \left(\log \frac{f_{j-1}(\mathbf{x}_{2j-2})}{f_j(\mathbf{x}_{2j})} \right)^{q_j} dx_{2j-1}\, dx_{2j},
$$

where the sequence of functions $f_j(\mathbf{x}_{2j})$, $j = 0, 1, 2, \ldots, n$, are defined by

$$f_0(\mathbf{x}_0) = 1, \quad f_1(x_1, x_2) = e^{x_1} + e^{x_2} - e^{x_1+x_2},$$

$$f_{j+1}(\mathbf{x}_{2j+2}) = e^{x_1+x_3+\cdots+x_{2j+1}} + e^{x_{2j+2}} f_j(\mathbf{x}_{2j}) - e^{x_1+x_3+\cdots+x_{2j+1}+x_{2j+2}}.$$

All these functions come from the transformations of

$$\log \frac{1 - t_{2j-1}}{1 - t_{2j}}, \quad j = 1, 2, \ldots, n.$$

From the relations

$$x_1 = \log \frac{1}{1 - t_1} \quad \text{and} \quad x_2 = \log \frac{t_2}{t_1},$$

we get immediately that

$$\log \frac{1 - t_1}{1 - t_2} = \log \frac{1}{f_1(\mathbf{x}_2)} \quad \text{and} \quad 1 - t_2 = e^{-x_1} f_1(\mathbf{x}_2).$$

Next, from the relations

$$x_3 + \log \frac{1}{1 - t_2} = \log \frac{1}{1 - t_3} \quad \text{and} \quad x_4 = \log \frac{t_4}{t_3},$$

along with the previous step, we get

$$\begin{aligned}
\log \frac{1 - t_3}{1 - t_4} &= \log \frac{1 - t_2}{e^{x_3} + e^{x_4}(1 - t_2) - e^{x_3+x_4}} \\
&= \log \frac{e^{-x_1} f_1(\mathbf{x}_2)}{e^{-x_1} f_2(\mathbf{x}_4)} \\
&= \log \frac{f_1(\mathbf{x}_2)}{f_2(\mathbf{x}_4)}
\end{aligned}$$

with

$$f_2(x_1, x_2, x_3, x_4) = e^{x_1+x_3} + e^{x_4} f_1(x_1, x_2) - e^{x_1+x_3+x_4}.$$

Also we have $1 - t_3 = e^{-x_3}(1 - t_2) = e^{-x_1-x_3} f_1(\mathbf{x}_2)$, and hence

$$1 - t_4 = e^{-x_1-x_3} f_2(\mathbf{x}_4).$$

Repeat such process to get

$$\log \frac{1 - t_{2j-1}}{1 - t_{2j}} = \log \frac{f_{j-1}(\mathbf{x}_{2j-2})}{f_j(\mathbf{x}_{2j})}, \quad j = 1, 2, \ldots, n.$$

Note that the condition $f_n(\mathbf{x}_{2n}) > 0$ implies that $f_j(\mathbf{x}_{2j}) > 0$ for $j = 1, 2, \ldots, n-1$ by the recursive formula of the sequence of the functions $f_j(\mathbf{x}_{2j})$ $(j = 1, 2, \ldots, n)$. Indeed, from the formula

$$f_n(\mathbf{x}_{2n}) = e^{x_1 + x_3 + \cdots + x_{2n-1}}(1 - e^{x_{2n}}) + e^{x_{2n}} f_{n-1}(\mathbf{x}_{2n-2}),$$

and the first term in the right is not positive since $1 - e^{x_{2n}} \le 0$, we conclude that $f_n(\mathbf{x}_{2n}) > 0$ implies $f_{n-1}(\mathbf{x}_{2n-2}) > 0$. Thus, by the multinomial theorem, the integral after transform is
(3.4.2)

$$\frac{1}{m!} \int_{\substack{x_j > 0, f_n(\mathbf{x}_{2n}) > 0 \\ j=1,2,\ldots,n}} \left(\log \frac{1}{f_n(\mathbf{x}_{2n})} \right)^m \prod_{j=1}^{n} \frac{1}{p_j! r_j!} x_{2j-1}^{p_j} x_{2j}^{r_j} \, dx_{2j-1} dx_{2j}.$$

Now, make another change of variables

$$x_{2j-1} = \log \frac{1 - u_{2j-1}}{1 - u_{2j}}, \quad x_{2j} = \log \frac{u_{2j+1}}{u_{2j}}, \quad j = 1, 2, \ldots, n,$$

so that the region of integration is transformed back to the simplex E^{2n}: $0 < u_1 < u_2 < \cdots < u_{2n-1} < u_{2n} < 1$. The difficult part is to find the transformation of $\log[1/f_n(\mathbf{x}_{2n})]$. For this, we shall prove that

$$\log \frac{1}{f_n(\mathbf{x}_{2n})} = \log \frac{1}{f_n(\sigma(\mathbf{x}_{2n}))} = \log \frac{u_2}{u_1} + \log \frac{u_4}{u_3} + \cdots + \log \frac{u_{2n}}{u_{2n-1}}.$$

Beginning with the last two relations

$$x_{2n-1} = \log \frac{1 - u_{2n-1}}{1 - u_{2n}} \quad \text{and} \quad x_{2n} = \log \frac{1}{u_{2n}},$$

we get

$$\log \frac{u_{2n}}{u_{2n-1}} = \log \frac{1}{f_1(x_{2n}, x_{2n-1})} \quad \text{and} \quad u_{2n-1} = e^{-x_{2n}} f_1(x_{2n}, x_{2n-1}).$$

Next from the relations

$$x_{2n-3} = \log \frac{1 - u_{2n-3}}{1 - u_{2n-2}} \quad \text{and} \quad x_{2n-2} + \log \frac{1}{u_{2n-1}} = \log \frac{1}{u_{2n-2}},$$

and the previous step, we get

$$\log \frac{u_{2n-2}}{u_{2n-3}} = \log \frac{u_{2n-1}}{e^{x_{2n-2}} + e^{x_{2n-3}} u_{2n-1} - e^{x_{2n-3} + x_{2n-2}}}$$

$$= \log \frac{e^{-x_{2n}} f_1(x_{2n}, x_{2n-1})}{e^{x_{2n-2}} + e^{x_{2n-3} - x_{2n}} f_1(x_{2n}, x_{2n-1}) - e^{x_{2n-2} + x_{2n-3}}}$$

$$= \log \frac{f_1(x_{2n}, x_{2n-1})}{f_2(x_{2n}, x_{2n-1}, x_{2n-2}, x_{2n-3})}$$

and $u_{2n-3} = e^{-x_{2n-2}-x_n} f_2(x_{2n}, x_{2n-1}, x_{2n-2}, x_{2n-3})$. Repeat this process to get

$$\log \frac{u_{2n}}{u_{2n-1}} = \log \frac{1}{f_1(x_{2n}, x_{2n-1})},$$

$$\log \frac{u_{2n-2}}{u_{2n-3}} = \log \frac{f_1(x_{2n}, x_{2n-1})}{f_2(x_{2n}, x_{2n-1}, x_{2n-2}, x_{2n-3})},$$

$$\vdots$$

$$\log \frac{u_2}{u_1} = \log \frac{f_{n-1}(x_{2n}, x_{2n-1}, \ldots, x_3)}{f_n(\sigma(\mathbf{x}_{2n}))},$$

and it follows from Lemma 3.4.4 that

$$\log \frac{u_2}{u_1} + \log \frac{u_4}{u_3} + \cdots + \log \frac{u_{2n}}{u_{2n-1}} = \log \frac{1}{f_n(\sigma(\mathbf{x}_{2n}))} = \log \frac{1}{f_n(\mathbf{x}_{2n})}.$$

This proves our assertions. Consequently, the second change of variables transform the integral (3.4.2) into

$$\sum_{|\mathbf{q}|=m} \int_{E_{2n}} \prod_{j=1}^{n} \frac{1}{p_j! q_j! r_j!}$$

$$\times \left(\log \frac{1-u_{2j-1}}{1-u_{2j}} \right)^{p_j} \left(\log \frac{u_{2j}}{u_{2j-1}} \right)^{q_j} \left(\log \frac{u_{2j+1}}{u_{2j}} \right)^{r_j} \frac{du_{2j-1} du_{2j}}{(1-u_{2j-1})u_{2j}},$$

which is the integral representation of

$$\sum_{|\mathbf{c}|=m} \zeta(\mathbf{k} + \mathbf{c}). \qquad \square$$

Here we mention an application of our theorem. The sum formula

$$\sum_{|\mathbf{a}|=m+n+1} \zeta(a_0, a_1, \ldots, a_m + 1) = \zeta(m+n+2), \quad m, n \geq 0,$$

is equivalent to the identity between two generating functions

$$\sum_{m=0}^{\infty} \sum_{n=0}^{\infty} \sum_{|\mathbf{a}|=m+n+1} \zeta(a_0, a_1, \ldots, a_m + 1) x^m y^n = \sum_{m=0}^{\infty} \sum_{n=0}^{\infty} \zeta(m+n+2) x^m y^n.$$

As the sum

$$\sum_{|\mathbf{a}|=m+n+1} \zeta(a_0, a_1, \ldots, a_m + 1)$$

has the integral representation

$$\frac{1}{m!n!} \int_{0<t_1<t_2<1} \left(\log \frac{1-t_1}{1-t_2}\right)^m \left(\log \frac{t_2}{t_1}\right)^n \frac{dt_1 dt_2}{(1-t_1)t_2},$$

so that the first generating function is

$$\int_{0<t_1<t_2<1} \left(\frac{1-t_1}{1-t_2}\right)^x \left(\frac{t_2}{t_1}\right)^y \frac{dt_1 dt_2}{(1-t_1)t_2}.$$

Under two consecutive transforms as given in the proof of Theorem 3.4.1, the above integral is transformed into

$$\int_{0<u_1<u_2<1} \left(\frac{u_2}{u_1}\right)^x \left(\frac{1}{u_2}\right)^y \frac{du_1 du_2}{(1-u_1)u_2},$$

which is equal

$$\sum_{k=1}^{\infty} \frac{1}{(k-x)(k-y)}.$$

The power series expansion

$$\frac{1}{(k-x)(k-y)} = \sum_{m=0}^{\infty} \sum_{n=0}^{\infty} k^{-(m+n+2)} x^m y^n$$

then leads to our identity. Here we give the vector version of the above as an application of our theorem. Consider the power series with $n+1$ variables given by

$$\sum_{1\leq k_1<k_2<\cdots<k_n} \prod_{j=1}^{n} \frac{1}{(k_j-x_j)(k_j-z)}.$$

It is easy to see that the above has the power series expansion as

$$\sum_{\substack{m_i=0 \\ 1\leq i\leq 2n}}^{\infty} \sum_{1\leq k_1<\cdots<k_n} \left(\prod_{j=1}^{n} k_j^{-(m_{2j-1}+m_{2j}+2)} x_j^{m_{2j-1}}\right) z^{m_2+m_4+\cdots+m_{2n}},$$

or

$$\sum_{\substack{m_i=0 \\ 1\leq i\leq 2n}}^{\infty} \zeta(m_1+m_2+2,\ldots,m_{2n-1}+m_{2n}+2) x_1^{m_1} x_2^{m_3} \cdots x_n^{m_{2n-1}} z^{m_2+m_4+\cdots+m_{2n}}.$$

On the other hand, the given series comes from the evaluation of the integral

$$\int_{E_{2n}} \prod_{j=1}^{n} \left(\frac{u_{2j}}{u_{2j-1}}\right)^{z} \left(\frac{u_{2j+1}}{u_{2j}}\right)^{x_j} \frac{du_{2j-1}du_{2j}}{(1-u_{2j-1})u_{2j}}.$$

Such an integral is just the image of the integral

$$\int_{E_{2n}} \prod_{j=1}^{n} \left(\frac{1-t_{2j-1}}{1-t_{2j}}\right)^{z} \left(\frac{t_{2j}}{t_{2j-1}}\right)^{x_j} \frac{dt_{2j-1}dt_{2j}}{(1-t_{2j-1})t_{2j}}$$

under the two consecutive transforms given in Theorem 3.4.1, also it is just the sum of the generating function

$$\sum_{\substack{m_i=0 \\ 1\leq i\leq 2n}} \sum_{\substack{|\boldsymbol{\alpha}_j|=m_{2j-1}+m_{2j}+1 \\ 1\leq j\leq n}} \zeta(\boldsymbol{\alpha}_1,\boldsymbol{\alpha}_2,\ldots,\boldsymbol{\alpha}_n)x_1^{m_1}x_2^{m_3}\cdots x_n^{m_{2n-1}}z^{m_2+m_4+\cdots+m_{2n}}$$

with $\boldsymbol{\alpha}_j = (\alpha_{j0},\alpha_{j1},\ldots,\alpha_{j,m_{2j}}+1)$. Consequently, we prove the following.

Corollary 3.4.5. *For any positive integer n, we have for $|x_j| < 1$, $1 \leq j \leq n$, and $|z| < 1$,*

$$\sum_{\substack{m_i=0 \\ 1\leq i\leq 2n}}^{\infty} \zeta(m_1+m_2+2,\ldots,m_{2n-1}+m_{2n}+2)$$

$$\times \left(\prod_{j=1}^{n} x_j^{m_{2j-1}}\right) z^{m_2+m_4+\cdots+m_{2n}}$$

$$= \sum_{\substack{m_i=0 \\ 1\leq i\leq 2n}}^{\infty} \sum_{\substack{|\boldsymbol{\alpha}_j|=m_{2j-1}+m_{2j}+1 \\ 1\leq j\leq n}} \zeta(\boldsymbol{\alpha}_1,\boldsymbol{\alpha}_2,\ldots,\boldsymbol{\alpha}_n)$$

$$\times x_1^{m_1}x_2^{m_3}\cdots x_n^{m_{2n-1}}z^{m_2+m_4+\cdots+m_{2n}}$$

$$= \sum_{1\leq k_1<k_2<\cdots<k_n} \prod_{j=1}^{n} \frac{1}{(k_j-x_j)(k_j-z)}.$$

Here $\boldsymbol{\alpha}_j = (\alpha_{j0},\alpha_{j1},\ldots,\alpha_{j,m_{2j}}+1)$ and $|\boldsymbol{\alpha}_j| = \alpha_{j0}+\alpha_{j1}+\cdots+\alpha_{j,m_{2j}}$, $j=1,2,\ldots,n$.

3.5   Exercises

1. Deduce a shuffle relation by considering the integral

$$\frac{1}{k!r!} \iint_{R_1 \times R_2} \left(\mu \log \frac{1}{1-t_1} + \log \frac{1}{1-u_1} \right)^k$$

$$\times \left(\log \frac{t_2}{t_1} + \lambda \log \frac{u_2}{u_1} \right)^r \frac{dt_1 dt_2}{(1-t_1)t_2} \frac{du_1 du_2}{(1-u_1)u_2}$$

with $R_1 : 0 < t_1 < t_2 < 1$ and $R_2 : 0 < u_1 < u_2 < 1$.

2. Prove the identity

$$(k+1) \left\{ E(\mu+1, \lambda+1) - E(\mu+1, 1) \right\}$$

$$+ \mu^k \lambda^{r+2} \left\{ E\left(\frac{\mu+1}{\mu}, \frac{\lambda+1}{\lambda} \right) - E\left(\frac{\mu+1}{\mu}, 1 \right) \right\}$$

$$- \mu \left\{ \frac{\partial}{\partial \mu} E(\mu+1, \lambda+1) - \frac{\partial}{\partial \mu} E(\mu+1, 1) \right\}$$

$$+ \mu^{k-1} \lambda^{r+2} \left\{ \frac{\partial}{\partial \mu} E\left(\frac{\mu+1}{\mu}, \frac{\lambda+1}{\lambda} \right) - \frac{\partial}{\partial \mu} E\left(\frac{\mu+1}{\mu}, 1 \right) \right\}$$

$$= \left\{ (k+1)(\mu+1)^k + \lambda^{r+2} \left[(\mu+1)^k + (k+2) \frac{(\mu+1)^k - 1}{\mu} \right] \right\}$$

$$\times \zeta(\{1\}^{k+1}, r+3)$$

$$+ \sum_{j=0}^{k} \sum_{\ell=0}^{r} (k-j+1) \mu^j \lambda^{\ell+1} \zeta(\{1\}^j, r-\ell+2) \zeta(\{1\}^{k-j}, \ell+2).$$

3. Prove that when r is even,

$$\sum_{p+q=k} 2^p \sum_{|\alpha|=q+r+3} 2^{\alpha_{q+1}} \zeta(\{1\}^p, \alpha_0, \alpha_1, \ldots, \alpha_{q+1}+1)$$

$$= \sum_{j=0}^{k} \sum_{\substack{\ell=0 \\ \ell : \text{even}}} \zeta(\{1\}^j, r-\ell+2) \zeta(\{1\}^{k-j}, \ell+2).$$

CHAPTER 4

Euler Decomposition Theorem

The well-known Euler decomposition theorem expressed a product of a pair of single zeta values as a sum of double Euler sums, which is employed to build up a weighted sum formula among Euler double sums.

Through shuffle products of integrals over the product of two simplices of dimension two, we are able to produce several analogues of Euler decomposition theorem, with applications in establishing shuffle relations as well as combinatorial identities.

Some shuffle relations we obtained have certain restrictions on the depth or the weight, we develop systematic ways to overcome such difficulty.

4.1   A Shuffle Relation with Two Parameters

Let

$$R_1: \ 0 < t_1 < t_2 < 1 \ \text{ and } \ R_2: \ 0 < u_1 < u_2 < 1$$

be simplices in two dimensional Euclidean space. For a pair of integers $k, r \geq 0$, we consider the integral

$$\frac{1}{k!r!} \iint_{R_1 \times R_2} \left(\mu \log \frac{1}{1-t_1} + \log \frac{1}{1-u_2} \right)^k$$

$$\times \left(\log \frac{1}{t_2} + \lambda \log \frac{1}{u_2} \right)^r \frac{dt_1 dt_2}{(1-t_1)t_2} \frac{du_1 du_2}{(1-u_1)u_2},$$

where μ, λ are complex numbers and $\mu\lambda \neq 0$. Substituting the integrand by its binomial expansion

$$\sum_{j=0}^{k} \sum_{\ell=0}^{r} \binom{k}{j} \binom{r}{\ell} \mu^j \lambda^\ell \left(\log \frac{1}{1-t_1} \right)^j \left(\log \frac{1}{t_2} \right)^{r-\ell}$$

$$\times \left(\log \frac{1}{1-u_2} \right)^{k-j} \left(\log \frac{1}{u_2} \right)^\ell,$$

we see immediately that the value of the integral is

$$\sum_{j=0}^{k} \sum_{\ell=0}^{r} \mu^j \lambda^\ell (k-j+1)\zeta(\{1\}^j, r-\ell+2)\zeta(\{1\}^{k-j}, \ell+2).$$

In the following, we shall decompose the region of integration $R_1 \times R_2$ into 6 simplices obtaining from all possible interlacings of variables t_1, t_2 and u_1, u_2, so that the value of the integral is equal to the sum of the values over these simplices. This is equivalent to carry out the shuffle process. These simplces are as follows:

$$D_1: \ 0 < t_1 < t_2 < u_1 < u_2 < 1,$$
$$D_2: \ 0 < u_1 < u_2 < t_1 < t_2 < 1,$$
$$D_3: \ 0 < t_1 < u_1 < t_2 < u_2 < 1,$$
$$D_4: \ 0 < t_1 < u_1 < u_2 < t_2 < 1,$$
$$D_5: \ 0 < u_1 < t_1 < u_2 < t_2 < 1 \text{ and}$$
$$D_6: \ 0 < u_1 < t_1 < t_2 < u_2 < 1.$$

1. On the simplex: $D_1: \ 0 < t_1 < t_2 < u_1 < u_2 < 1$, we expand the factors

$$\left(\mu \log \frac{1}{1-t_1} + \log \frac{1}{1-u_2} \right)^k \quad \text{and} \quad \left(\log \frac{1}{t_2} + \lambda \log \frac{1}{u_2} \right)^r$$

as

$$\sum_{a+b+c=k} \frac{k!}{a!b!c!} \left[(\mu + 1) \log \frac{1}{1-t_1} + \log \frac{1-t_1}{1-t_2} \right]^a$$
$$\times \left(\log \frac{1-t_2}{1-u_1} \right)^b \left(\log \frac{1-u_1}{1-u_2} \right)^c$$

and

$$\sum_{m+n+\ell=r} (\lambda + 1)^\ell \frac{r!}{m!n!\ell!} \left(\log \frac{u_1}{t_2} \right)^m \left(\log \frac{u_2}{u_1} \right)^n \left(\log \frac{1}{u_2} \right)^\ell.$$

In terms of multiple zeta values, the value of integration over D_1 is

$$\sum_{a+b+c=k} \left(\sum_{d=0}^{a} (\mu + 1)^d \right) \sum_{m+n+\ell=r} (\lambda + 1)^\ell$$
$$\times \sum_{\substack{|\alpha|=b+m+1 \\ |\beta|=c+n+1}} \zeta(\{1\}^a, \alpha_0 + 1, \ldots, \alpha_b, \beta_0, \ldots, \beta_c + 1 + \ell)$$

or

$$\sum_{a+b+c=k} (\mu + 1)^a \sum_{m+n+\ell=r} (\lambda + 1)^\ell$$
$$\times \sum_{\substack{|\alpha|=b+m+2 \\ |\beta|=c+n+1}} \zeta(\{1\}^a, \alpha_0, \ldots, \alpha_b, \beta_0, \ldots, \beta_c + 1 + \ell)$$

by an elementary consideration. Combining together α and β as a simple dummy variable, we rewrite the value as

(4.1.1)

$$\sum_{p+q=k} (\mu + 1)^p (q + 1) \sum_{\ell=0}^{r} (\lambda + 1)^{r-\ell}$$
$$\times \sum_{|\alpha|=q+\ell+3} \zeta(\{1\}^p, \alpha_0, \ldots, \alpha_{q+1} + r - \ell + 1)$$
$$- \sum_{a+b+c=k} (\mu + 1)^a \sum_{\ell=0}^{r} (\lambda + 1)^{r-\ell}$$
$$\times \sum_{|\beta|=c+\ell+2} \zeta(\{1\}^{a+b+1}, \beta_0, \ldots, \beta_c + r - \ell + 1).$$

In our new notation for double weighted sums, the first term is

$$(k+1)\frac{1}{\lambda}\left\{E(\mu+1,\lambda+1)-E(\mu+1,1)\right\}$$

$$-\frac{\mu+1}{\lambda}\left\{\frac{\partial}{\partial\mu}E(\mu+1,\lambda+1)-\frac{\partial}{\partial\mu}E(\mu+1,1)\right\}$$

$$-\sum_{p+q=k}(\mu+1)^p(q+1)(\lambda+1)^{r+1}\zeta(\{1\}^{k+1},r+3).$$

2. In the same manner, the value of integration over D_3 : $0 < t_1 < u_1 < t_2 < u_2 < 1$, in terms of multiple zeta values, is given by

$$\sum_{a+b+c=k}(\mu+1)^a(b+1)\sum_{\ell=0}^{r}(\lambda+1)^{r-\ell}$$

$$\times\sum_{|\boldsymbol{\alpha}|=c+\ell+1}\zeta(\{1\}^{a+b+1},\alpha_0+1,\ldots,\alpha_c+r-\ell+1).$$

Note that for $c \geq 1$,

$$\sum_{|\boldsymbol{\alpha}|=c+\ell+1}\zeta(\{1\}^{a+b+1},\alpha_0+1,\ldots,\alpha_c+r-\ell+1)$$

$$(4.1.2)\qquad =\sum_{|\boldsymbol{\alpha}|=c+\ell+2}\zeta(\{1\}^{a+b+1},\alpha_0,\ldots,\alpha_c+r-\ell+1)$$

$$-\sum_{|\boldsymbol{\alpha}|=c+\ell+1}\zeta(\{1\}^{a+b+2},\alpha_0,\ldots,\alpha_{c-1}+r-\ell+1).$$

Therefore, the value is equal to

$$\sum_{a+b+c=k}(\mu+1)^a\sum_{\ell=0}^{r}(\lambda+1)^{r-\ell}\sum_{|\boldsymbol{\alpha}|=c+\ell+2}\zeta(\{1\}^{a+b+1},\alpha_0,\ldots,\alpha_c+r-\ell+1).$$

Such a value will be canceled with the negative term as shown in (4.1.1) comes form the value of integration over D_1.

3. The value of integration over D_6 : $0 < u_1 < t_1 < t_2 < u_2 < 1$, in terms of multiple zeta values, is given by

$$\sum_{a+b+c=k}(\mu+1)^a(a+1)\sum_{\ell=0}^{r}(\lambda+1)^{r-\ell}$$

$$\times\sum_{|\boldsymbol{\alpha}|=q+\ell+1}\zeta(\{1\}^{a+b+1},\alpha_0+1,\ldots,\alpha_q+r-\ell+1).$$

With the same reason as shown in (4.1.2), it is equal to

$$\sum_{p+q=k}(\mu+1)^p(p+1)\sum_{\ell=0}^{r}(\lambda+1)^{r-\ell}\sum_{|\alpha|=c+\ell+2}\zeta(\{1\}^{p+1},\alpha_0,\ldots,\alpha_c+r-\ell+1).$$

In terms of our notation, it is

$$\frac{1}{\lambda}\left\{\frac{\partial}{\partial\mu}E(\mu+1,\lambda+1)-\frac{\partial}{\partial\mu}E(\mu+1,1)\right\}$$

$$-\sum_{p=0}^{k}(\mu+1)^p(p+1)(\lambda+1)^{r+1}\zeta(\{1\}^{k+1},r+3)$$

$$+(k+1)(\mu+1)^k\frac{1}{\lambda}\left\{(\lambda+1)^{r+2}-1\right\}\zeta(\{1\}^{k+1},r+3).$$

4. Integrations over D_4 and D_5, only produce constant multiples of $\zeta(\{1\}^{k+1},r+3)$. Their values are

$$\sum_{p+q=k}(\mu+1)^p(q+1)\left\{(\lambda+1)^{r+1}-\lambda^{r+1}\right\}\zeta(\{1\}^{k+1},r+3)$$

and

$$\sum_{p+q=k}(\mu+1)^p(p+1)\left\{(\lambda+1)^{r+1}-\lambda^{r+1}\right\}\zeta(\{1\}^{k+1},r+3),$$

respectively. The total is

$$(k+2)\frac{(\mu+1)^{k+1}-1}{\mu}\left\{(\lambda+1)^{r+1}-\lambda^{r+1}\right\}\zeta(\{1\}^{k+1},r+3).$$

5. Now it remains to evaluate the integration over $D_2 : 0 < u_1 < u_2 < t_1 < t_2 < 1$. In terms of multiple zeta values, the value of integration over D_2 is

$$\mu^k\lambda^{r+1}\sum_{p+q=k}\left(\frac{\mu+1}{\mu}\right)^p(p+1)\sum_{\ell=0}^{r}\left\{\left(\frac{\lambda+1}{\lambda}\right)^{r-\ell+1}-1\right\}$$

$$\times\sum_{|\alpha|=q+\ell+1}\zeta(\{1\}^p,\alpha_0+1,\ldots,\alpha_q,r-\ell+2),$$

or

$$\mu^k \lambda^{r+1} \sum_{p+q=k} \left(\frac{\mu+1}{\mu}\right)^p (p+1) \sum_{\ell=0}^{r} \left\{ \left(\frac{\lambda+1}{\lambda}\right)^{r-\ell+1} - 1 \right\}$$

$$\times \sum_{|\boldsymbol{\alpha}|=q+\ell+2} \zeta(\{1\}^p, \alpha_0, \ldots, \alpha_q, r-\ell+2)$$

$$-\mu^k \lambda^{r+1} \sum_{\substack{p+q=k \\ p \le k-1}} \left(\frac{\mu+1}{\mu}\right)^p (p+1) \sum_{\ell=0}^{r} \left\{ \left(\frac{\lambda+1}{\lambda}\right)^{r-\ell+1} - 1 \right\}$$

$$\times \sum_{|\boldsymbol{\alpha}|=q+\ell+1} \zeta(\{1\}^{p+1}, \alpha_0, \ldots, \alpha_{q-1}, r-\ell+2),$$

or

$$\mu^k \lambda^{r+1} \sum_{p+q=k} \left(\frac{\mu+1}{\mu}\right)^p \sum_{\ell=0}^{r} \left\{ \left(\frac{\lambda+1}{\lambda}\right)^{r-\ell+1} - 1 \right\}$$

$$\times \sum_{|\boldsymbol{\alpha}|=q+\ell+2} \zeta(\{1\}^p, \alpha_0, \ldots, \alpha_q, r-\ell+2)$$

$$+\mu^{k-1} \lambda^{r+1} \sum_{\substack{p+q=k \\ p \le k-1}} \left(\frac{\mu+1}{\mu}\right)^p (p+1) \sum_{\ell=0}^{r} \left\{ \left(\frac{\lambda+1}{\lambda}\right)^{r-\ell+1} - 1 \right\}$$

$$\times \sum_{|\boldsymbol{\alpha}|=q+\ell+1} \zeta(\{1\}^{p+1}, \alpha_0, \ldots, \alpha_{q-1}, r-\ell+2).$$

In new notation of double weighted sums, it is equal to

$$\mu^k \lambda^{r+1} \left\{ E\left(\frac{\mu+1}{\mu}, \frac{\lambda+1}{\lambda}\right) - E\left(\frac{\mu+1}{\mu}, 1\right) \right\}$$

$$+ \mu^{k-1} \lambda^{r+1} \left\{ \frac{\partial}{\partial \mu} E\left(\frac{\mu+1}{\mu}, \frac{\lambda+1}{\lambda}\right) - \frac{\partial}{\partial \mu} E\left(\frac{\mu+1}{\mu}, 1\right) \right\}$$

$$- \left\{ (\mu+1)^{k+1} - \mu^{k+1} \right\} \left\{ \frac{(\lambda+1)^{r+2}}{\lambda} - \lambda^{r+1} \right\} \zeta(\{1\}^{k+1}, r+3)$$

$$- \mu^{k-1} \sum_{p=0}^{k-1} \left(\frac{\mu+1}{\mu}\right)^p (p+1) \left\{ \frac{(\lambda+1)^{r+2}}{\lambda} - \lambda^{r+1} \right\} \zeta(\{1\}^{k+1}, r+3).$$

Now we can write down the final identity after a multiplication of λ to the values.

$$
\begin{aligned}
(4.1.3) \quad & (k+1)\left\{E(\mu+1,\lambda+1) - E(\mu+1,1)\right\} \\
& + \mu^k \lambda^{r+2}\left\{E\left(\frac{\mu+1}{\mu},\frac{\lambda+1}{\lambda}\right) - E\left(\frac{\mu+1}{\mu},1\right)\right\} \\
& - \mu\left\{\frac{\partial}{\partial\mu}E(\mu+1,\lambda+1) - \frac{\partial}{\partial\mu}E(\mu+1,1)\right\} \\
& + \mu^{k-1}\lambda^{r+2}\left\{\frac{\partial}{\partial\mu}E\left(\frac{\mu+1}{\mu},\frac{\lambda+1}{\lambda}\right) - \frac{\partial}{\partial\mu}E\left(\frac{\mu+1}{\mu},1\right)\right\} \\
& = \left\{(k+1)(\mu+1)^k + \lambda^{r+2}\left[(\mu+1)^k + (k+2)\frac{(\mu+1)^k-1}{\mu}\right]\right\} \\
& \quad \times \zeta(\{1\}^{k+1},r+3) \\
& + \sum_{j=0}^{k}\sum_{\ell=0}^{r}(k-j+1)\mu^j\lambda^{\ell+1}\zeta(\{1\}^j,r-\ell+2)\zeta(\{1\}^{k-j},\ell+2).
\end{aligned}
$$

Here are a few direct consequences of (4.1.3) obtained by specifying the values of μ and λ.

1. Letting $\mu = \lambda = 1$, we get

$$
\begin{aligned}
(4.1.4) \quad & \sum_{p+q=k} 2^p \sum_{|\alpha|=q+r+3}(2^{\alpha_{q+1}}-1)\zeta(\{1\}^p,\alpha_0,\alpha_1,\ldots,\alpha_{q+1}+1) \\
& = (2^{k+1}-1)\zeta(\{1\}^{k+1},r+3) \\
& + \frac{1}{2}\sum_{j=0}^{k}\sum_{\ell=0}^{r}\zeta(\{1\}^j,r-\ell+2)\zeta(\{1\}^{k-j},\ell+2).
\end{aligned}
$$

2. Let $\mu = -1$, $\lambda = 1$ and suppose that k is even,

$$
\begin{aligned}
(4.1.5) \quad & \sum_{|\alpha|=k+r+3}(2^{\alpha_{k+1}}-1)\zeta(\alpha_0,\alpha_1,\ldots,\alpha_{k+1}+1) \\
& = \frac{1}{2}\sum_{j=0}^{k}\sum_{\ell=0}^{r}(-1)^j\zeta(\{1\}^j,r-\ell+2)\zeta(\{1\}^{k-j},\ell+2) \\
& + \zeta(\{1\}^{k+1},r+3).
\end{aligned}
$$

On the other hand, when k is odd, the identity is

$$k \sum_{|\boldsymbol{\alpha}|=k+r+3} (2^{\alpha_{k+1}} - 1)\zeta(\alpha_0, \alpha_1, \ldots, \alpha_{k+1} + 1)$$

$$+ 2 \sum_{|\boldsymbol{\alpha}|=k+r+2} (2^{\alpha_k} - 1)\zeta(1, \alpha_0, \alpha_1, \ldots, \alpha_k + 1)$$

$$= (k + 2)\zeta(\{1\}^{k+1}, r + 3)$$

$$+ \sum_{j=0}^{k}\sum_{\ell=0}^{r}(k - j + 1)(-1)^{j}\zeta(\{1\}^{j}, r - \ell + 2)\zeta(\{1\}^{k-j}, \ell + 2).$$

3. Let $\mu = 1$, $\lambda = -1$ and suppose that r is even,

$$\sum_{p+q=k} 2^p \sum_{|\boldsymbol{\alpha}|=q+r+3} \zeta(\{1\}^{p}, \alpha_0, \alpha_1, \ldots, \alpha_{q+1} + 1)$$

(4.1.6) $$= - (2^{k+1} - 1)\zeta(\{1\}^{k+1}, r + 3)$$

$$+ \frac{1}{2}\sum_{j=0}^{k}\sum_{\ell=0}^{r}(-1)^{\ell}\zeta(\{1\}^{j}, r - \ell + 2)\zeta(\{1\}^{k-j}, \ell + 2).$$

On the other hand, when r is odd, the identity is

$$\sum_{p+q=k} 2^p q \sum_{|\boldsymbol{\alpha}|=q+r+3} \zeta(\{1\}^{p}, \alpha_0, \ldots, \alpha_{q+1} + 1)$$

$$= (2^{k+1} - k + 2)\zeta(\{1\}^{k+1}, r + 3)$$

$$+ \sum_{j=0}^{k}\sum_{\ell=0}^{r}(k - j + 1)(-1)^{\ell}\zeta(\{1\}^{j}, r - \ell + 2)\zeta(\{1\}^{k-j}, \ell + 2).$$

With identities (4.1.4), (4.1.5) and (4.1.6), we get the value of $E(2, 2)$ as follows.

Corollary A. *When r is even,*

$$\sum_{p+q=k} 2^p \sum_{|\boldsymbol{\alpha}|=q+r+3} 2^{\alpha_{q+1}}\zeta(\{1\}^{p}, \alpha_0, \alpha_1, \ldots, \alpha_{q+1} + 1)$$

$$= \sum_{j=0}^{k}\sum_{\substack{\ell=0 \\ \ell:\text{even}}}^{r} \zeta(\{1\}^{j}, r - \ell + 2)\zeta(\{1\}^{k-j}, \ell + 2).$$

Corollary B. *When r is odd*

$$\sum_{p+q=k} 2^p q \sum_{|\alpha|=q+r+3} 2^{\alpha_{q+1}} \zeta(\{1\}^p, \alpha_0, \alpha_1, \ldots, \alpha_{q+1}+1)$$

$$+ \sum_{p+q=k} 2^p p \sum_{|\alpha|=q+r+3} (2^{\alpha_{q+1}} - 1) \zeta(\{1\}^p, \alpha_0, \alpha_1, \ldots, \alpha_{q+1}+1)$$

$$= (k+1)\left\{2^{k+1} - 2\right\} \zeta(\{1\}^{k+1}, r+3)$$

$$+ 2 \sum_{j=0}^{k} \sum_{\substack{\ell=0 \\ \ell:\text{odd}}}^{r} j\zeta(\{1\}^j, r-\ell+2)\zeta(\{1\}^{k-j}, \ell+2).$$

4. Let $\mu = \lambda = -1$ and suppose that $k+r$ is even,

$$\sum_{|\alpha|=k+r+3} \zeta(\alpha_0, \alpha_1, \ldots, \alpha_{k+1}+1) + (-1)^r \zeta(\{1\}^{k+1}, r+3)$$

$$= \frac{1}{2} \sum_{j=0}^{k} \sum_{\ell=0}^{r} (-1)^{j+\ell} \zeta(\{1\}^j, r-\ell+2)\zeta(\{1\}^{k-j}, \ell+2).$$

One the other hand, when $k+r$ is odd, the identity is

$$k \sum_{|\alpha|=k+r+3} \zeta(\alpha_0, \alpha_1, \ldots, \alpha_{k+1}+1)$$

$$+ 2 \sum_{|\alpha|=k+r+2} \zeta(1, \alpha_0, \ldots, \alpha_k+1)$$

$$= (-1)^{r+1}(k+2)\zeta(\{1\}^{k+1}, r+3)$$

$$+ \sum_{j=0}^{k} \sum_{\ell=0}^{r} (k-j+1)(-1)^{j+\ell} \zeta(\{1\}^j, r-\ell+2)\zeta(\{1\}^{k-j}, \ell+2)$$

or

$$k\zeta(k+r+4) + 2 \sum_{|\alpha|=k+2} \zeta(\alpha_0, \alpha_1+r+2)$$

$$= (-1)^{r+1}(k+2)\zeta(\{1\}^{k+1}, r+3)$$

$$+ \sum_{j=0}^{k} \sum_{\ell=0}^{r} (k-j+1)(-1)^{j+\ell} \zeta(\{1\}^j, r-\ell+2)\zeta(\{1\}^{k-j}, \ell+2)$$

by the restricted sum formula.

5. Setting $\mu = 0$ and $\lambda = 1$, we get

$$\sum_{p+q=k} \sum_{|\alpha|=q+r+3} (2^{\alpha_{q+1}} - 1)\zeta(\{1\}^p, \alpha_0, \ldots, \alpha_{q+1} + 1)$$

$$+ \sum_{|\alpha|=r+2} (2^{\alpha_1} - 1)\zeta(\{1\}^k, \alpha_0, \alpha_1 + 1)$$

$$= (k+2)\zeta(\{1\}^{k+1}, r+3) + \sum_{\ell=0}^{r} \zeta(r - \ell + 2)\zeta(\{1\}^k, \ell + 2),$$

or

$$\sum_{p+q=k} (1 + \delta_{pk}) \sum_{|\alpha|=q+r+3} (2^{\alpha_{q+1}} - 1)\zeta(\{1\}^p, \alpha_0, \ldots, \alpha_{q+1} + 1)$$

$$= (k+2)\zeta(\{1\}^{k+1}, r+3) + \sum_{\ell=0}^{r} \zeta(r - \ell + 2)\zeta(\{1\}^k, \ell + 2).$$

6. Setting $\mu = 0$ and $\lambda = -1$, we get the identity

$$\sum_{p+q=k} \{1 + (-1)^r \delta_{pk}\} \sum_{|\alpha|=q+r+3} \zeta(\{1\}^p, \alpha_0, \ldots, \alpha_q, \alpha_{q+1} + 1)$$

$$= -\{(-1)^r(k+1) + 1\} \zeta(\{1\}^{k+1}, r+3)$$

$$+ \sum_{\ell=0}^{r} (-1)^\ell \zeta(r - \ell + 2)\zeta(\{1\}^k, \ell + 2).$$

From 5 and 6, we obtain the value of $E(1,2)$ as follows.

Corollary C. *For a pair of nonnegative integers k and r, when r is even, we have*

$$\sum_{p+q=k} (1 + \delta_{pk}) \sum_{|\alpha|=q+r+3} 2^{\alpha_{q+1}} \zeta(\{1\}^p, \alpha_0, \ldots, \alpha_q, \alpha_{q+1} + 1)$$

$$= 2 \sum_{\substack{\ell=0, \\ \ell:\text{even}}}^{r} \zeta(r - \ell + 2)\zeta(\{1\}^k, \ell + 2).$$

Also, when r is odd, we have

$$\sum_{\substack{p+q=k,\, |\alpha|=q+r+3 \\ p\leq k-1}} 2^{\alpha_{q+1}} \zeta(\{1\}^p, \alpha_0, \ldots, \alpha_q, \alpha_{q+1}+1)$$

$$+ 2 \sum_{|\alpha|=r+3} (2^{\alpha_1}-1)\zeta(\{1\}^k, \alpha_0, \alpha_1+1)$$

$$= 2\zeta(\{1\}^{k+1}, r+3) + 2 \sum_{\substack{\ell=0, \\ \ell:\text{even}}}^{r} \zeta(r-\ell+2)\zeta(\{1\}^k, \ell+2).$$

4.2 Integrals with Three Factors

In the following, we are going to produce shuffle relations from integrals of the form

(4.2.1)

$$\frac{1}{k!q!r!} \iint_{R_1 \times R_2} \left(\xi \log\frac{1}{1-t_1} + \log\frac{1}{1-u_1} \right)^k \left(\mu \log\frac{t_2}{t_1} + \log\frac{u_2}{u_1} \right)^q$$

$$\times \left(\log\frac{1}{t_2} + \lambda \log\frac{1}{u_2} \right)^r \frac{dt_1 dt_2}{(1-t_1)t_2} \frac{du_1 du_2}{(1-u_1)u_2},$$

where R_1 is the simplex defined by $0 < t_1 < t_2 < 1$ and R_2 is the simplex defined by $0 < u_1 < u_2 < 1$. First we note that for nonnegative integers m, n and ℓ,

$$\frac{1}{m!n!\ell!} \int_{0<t_1<t_2<1} \left(\log\frac{1}{1-t_1} \right)^m \left(\log\frac{t_2}{t_1} \right)^n \left(\log\frac{1}{t_2} \right)^\ell \frac{dt_1 dt_2}{(1-t_1)t_2}$$

$$= \zeta(\{1\}^m, n+\ell+2)$$

since the integral comes from the iterated integral

$$\int_{0<t_1<\cdots<t_{m+n+\ell+2}<1} \left(\prod_{j=1}^m \frac{dt_j}{1-t_j} \right) \frac{dt_{m+1}}{1-t_{m+1}}$$

$$\times \left(\prod_{k=m+2}^{m+n+1} \frac{dt_k}{t_k} \right) \frac{dt_{m+n+2}}{t_{m+n+2}} \left(\prod_{p=m+n+3}^{m+n+\ell+2} \frac{dt_p}{t_p} \right).$$

Therefore, the value of the integral (4.2.1) is

$$\sum_{j=0}^k \sum_{m=0}^q \sum_{\ell=0}^r \xi^j \mu^m \lambda^\ell \zeta(\{1\}^j, m+r-\ell+2)\zeta(\{1\}^{k-j}, q-m+\ell+2).$$

Two basic identities play important roles during our shuffle process.

Proposition 4.2.1. *For nonnegative integers q, r and a complex number λ, we have*

$$\sum_{h+\ell=r} \binom{q+h}{q} (\lambda+1)^\ell = \frac{(q+r+1)!}{q!r!} \int_0^1 (\lambda+1-\lambda x)^r x^q dx$$

$$= \sum_{\ell=0}^r \binom{q+r+1}{\ell} \lambda^{r-\ell}.$$

Proof. Rewrite the sum as

$$(\lambda+1)^r \sum_{h=0}^r \binom{q+h}{q} \left(\frac{1}{\lambda+1}\right)^h,$$

so that it is equal to the coefficient of x^q in the polynomial in x of

$$(\lambda+1)^r \sum_{h=0}^r \left(x + \frac{1}{\lambda+1}\right)^{q+h},$$

or

$$(\lambda+1)^r \left\{ \left(x + \frac{1}{\lambda+1}\right)^{q+r+1} - \left(x + \frac{1}{\lambda+1}\right)^q \right\} \left(x - \frac{\lambda}{\lambda+1}\right)^{-1}.$$

Express both $\left(x + \frac{1}{\lambda+1}\right)^{q+r+1}$ and $\left(x + \frac{1}{\lambda+1}\right)^q$ as polynomials in $x - \frac{\lambda}{\lambda+1}$ via binomial expansions, so the coefficient of x^q in the resulting polynomial is

$$(\lambda+1)^r \sum_{\ell=q+1}^{q+r+1} \binom{q+r+1}{\ell} \binom{\ell-1}{q} \left(-\frac{\lambda}{\lambda+1}\right)^{\ell-q-1},$$

or

$$\frac{(q+r+1)!(\lambda+1)^r}{q!r!} \sum_{\ell=0}^r \binom{r}{\ell} \left(-\frac{\lambda}{\lambda+1}\right)^\ell \frac{1}{\ell+q+1}.$$

This is precisely equal to the value of the integral

$$\frac{(q+r+1)!}{q!r!} \int_0^1 (\lambda+1-\lambda x)^r x^q dx.$$

The second part is trivial so we omit the proof.   □

Proposition 4.2.2. *For a pair of nonnegative integers q, r, and complex numbers μ, λ, we have*

$$\sum_{m+n=q} (\mu+1)^m \sum_{h+\ell=r} \binom{n+h}{n} (\lambda+1)^\ell$$

$$= \frac{(q+r+2)!}{q!r!} \int_0^1 \int_0^y [(\mu+1)(1-y)+x]^q [(\lambda+1)y - \lambda x]^r \, dx dy.$$

Proof. The inner sum is equal to

$$\frac{(n+r+1)!}{n!r!} \int_0^1 (\lambda+1-\lambda x)^r x^n dx$$

or

$$(r+1)\binom{n+r+1}{r+1} \int_0^1 (\lambda+1-\lambda x)^r x^n dx$$

by our previous proposition. Next we have to find the sum of

$$\sum_{m+n=q} (\mu+1)^m \binom{n+r+1}{r+1} x^n$$

or

$$(\mu+1)^q \sum_{n=0}^q \binom{n+r+1}{r+1} \left(\frac{x}{\mu+1} \right)^n.$$

Again by Proposition 4.2.1, it is equal to

$$\frac{(q+r+2)!}{q!(r+1)!} \int_0^1 [(\mu+1)(1-y)+yx]^q y^{r+1} dy,$$

so that the sum is equal to

$$\frac{(q+r+2)!}{q!r!} \int_0^1 \int_0^1 [(\mu+1)(1-y)+yx]^q [(\lambda+1)-\lambda x]^r y^{r+1} dx dy.$$

A simple change of variables yields our assertion.   □

For convenience, we let

$$I_1(m,n;\lambda) = \frac{(m+n+1)!}{m!n!} \int_0^1 (\lambda+1-\lambda x)^n x^m dx$$

so that

$$\lambda^m I_1\left(m, n; \frac{1}{\lambda} \right) = \frac{(m+n+1)!}{m!n!} \int_0^1 (\lambda+1-x)^n x^m dx.$$

On the other hand, we let

$$
I_2(q, r; \mu, \lambda)
$$
$$
= \frac{(q+r+2)!}{q!r!} \int_0^1 \int_0^y [(\mu+1)(1-y)+x]^q [(\lambda+1)y - \lambda x]^r \, dx \, dy.
$$

First, we consider the case $k = 0$. The shuffle relation we are going to see is a weighted sum formula on Euler double sums with weights in terms of I_1 and I_2.

Theorem 4.2.3. *For a pair of nonnegative integers q, r, we have*

$$
\sum_{m+n=q} \sum_{h+\ell=r} \zeta(m+h+2, n+\ell+2) \left\{ \mu^m I_1(n, \ell; \lambda) + \mu^n \lambda^r I_1\left(n, \ell; \frac{1}{\lambda}\right) \right\}
$$
$$
+ \sum_{m+n=q} \zeta(m+1, n+r+3)(\mu^m+1) \left\{ I_2(n, r; \mu, \lambda) + \mu^n \lambda^r I_2\left(n, r; \frac{1}{\mu}, \frac{1}{\lambda}\right) \right\}
$$
$$
= \sum_{m=0}^q \sum_{\ell=0}^r \mu^m \lambda^\ell \zeta(m+r-\ell+2)\zeta(q-m+\ell+2).
$$

Proof. Consider the integral

$$
\frac{1}{q!r!} \iint_{R_1 \times R_2} \left(\mu \log \frac{t_2}{t_1} + \log \frac{u_2}{u_1} \right)^q \left(\log \frac{1}{t_2} + \lambda \log \frac{1}{u_2} \right)^r
$$
$$
\times \frac{dt_1 dt_2}{(1-t_1)t_2} \frac{du_1 du_2}{(1-u_1)u_2},
$$

where R_1 is the simplex defined by $0 < t_1 < t_2 < 1$ and R_2 is the simplex defined by $0 < u_1 < u_2 < 1$. Once we expand the factors

$$
\left(\mu \log \frac{t_2}{t_1} + \log \frac{u_2}{u_1} \right)^q \quad \text{and} \quad \left(\log \frac{1}{t_2} + \lambda \log \frac{1}{u_2} \right)^r
$$

by their binomial expansions

$$
\sum_{m=0}^q \binom{q}{m} \left(\mu \log \frac{t_2}{t_1} \right)^m \left(\log \frac{u_2}{u_1} \right)^{q-m}
$$

and

$$
\sum_{\ell=0}^r \binom{r}{\ell} \left(\log \frac{1}{t_2} \right)^{r-\ell} \left(\lambda \log \frac{1}{u_2} \right)^\ell,
$$

we see immediately that the value of the integral is

$$\sum_{m=0}^{q}\sum_{\ell=0}^{r}\mu^m\lambda^\ell\zeta(m+r-\ell+2)\zeta(q-m+\ell+2).$$

As a replacement of shuffle process, we decompose $R_1 \times R_2$ into 6 simplices obtaining from all possible interlacings of t_1, t_2 and u_1, u_2. Here are these simplices:

$$D_1 : \ 0 < t_1 < t_2 < u_1 < u_2 < 1,$$
$$D_2 : \ 0 < u_1 < u_2 < t_1 < t_2 < 1,$$
$$D_3 : \ 0 < t_1 < u_1 < t_2 < u_2 < 1,$$
$$D_4 : \ 0 < t_1 < u_1 < u_2 < t_2 < 1,$$
$$D_5 : \ 0 < u_1 < t_1 < u_2 < t_2 < 1 \text{ and}$$
$$D_6 : \ 0 < u_1 < t_1 < t_2 < u_2 < 1.$$

We need to evaluate the integrations over D_j $(j = 1, 2, \ldots, 6)$ one by one. However, the integration over D_2 is similar to the integration over D_1 and the integrations over D_3, D_4, D_5 and D_6 are similar to one another. So we only evaluate the integrations over D_1 and D_3.

On the simplex $D_1 : \ 0 < t_1 < t_2 < u_1 < u_2 < 1$, we expand the factor

$$\left(\mu\log\frac{t_2}{t_1} + \log\frac{u_2}{u_1}\right)^q$$

into

$$\sum_{m+n=q}\binom{q}{m}\mu^m\left(\log\frac{t_2}{t_1}\right)^m\left(\log\frac{u_2}{u_1}\right)^n$$

and expand the factor

$$\left(\log\frac{1}{t_2} + \lambda\log\frac{1}{u_2}\right)^r = \left[\log\frac{u_1}{t_2} + \log\frac{u_2}{u_1} + (\lambda+1)\log\frac{1}{u_2}\right]^r$$

into

$$\sum_{h+j+\ell=r}\frac{r!}{h!j!\ell!}\left(\log\frac{u_1}{t_2}\right)^h\left(\log\frac{u_2}{u_1}\right)^j\left[(\lambda+1)\log\frac{1}{u_2}\right]^\ell$$

so that the value of the integration over D_1 is

$$\sum_{m+n=q}\mu^m\sum_{h+j+\ell=r}\zeta(m+h+2, n+j+\ell+2)\binom{n+j}{j}(\lambda+1)^\ell.$$

By Proposition 4.2.1, the weight

$$\sum_{j+\ell=p} \binom{n+j}{j} (\lambda+1)^\ell = I_1(n,p;\lambda).$$

So the value is

$$\sum_{m+n=q} \mu^m \sum_{h+\ell=r} \zeta(m+h+2,n+\ell+2) I_1(n,\ell;\lambda).$$

In the same manner, we obtain the value of the integration over D_2 given by

$$\sum_{m+n=q} \mu^n \sum_{h+\ell=r} \zeta(m+h+2,n+\ell+2) \left\{ \lambda^r I_1\left(n,\ell;\frac{1}{\lambda}\right) \right\}.$$

On the simplex $D_3 :\ 0 < t_1 < u_1 < t_2 < u_2 < 1$, two factors in the integrand are substituted by

$$\sum_{a+b+c=q} \frac{q!}{a!b!c!} \left(\mu \log \frac{u_1}{t_1}\right)^a \left[(\mu+1)\log\frac{t_2}{u_1}\right]^b \left(\log\frac{u_2}{t_2}\right)^c$$

and

$$\sum_{h+\ell=r} \binom{r}{\ell} \left(\log\frac{u_2}{t_2}\right)^h \left((\lambda+1)\log\frac{1}{u_2}\right)^\ell$$

respectively.

Therefore the value of the integration over D_3 is

$$\sum_{a+b+c=q} \mu^a (\mu+1)^b \sum_{h+\ell=r} \zeta(a+1,b+c+r+3) \binom{c+h}{c}(\lambda+1)^\ell.$$

By Proposition 4.2.1, the above value is

$$\sum_{m+n=q} \mu^m \zeta(m+1,n+r+3) I_2(n,r;\mu,\lambda).$$

In the similar way, we obtain the values of the integrations over D_4, D_5 and D_6 as

$$\sum_{m+n=q} \mu^m \zeta(m+1,n+r+3) \mu^n \lambda^r I_2\left(n,r;\frac{1}{\mu},\frac{1}{\lambda}\right),$$

$$\sum_{m+n=q} \zeta(m+1, n+r+3)\mu^n \lambda^r I_2\left(n, r; \frac{1}{\mu}, \frac{1}{\lambda}\right)$$

and

$$\sum_{m+n=q} \zeta(m+1, n+r+3)I_2\left(n, r; \mu, \lambda\right).$$

$\qquad\qquad\qquad\qquad\qquad\qquad\qquad\qquad\qquad\qquad\qquad\qquad$ □

Here are some immediate consequences of Theorem 4.2.3 obtained by specifying the values of μ and λ.

(1) When $\mu = 1$ and $\lambda = 1$, the shuffle relation is

$$\sum_{m+n=q}\sum_{h+\ell=r} \zeta(m+h+2, n+\ell+2)\sum_{p=0}^{\ell}\binom{n+\ell+1}{p}$$

$$+\, 2\sum_{m+n=q} \zeta(m+1, n+r+2)\sum_{a+b=n} 2^a\sum_{\ell=0}^{r}\binom{r+b+1}{\ell}$$

$$=\frac{1}{2}\sum_{m=0}^{q}\sum_{\ell=0}^{r}\zeta(m+r-\ell+2)\zeta(q-m+\ell+2).$$

(2) When $\mu = 1$ and $\lambda = -1$ with r even, the shuffle relation is

$$\sum_{m+n=q}\sum_{h+\ell=r} \zeta(m+h+2, n+\ell+2)\binom{n+\ell}{n}$$

$$+\, 2\sum_{m+n=q} \zeta(m+1, n+r+3)\sum_{a=0}^{n}\binom{n+r+1}{a}$$

$$=\frac{1}{2}\sum_{m=0}^{q}\sum_{\ell=0}^{r}(-1)^\ell\zeta(m+r-\ell+2)\zeta(q-m+\ell+2).$$

(3) When $\mu = -1$ and $\lambda = 1$ with q even, the shuffle relation is

$$\sum_{m+n=q}(-1)^m\sum_{h+\ell=r} \zeta(m+h+2, n+\ell+2)\sum_{p=0}^{\ell}\binom{n+\ell+1}{p}$$

$$+\, 2\sum_{\substack{m+n=q\\ n:\text{even}}} \zeta(m+1, n+r+3)\sum_{\ell=0}^{r}\binom{n+r+1}{\ell}$$

$$=\frac{1}{2}\sum_{m=0}^{q}\sum_{\ell=0}^{r}(-1)^m\zeta(m+j-\ell+2)\zeta(q-m+\ell+2).$$

(4) When $\mu = \lambda = -1$ and $q + r$ is even, the shuffle relation is

$$\sum_{m+n=q} (-1)^m \sum_{h+\ell=r} \zeta(m+h+2, n+\ell+2)\binom{n+\ell}{\ell}$$

$$+ 2 \sum_{\substack{m+n=q \\ n:\text{even}}} \zeta(m+1, n+r+3)\binom{n+r}{n}$$

$$= \frac{1}{2} \sum_{m=0}^{q} \sum_{\ell=0}^{r} (-1)^{m+\ell}\zeta(m+r-\ell+2)\zeta(q-m+\ell+2).$$

4.3   Generalizations of Euler Decomposition Theorem

For any nonnegative integer r, the well-known Euler decomposition theorem for a product of two zeta values asserted that for $0 \leq \ell \leq r$,

$$\sum_{|\alpha|=r+3} \zeta(\alpha_1, \alpha_2 + 1)\left\{\binom{\alpha_2}{\ell+1} + \binom{\alpha_2}{r-\ell+1}\right\} = \zeta(\ell+2)\zeta(r-\ell+2).$$

Just sum over all ℓ with $0 \leq \ell \leq r$, it follows the weighted Euler sum formula

$$\sum_{|\alpha|=r+3} 2^{\alpha_2}\zeta(\alpha_1, \alpha_2 + 1) = \frac{r+5}{2}\zeta(r+4)$$

which was proved by Y. Ohno and W. Zudilin.

In the following, we shall separate the coefficients of $\mu^m\lambda^\ell$ on both sides of the shuffle relation in Theorem 13.3. As a result, we obtain a decomposition theorem for product $\zeta(m+r-\ell+2)\zeta(q-m+\ell+2)$. Of course, such decomposition is a generalization of Euler decomposition theorem.

Theorem 4.3.1. *For all nonnegative integers q, r, m and ℓ with $0 \leq m \leq$*

$q,\ 0 \le \ell \le r$, we have

$$\sum_{h+n=r} \zeta(m+h+2, q-m+n+2)\binom{q-m+n+1}{n-\ell}$$

$$+ \sum_{h+n=r} \zeta(q-m+h+2, m+n+2)\binom{m+n+1}{\ell-h}$$

$$+ \sum_{a+b=q} \zeta(a+1, b+r+3) \sum_{g=0}^{b} \left\{ \binom{r+g+1}{r-\ell}\binom{b-g}{m-a} \right.$$

$$+ \binom{r+g+1}{\ell}\binom{b-g}{q-m} + \binom{r+g+1}{\ell}\binom{b-g}{b-m}$$

$$\left. + \binom{r+g+1}{r-\ell}\binom{b-g}{m} \right\}$$

$$= \zeta(m+r-\ell+2)\zeta(q-m+\ell+2).$$

Proof. First, we separate the coefficient of $\mu^m \lambda^\ell$ from the value obtained from integration over D_1 given by

$$\sum_{a+b=q} \mu^a \sum_{h+n=r} \zeta(a+h+2, b+n+2)\frac{(b+n+1)!}{b!n!} \int_0^1 (\lambda+1-\lambda x)^n x^b dx.$$

The coefficient of $\mu^m \lambda^\ell$ on the above sum is zero unless $a = m$ and $n \ge \ell$. When $a = m$ and $n \ge \ell$, the coefficient of $\mu^m \lambda^\ell$ is

$$\sum_{h+n=r} \zeta(m+h+2, q-m+n+2)\frac{(q-m+n+1)!}{(q-m)!n!}\binom{n}{\ell} \int_0^1 (1-x)^\ell x^{q-m} dx$$

or

$$\sum_{h+n=r} \zeta(m+h+2, q-m+n+2)\binom{q-m+n+1}{n-\ell}.$$

In the same way, we get the coefficient of $\mu^m \lambda^\ell$ in the value obtained from integration over D_2 as

$$\sum_{h+n=r} \zeta(q-m+h+2, m+n+2)\binom{m+n+1}{\ell-h}.$$

Next, we proceed to separate the coefficient of $\mu^m \lambda^\ell$ from the value obtained from integration over D_3 given by

$$\sum_{a+b=q} \mu^a \zeta(a+1, b+r+3) \frac{(b+r+2)!}{b!r!}$$

$$\times \int_0^1 \int_0^y [(\mu+1)(1-y)+x]^b [(\lambda+1)y - \lambda x]^r \, dx \, dy.$$

The coefficient of $\mu^m \lambda^\ell$ is zero unless $a \le m$. When $a \le m$, the coefficient is given by

$$\sum_{a+b=q} \zeta(a+1, b+r+3) \frac{(b+r+2)!}{b!r!} \binom{b}{m-a} \binom{r}{\ell} I$$

with

$$I = \int_0^1 \int_0^y (1-y+x)^{q-m} (1-y)^{m-a} (y-x)^\ell y^{r-\ell} \, dx \, dy.$$

The above integral is equal to

$$\sum_{g=0}^{q-m} \binom{q-m}{g} \int_0^1 (1-y)^{b-g} y^{r-\ell} \, dy \int_0^y x^g (y-x)^\ell \, dx$$

and its value is

$$\sum_{g=0}^{q-m} \binom{q-m}{g} \frac{g! \ell!}{(g+\ell+1)!} \frac{(r+g+1)!(b-g)!}{(r+b+2)!}.$$

Therefore the coefficient of $\mu^m \lambda^\ell$ in the value obtained from integration over D_3 is

$$\sum_{a+b=q} \zeta(a+1, b+r+3) \sum_{g=0}^{q-m} \binom{r+g+1}{r-\ell} \binom{b-g}{m-a}.$$

In the same manner, we separate the coefficients of $\mu^m \lambda^\ell$ from integrations over D_4, D_5 and D_6. Here are the exact values:

$$\sum_{a+b=q} \zeta(a+1, b+r+3) \sum_{g=0}^{m-a} \binom{r+g+1}{\ell} \binom{b-g}{q-m}$$

$$\sum_{a+b=q} \zeta(a+1, b+r+3) \sum_{g=0}^{m} \binom{r+g+1}{\ell} \binom{b-g}{b-m}$$

and
$$\sum_{a+b=q} \zeta(a+1, b+r+3) \sum_{g=0}^{b-m} \binom{r+g+1}{r-\ell}\binom{b-g}{m}.$$

Note that all the upper bounds in the inner summations can be adjusted automatically. So it is harmless to write all the bounds as b.

$\square$

Not only the Euler decomposition theorem can be extended to a product of the form
$$\zeta(m+r-\ell+2)\zeta(q-m+\ell+2)$$

as we have done before, but also it can be extended to a product of two multiple zeta values of the form $\zeta(\{1\}^m, n+2)$. Indeed, we have for non-negative integers k, r and j, ℓ with $0 \le j \le k$, $0 \le \ell \le r$ that

$$\sum_{p+d=k}\binom{p}{j}\sum_{|\alpha|=d+r+3}\zeta(\{1\}^p, \alpha_0, \ldots, \alpha_d, \alpha_{d+1}+1)\binom{\alpha_{d+1}}{\ell+1}$$
$$+ \sum_{p+d=k}\binom{p}{k-j}\sum_{|\alpha|=d+r+3}\zeta(\{1\}^p, \alpha_0, \ldots, \alpha_d, \alpha_{d+1}+1)\binom{\alpha_{d+1}}{r-\ell+1}$$
$$= \zeta(\{1\}^j, r-\ell+2)\zeta(\{1\}^{k-j}, \ell+2).$$

In an attempt to obtain the decomposition for the product
$$\zeta(\{1\}^j, m+r-\ell+2)\zeta(\{1\}^{k-j}, q-m+\ell+2)$$

with three free indices k, q, r and fixed indices j, m, ℓ with $0 \le j \le k$, $0 \le m \le q$, $0 \le \ell \le r$, we begin with a more general integral

$$\frac{1}{k!q!r!}\iint_{R_1\times R_2}\left(\xi\log\frac{1}{1-t_1}+\log\frac{1}{1-u_1}\right)^k\left(\mu\log\frac{t_2}{t_1}+\log\frac{u_2}{u_1}\right)^q$$
$$\times \left(\log\frac{1}{t_2}+\lambda\log\frac{1}{u_2}\right)^r\frac{dt_1dt_2}{(1-t_1)t_2}\frac{du_1du_2}{(1-u_1)u_2}.$$

Of course, we already know the value of the above integral is

$$\sum_{j=0}^{k}\sum_{m=0}^{q}\sum_{\ell=0}^{r}\xi^j\mu^m\lambda^\ell\zeta(\{1\}^j, m+r-\ell+2)\zeta(\{1\}^{k-j}, q-m+\ell+2).$$

With the same shuffle process as described and carried out in the previous section, we get the following theorems.

Theorem 4.3.2. *For nonnegative integers* k, q, r *and nonzero complex numbers* ξ, μ, λ, *we have for* $w = k + q + r + 4$ *that*

$$
\sum_{p+c+d=k} (\xi+1)^p \sum_{m+n=q} \mu^m
$$

$$
\times \sum_{\substack{|\boldsymbol{\alpha}|=c+m+1 \\ |\boldsymbol{\beta}|=n+r+d+2}} \zeta(\{1\}^p, \alpha_0, \ldots, \alpha_c + \beta_0, \ldots, \beta_d, \beta_{d+1} + 1)
$$

$$
\times I_1(\beta_{d+1} - n - 1, n; \lambda)
$$

$$
+ \sum_{p+c+d=k} (\xi+1)^p \xi^{c+d} \sum_{m+n=q} \mu^n
$$

$$
\times \sum_{\substack{|\boldsymbol{\alpha}|=c+m+1 \\ |\boldsymbol{\beta}|=n+r+d+2}} \zeta(\{1\}^p, \alpha_0, \ldots, \alpha_c + \beta_0, \ldots, \beta_d, \beta_{d+1} + 1)
$$

$$
\times \left\{ \lambda^r I_1\left(\beta_{d+1} - n - 1, n; \frac{1}{\lambda}\right) \right\}
$$

$$
+ \sum_{p+d=k} (\xi+1)^p \sum_{m+n=q} \mu^m \sum_{\substack{|\boldsymbol{\alpha}|=w-p-1 \\ \alpha_{d+1}=n+r+2}} \zeta(\{1\}^p, \alpha_0, \ldots, \alpha_d, \alpha_{d+1} + 1)
$$

$$
\times \left\{ I_2(n, r; \mu, \lambda) + \mu^n \lambda^r I_2\left(n, r; \frac{1}{\mu}, \frac{1}{\lambda}\right) \right\}
$$

$$
+ \sum_{p+d=k} (\xi+1)^p \xi^d \sum_{m+n=q} \sum_{\substack{|\boldsymbol{\alpha}|=w-p-1 \\ \alpha_{d+1}=n+r+2}} \zeta(\{1\}^p, \alpha_0, \ldots, \alpha_d, \alpha_{d+1} + 1)
$$

$$
\times \left\{ I_2(n, r; \mu, \lambda) + \mu^n \lambda^r I_2\left(n, r; \frac{1}{\mu}, \frac{1}{\lambda}\right) \right\}
$$

$$
= \sum_{j=0}^{k} \sum_{m=0}^{q} \sum_{\ell=0}^{r} \xi^j \mu^m \lambda^\ell \zeta(\{1\}^j, m + r - \ell + 2) \zeta(\{1\}^{k-j}, q - m + \ell + 2)
$$

where for integers $m, n \geq 0$,

$$
I_1(m, n; \lambda) = \frac{(m+n+1)!}{m! n!} \int_0^1 (\lambda + 1 - \lambda x)^m x^n dx
$$

and

$$I_2(m, n; \mu, \lambda)$$
$$= \frac{(m+n+2)!}{m!n!} \int_0^1 \int_0^y \left[(\mu+1)(1-y) + x \right]^m \left[(\lambda+1)y - \lambda x \right]^n dx dy.$$

Theorem 4.3.3. *For integers* $k, q, r \geq 0$ *and* j, m, ℓ *with* $0 \leq j \leq k$, $0 \leq m \leq q$, $0 \leq \ell \leq r$, *we have for* $w = k + q + r + 4$ *that*

$$\sum_{p+c+d=k} \binom{p}{j} \sum_{n=\ell}^{r} \sum_{\substack{|\boldsymbol{\alpha}|=c+m+1, \\ |\boldsymbol{\beta}|=d+q-m+r+2, \\ \beta_{d+1}=q-m+n+1}} \zeta(\{1\}^p, \alpha_0, \ldots, \alpha_c + \beta_0, \ldots, \beta_d, \beta_{d+1}+1)$$

$$\times \binom{\beta_{d+1}}{n-\ell}$$

$$+ \sum_{p+c+d=k} \binom{p}{k-j} \sum_{n=0}^{\ell} \sum_{\substack{|\boldsymbol{\alpha}|=c+q-m+1, \\ |\boldsymbol{\beta}|=d+m+r+2, \\ \beta_{d+1}=m+n+1}} \zeta(\{1\}^p, \alpha_0, \ldots, \alpha_c + \beta_0, \ldots, \beta_{d+1}+1)$$

$$\times \binom{\beta_{d+1}}{\ell-n}$$

$$+ \sum_{p+d=k} \binom{p}{j} \sum_{b=0}^{q} \sum_{\substack{|\boldsymbol{\alpha}|=w-p-1, \\ \beta_{d+1}=r+b+2}} \zeta(\{1\}^p, \alpha_0, \ldots, \alpha_d, \alpha_{d+1}+1)$$

$$\times \sum_{g=0}^{b} \left\{ \binom{r+g+1}{r-\ell} \binom{b-g}{b+m-q} + \binom{r+g+1}{\ell} \binom{b-g}{q-m} \right\}$$

$$+ \sum_{p+d=k} \binom{p}{k-j} \sum_{b=0}^{q} \sum_{\substack{|\boldsymbol{\alpha}|=w-p-1, \\ \alpha_{d+1}=r+b+2}} \zeta(\{1\}^p, \alpha_0, \ldots, \alpha_d, \alpha_{d+1}+1)$$

$$\times \sum_{g=0}^{b} \left\{ \binom{r+g+1}{\ell} \binom{b-g}{b-m} + \binom{r+g+1}{r-\ell} \binom{b-g}{m} \right\}$$

$$= \zeta(\{1\}^j, m+r-\ell+2)\zeta(\{1\}^{k-j}, q-m+\ell+2).$$

Restrict $q = 0$ or $r = 0$ leads to the following two special cases.

Corollary 4.3.4. *For a pair of nonnegative integers* k, r *and integers* j, ℓ

with $0 \leq j \leq k$, $0 \leq \ell \leq r$, we have

$$\sum_{p+q=k} \binom{p}{j} \sum_{|\boldsymbol{\alpha}|=q+r+3} \zeta\left(\{1\}^p, \alpha_0, \ldots, \alpha_q, \alpha_{q+1}+1\right) \binom{\alpha_{q+1}}{\ell+1}$$

$$+ \sum_{p+q=k} \binom{p}{k-j} \sum_{|\boldsymbol{\alpha}|=q+r+3} \zeta\left(\{1\}^p, \alpha_0, \ldots, \alpha_q, \alpha_{q+1}+1\right) \binom{\alpha_{q+1}}{r-\ell+1}$$

$$= \zeta(\{1\}^j, r-\ell+2)\zeta(\{1\}^{k-j}, \ell+2).$$

Corollary 4.3.5. *For a pair of nonnegative integers k, q and integers j, m with $0 \leq j \leq k$, $0 \leq m \leq q$, we have*

$$\sum_{p+d=k} \binom{p}{j} \sum_{|\boldsymbol{\alpha}|=d+m+2} \zeta\left(\{1\}^p, \alpha_0, \ldots, \alpha_d, q-m+2\right)$$

$$+ \sum_{p+d=k} \binom{p}{k-j} \sum_{|\boldsymbol{\alpha}|=d+q-m+2} \zeta\left(\{1\}^p, \alpha_0, \ldots, \alpha_d, m+2\right)$$

$$+ \sum_{p+d=k} \binom{p}{j} \sum_{b=0}^{q} \sum_{\substack{|\boldsymbol{\alpha}|=q+d+3 \\ \alpha_{d+1}=b+2}} \zeta\left(\{1\}^p, \alpha_0, \ldots, \alpha_d, \alpha_{d+1}+1\right)$$

$$\times \sum_{g=0}^{b} \left\{ \binom{b-g}{q-m} + \binom{b-g}{b+m-q} \right\}$$

$$+ \sum_{p+d=k} \binom{p}{k-j} \sum_{b=0}^{q} \sum_{\substack{|\boldsymbol{\alpha}|=q+d+3 \\ \alpha_{d+1}=b+2}} \zeta\left(\{1\}^p, \alpha_0, \ldots, \alpha_d, \alpha_{d+1}+1\right)$$

$$\times \sum_{g=0}^{b} \left\{ \binom{b-g}{b-m} + \binom{b-g}{m} \right\}$$

$$= \zeta(\{1\}^j, m+2)\zeta(\{1\}^{k-j}, q-m+2).$$

Some applications in combinatorics concerning sums of binomial coefficients are worth to mention here.

Proposition 4.3.6. *For a pair of positive integers q, r and an integer ℓ with $0 \leq \ell \leq r$, then*

$$\sum_{h+n=r} \binom{q+h}{q}\binom{n}{\ell} = \binom{q+r+1}{r-\ell}.$$

Proof. For any complex number λ, by Proposition 4.2.1, we have

$$\sum_{h+n=r} \binom{q+h}{q} (\lambda + 1)^n = \frac{(q+r+1)!}{q!r!} \int_0^1 (\lambda + 1 - \lambda x)^r x^q dx.$$

The coefficient of λ^ℓ in the left of above is

$$\sum_{h+n=r} \binom{q+h}{q} \binom{n}{\ell}$$

while the coefficient on λ^ℓ in the right of the above identity is

$$\frac{(q+r+1)!}{q!r!} \binom{r}{\ell} \int_0^1 (1-x)^\ell x^q dx,$$

or

$$\binom{q+r+1}{r-\ell}.$$

Thus our assertion follows. $\qquad\square$

Remark 4.3.7. If we substitute ℓ by $r - \ell$, we obtain the identity

$$\sum_{h+n=r} \binom{q+h}{q} \binom{n}{\ell - h} = \binom{q+r+1}{\ell}.$$

Proposition 4.3.8. *For a pair of positive integers q, r and integers m, ℓ with $0 \le m \le q$, $0 \le \ell \le r$, we have*

$$\sum_{a+b=q} \binom{a}{m} \sum_{h+n=r} \binom{b+h}{b} \binom{n}{\ell} = \sum_{g=0}^{q-m} \binom{r+g+1}{r-\ell} \binom{q-g}{m}.$$

Proof. For complex numbers μ and λ, consider the polynomial function in μ and λ defined as

$$\sum_{a+b=q} (\mu + 1)^a \sum_{h+n=r} \binom{b+h}{b} (\lambda + 1)^n.$$

The coefficient of $\mu^m \lambda^\ell$ of the above polynomial is given by

$$\sum_{a+b=q} \binom{a}{m} \sum_{h+n=r} \binom{b+h}{b} \binom{n}{\ell}.$$

On the other hand, by Proposition 4.2.2, the polynomial is equal to

$$\frac{(q+r+2)!}{q!r!} \int_0^1 \int_0^y [(\mu+1)(1-y)+x]^q [(\lambda+1)y - \lambda x]^r \, dx dy$$

and its coefficient of $\mu^m \lambda^\ell$ is

$$\sum_{g=0}^{q-m} \frac{(q+r+2)!}{q!r!} \binom{q}{m} \binom{r}{\ell} \binom{q-m}{g} \int_0^1 (1-y)^{q-g} \, y^{r-\ell} dy \int_0^y (y-x)^\ell x^g dx$$

or

$$\sum_{g=0}^{q-m} \binom{r+g+1}{r-\ell} \binom{q-g}{m}.$$

Therefore our assertion follows. □

In the same manner, we have the following proposition.

Proposition 4.3.9. *For a pair of positive integers q, r and integers m, ℓ with $0 \le m \le q$, $0 \le \ell \le r$, we have*

$$\sum_{a+b=q} \binom{a}{m-b} \sum_{h+n=r} \binom{b+h}{b} \binom{n}{\ell-h} = \sum_{g=0}^m \binom{r+g+1}{\ell} \binom{q-g}{q-m}.$$

4.4 Applications of the Decomposition Theorem

Recall that the hypergeometric distribution in the theory of probability asserted that for all positive integers G, B and n.

$$\sum_{g=0}^n \binom{G}{g} \binom{B}{n-g} = \binom{G+B}{n}.$$

The following are trivial consequences of the above.

Lemma 4.4.1. *For a pair of nonnegative integers k, p with $0 \le p \le k$, we have*

$$\sum_{j=0}^k \binom{p}{j} \binom{k}{j} = \binom{k+p}{k}$$

and

$$\sum_{j=0}^k \binom{p}{k-j} \binom{k}{j} = \binom{k+p}{k}.$$

Lemma 4.4.2. *For a pair of nonnegative integers r, g, we have*

$$\sum_{\ell=0}^{r} \binom{r+g+1}{r-\ell} \binom{r}{\ell} = \binom{2r+g+1}{r}$$

and

$$\sum_{\ell=0}^{r} \binom{r+g+1}{\ell} \binom{r}{\ell} = \binom{2r+g+1}{r}.$$

Lemma 4.4.3. *For nonnegative integers r, b, n, h with $n + h = r$, we have*

$$\sum_{\ell=0}^{r} \binom{b+n+1}{n-\ell} \binom{r}{\ell} = \binom{b+r+n+1}{n}$$

and

$$\sum_{\ell=0}^{r} \binom{b+n+1}{\ell-h} \binom{r}{\ell} = \binom{b+r+n+1}{n}.$$

Proof. The upper limit of the first summation can be changed to n, so the first assertion follows from the hypergeometric distribution. Let $\ell' = \ell - h$ be a new dummy variable in the second summation. Then it is equal to

$$\sum_{\ell'=0}^{n} \binom{b+n+1}{\ell'} \binom{r}{\ell'+h}$$

or

$$\sum_{\ell'=0}^{n} \binom{b+n+1}{\ell'} \binom{r}{n-\ell'},$$

hence the second assertion follows. $\qquad\square$

Based on the previous lemmas and the decomposition theorem of $\zeta(\{1\}^j, m + r - \ell + 2)\zeta(\{1\}^{k-j}, q - m + \ell + 2)$, we get the following theorem.

Theorem 4.4.4. *For nonnegative integers k, q, r, we have*

$$\sum_{p+c+d=k} \binom{k+p}{k} \sum_{m=0}^{q} \sum_{n=0}^{r}$$

$$\times \sum_{\substack{|\alpha|=q-m+c+1 \\ |\beta|=d+m+r+2 \\ \beta_{d+1}=m+n+1}} \zeta\left(\{1\}^p, \alpha_0, \ldots, \alpha_c + \beta_0, \ldots, \beta_d, \beta_{d+1} + 1\right)$$

$$\times \binom{r + \beta_{d+1}}{n}\binom{q}{m}$$

$$+ \sum_{p+d=k} \binom{k+p}{k} \sum_{b=0}^{q} \sum_{\substack{|\alpha|=q+r+d+3 \\ \alpha_{d+1}=r+b+2}} \zeta\left(\{1\}^p, \alpha_0, \ldots, \alpha_d, \alpha_{d+1}+1\right)$$

$$\times \sum_{g=0}^{b} \binom{2r+g+1}{r}\left\{\binom{q+b-g}{q} + \binom{q+b-g}{b}\right\}$$

$$= \frac{1}{2} \sum_{j=0}^{k} \sum_{m=0}^{q} \sum_{\ell=0}^{r} \binom{k}{j}\binom{q}{m}\binom{r}{\ell}$$

$$\times \zeta(\{1\}^j, m+r-\ell+2)\zeta(\{1\}^{k-j}, q-m+\ell+2).$$

The following lemmas are used to deal with alternating sums.

Lemma 4.4.5. *For a pair of nonnegative integers* r, g, *we have*

$$\sum_{\ell=0}^{r} \binom{r+g+1}{\ell}(-1)^\ell = (-1)^r \binom{r+g}{r}$$

and

$$\sum_{\ell=0}^{r} \binom{r+g+1}{r-\ell}(-1)^\ell = \binom{r+g}{r}.$$

Proof. By Proposition 4.2.1, we have

$$\sum_{n+\ell=r} \binom{g+n}{g}(\lambda+1)^\ell = \frac{(r+g+1)!}{g!r!} \int_0^1 (\lambda+1-\lambda x)^r x^g \, dx$$

$$= \sum_{\ell=0}^{r} \binom{r+g+1}{\ell}\lambda^{r-\ell}.$$

Setting $\lambda = -1$, we get our first assertion since the left hand side is equal to

$$\binom{r+g}{g}. \qquad \qquad \square$$

In the same way, we prove the following lemma.

Lemma 4.4.6. *For nonnegative integers* m, n, h, r *with* $r = n+h$, *we have*

$$\sum_{\ell=0}^{r} \binom{m+n+1}{n-\ell}(-1)^\ell = \binom{m+n}{n}$$

and

$$\sum_{\ell=0}^{r} \binom{m+n+1}{\ell-h}(-1)^{\ell} = (-1)^r \binom{m+n}{n}.$$

Based on Lemmas 4.4.5 and 4.4.6, we obtain the following theorems.

Theorem 4.4.7. *For nonnegative integers k, q, r with $k + r$ even, then we have for $w = k + q + r + 4$,*

$$\sum_{c+d=k} \sum_{\substack{|\boldsymbol{\alpha}|=q-m+c+1 \\ |\boldsymbol{\beta}|=d+m+r+2}} \zeta(\alpha_0,\ldots,\alpha_c+\beta_0,\ldots,\beta_d,\beta_{d+1}+1) \binom{\beta_{d+1}-1}{m}\binom{q}{m}$$

$$+\sum_{b=0}^{m} \sum_{\substack{|\boldsymbol{\alpha}|=w-1 \\ \alpha_{k+1}=r+2+b}} \zeta(\alpha_0,\ldots,\alpha_k,\alpha_{k+1}+1)$$

$$\times \sum_{g=0}^{b}\left\{\binom{r+g}{r}\binom{q+b-g}{b} + (-1)^r\binom{r+g}{r}\binom{q+b-g}{q}\right\}$$

$$= \frac{1}{2}\sum_{j=0}^{k}\sum_{m=0}^{q}\sum_{\ell=0}^{r}(-1)^{j+\ell}\binom{q}{m}$$

$$\times \zeta(\{1\}^j,m+r-\ell+2)\zeta(\{1\}^{k-j},q-m+\ell+2).$$

Theorem 4.4.8. *For nonnegative integers k, q, r with $q + r$ even, then we have for $w = k + q + r + 4$*

$$\sum_{p+c+d=k}\binom{k+p}{k}\sum_{\substack{|\boldsymbol{\alpha}|=q-m+c+1 \\ |\boldsymbol{\beta}|=d+m+r+2}}\zeta(\{1\}^p,\alpha_0,\ldots,\alpha_c+\beta_0,\ldots,\beta_d,\beta_{d+1}+1)$$

$$\times\binom{\beta_{d+1}-1}{m}(-1)^{m+r}$$

$$+\sum_{p+d=k}\binom{k+p}{k}\sum_{b=0}^{q}\sum_{\substack{|\boldsymbol{\alpha}|=d+q+r+3 \\ \alpha_{d+1}=r+b+2}}\zeta(\{1\}^p,\alpha_0,\ldots,\alpha_d,\alpha_{d+1}+1)$$

$$\times\binom{r+b}{r}\left[1+(-1)^{q-b}\right]$$

$$=\frac{1}{2}\sum_{j=0}^{k}\sum_{m=0}^{q}\sum_{\ell=0}^{r}(-1)^{m+\ell}\binom{k}{j}$$

$$\times\zeta(\{1\}^j,m+r-\ell+2)\zeta(\{1\}^{k-j},q-m+\ell+2).$$

Theorem 4.4.9. *For nonnegative integers* k, q, r *with* $k + q$ *even, then we have for* $w = k + q + r + 4$ *that*

$$
\sum_{c+d=k} \sum_{m=0}^{q} \sum_{n=0}^{r} \sum_{\substack{|\boldsymbol{\alpha}|=c+m+1 \\ |\boldsymbol{\beta}|=q-m+r+d+2 \\ \beta_{d+1}=q-m+n+1}} \zeta\left(\alpha_0, \ldots, \alpha_c + \beta_0, \ldots, \beta_d, \beta_{d+1} + 1\right)
$$

$$
\times \binom{\beta_{d+1} + r}{n} (-1)^m
$$

$$
+ \sum_{a+b=q} \sum_{\substack{|\boldsymbol{\alpha}|=w-1 \\ \alpha_{k+1}=r+b+2}} \zeta\left(\alpha_0, \ldots, \alpha_k, \alpha_{k+1} + 1\right) \binom{2r+b+1}{r} \{(-1)^a + (-1)^q\}
$$

$$
= \frac{1}{2} \sum_{j=0}^{k} \sum_{m=0}^{q} \sum_{\ell=0}^{r} (-1)^{j+m} \binom{r}{\ell} \zeta(\{1\}^j, m+r-\ell+2) \zeta(\{1\}^{k-j}, q-m+\ell+2).
$$

Theorem 4.4.10. *For nonnegative integers* k, q, r *with* $k + q + r$ *even, then we have for* $w = k + q + r + 4$ *that*

$$
\sum_{c+d=k} \sum_{m=0}^{q} \sum_{n=0}^{r} \sum_{\substack{|\boldsymbol{\alpha}|=c+m+1 \\ |\boldsymbol{\beta}|=q-m+r+d+2 \\ \beta_{d+1}=q-m+n+1}} \zeta\left(\alpha_0, \ldots, \alpha_c + \beta_0, \ldots, \beta_d, \beta_{d+1} + 1\right)
$$

$$
\times \binom{\beta_{d+1} - 1}{n} (-1)^m
$$

$$
+ \sum_{a+b=q} \sum_{\substack{|\boldsymbol{\alpha}|=w-1 \\ \alpha_{k+1}=r+b+2}} \zeta\left(\alpha_0, \ldots, \alpha_k, \alpha_{k+1} + 1\right) \binom{r+b}{r} \{(-1)^a + (-1)^{q+r}\}
$$

$$
= \frac{1}{2} \sum_{j=0}^{k} \sum_{m=0}^{q} \sum_{\ell=0}^{r} (-1)^{j+m+\ell} \zeta(\{1\}^j, m+r-\ell+2) \zeta(\{1\}^{k-j}, q-m+\ell+2).
$$

4.5   Applications of Another Decomposition Theorem

A lot of double weighted sum formulas had been produced from the decomposition theorem in Corollary 4.3.4. Presently, we shall use another decomposition theorem as given in Corollary 4.3.5 as follows

(4.5.1)

$$\sum_{p+d=k} \binom{p}{j} \sum_{|\boldsymbol{\alpha}|=d+m+2} \zeta(\{1\}^p, \alpha_0, \ldots, \alpha_d, q-m+2)$$

$$+ \sum_{p+d=k} \binom{p}{k-j} \sum_{|\boldsymbol{\alpha}|=d+q-m+2} \zeta(\{1\}^p, \alpha_0, \ldots, \alpha_d, m+2)$$

$$+ \sum_{p+d=k} \binom{p}{j} \sum_{b=0}^{q} \sum_{|\boldsymbol{\alpha}|=d+q-b+1} \zeta(\{1\}^p, \alpha_0, \ldots, \alpha_d, b+3)$$

$$\times \sum_{g=0}^{b} \left\{ \binom{b-g}{q-m} + \binom{b-g}{b+m-q} \right\}$$

$$+ \sum_{p+d=k} \binom{p}{k-j} \sum_{b=0}^{q} \sum_{|\boldsymbol{\alpha}|=d+q-b+1} \zeta(\{1\}^p, \alpha_0, \ldots, \alpha_d, b+3)$$

$$\times \sum_{g=0}^{b} \left\{ \binom{b-g}{b-m} + \binom{b-g}{m} \right\}$$

$$= \zeta(\{1\}^j, m+2)\zeta(\{1\}^{k-j}, q-m+2)$$

to produce weighted sum formulas with identities from the hypergeometric distribution in probability.

Theorem 4.5.1. *For a pair of nonnegative integers k, q, we have*

$$\sum_{p+d=k} \binom{k+p}{k} \sum_{|\boldsymbol{\alpha}|=q+d+3} \zeta(\{1\}^p, \alpha_0, \ldots, \alpha_d, \alpha_{d+1}+1)\binom{\alpha_{d+1}+q}{q+1}$$

$$= \frac{1}{2} \sum_{j=0}^{k} \sum_{m=0}^{q} \binom{k}{j}\binom{q}{m}\zeta(\{1\}^j, m+2)\zeta(\{1\}^{k-j}, q-m+2).$$

Proof. Multiplying both sides of (4.5.1) by $\binom{k}{j}\binom{q}{m}$ and then summing over all $0 \le j \le k$ and $0 \le m \le q$, we obtain

(4.5.2)

$$\sum_{p+d=k} \binom{p+k}{k} \sum_{|\alpha|=d+q+3} \zeta(\{1\}^p, \alpha_0, \ldots, \alpha_d, \alpha_{d+1}+1) \binom{q}{\alpha_{d+1}-1}$$

$$+ \sum_{p+d=k} \binom{p+k}{k} \sum_{b=0}^{q} \sum_{|\alpha|=d+q-b+1} \zeta(\{1\}^p, \alpha_0, \ldots, \alpha_d, b+3)$$

$$\times \sum_{g=0}^{b} \sum_{m=0}^{q} \left\{ \binom{b-g}{b-m}\binom{q}{m} + \binom{b-g}{m}\binom{q}{m} \right\}$$

$$= \frac{1}{2} \sum_{j=0}^{k} \sum_{m=0}^{q} \binom{k}{j}\binom{q}{m} \zeta(\{1\}^j, m+2) \zeta(\{1\}^{k-j}, q-m+2)$$

simply by the following well-known identities such as

$$\sum_{j=0}^{k} \binom{p}{j}\binom{k}{j} = \sum_{j=0}^{k} \binom{p}{j}\binom{k}{k-j} = \binom{k+p}{k}$$

and

$$\sum_{j=0}^{k} \binom{p}{k-j}\binom{k}{j} = \binom{k+p}{k}.$$

On the other hand, by the hypergeometric distribution from the theory of probability, we have

$$\sum_{m=0}^{q} \binom{b-g}{m}\binom{q}{m} = \sum_{m=0}^{q} \binom{b-g}{m}\binom{q}{q-m} = \binom{q+b-g}{q}$$

and

$$\sum_{m=0}^{q} \binom{b-g}{b-m}\binom{q}{m} = \sum_{m=0}^{b} \binom{b-g}{b-m}\binom{q}{m} = \binom{q+b-g}{b}.$$

Sum over g with $0 \le g \le b$ yields

$$\binom{q+b+1}{q+1} \quad \text{and} \quad \binom{q+b+1}{b+1} - \binom{q}{b+1} = \binom{q+b+1}{q} - \binom{q}{b+1},$$

respectively. Therefore, the second term on the left of (16.2) is rewritten

as

$$\sum_{p+d=k} \binom{p+k}{k} \sum_{b=0}^{q} \sum_{|\boldsymbol{\alpha}|=d+q-b+1} \zeta(\{1\}^p, \alpha_0, \ldots, \alpha_d, b+3)$$
$$\times \left\{ \binom{q+b+2}{q+1} - \binom{q}{b+1} \right\}.$$

Let $b+2$ be a new dummy variable α_{d+1}. After a cancellation with the first term on the left of (4.5.2), we obtain our assertion. $\square$

With the same manner, we obtain the following theorems.

Theorem 4.5.2. *For a pair of nonnegative integers k, q with k even, then we have*

$$\sum_{|\boldsymbol{\alpha}|=k+q+3} \zeta(\alpha_0, \ldots, \alpha_k, \alpha_{k+1}+1) \binom{\alpha_{k+1}+q}{q+1}$$
$$= \frac{1}{2} \sum_{j=0}^{k} \sum_{m=0}^{q} (-1)^j \binom{q}{m} \zeta(\{1\}^j, m+2) \zeta(\{1\}^{k-j}, q-m+2).$$

Theorem 4.5.3. *For a pair of nonnegative integers k, q, we have*

$$\sum_{p+d=k} 2^p \sum_{|\boldsymbol{\alpha}|=q+r+3} \zeta(\{1\}^p, \alpha_0, \ldots, \alpha_d, \alpha_{d+1}+1) \binom{\alpha_{d+1}+q}{q+1}$$
$$= \frac{1}{2} \sum_{j=0}^{k} \sum_{m=0}^{q} \binom{q}{m} \zeta(\{1\}^j, m+2) \zeta(\{1\}^{k-j}, q-m+2).$$

Remark 4.5.4. A counterpart of Theorem 4.5.2 when k is odd can be obtained by multiplying (4.5.1) by $(-1)^j (j+1) \binom{q}{m}$ and then summing over all $0 \le j \le k$ and $0 \le m \le q$. The result is

$$k \sum_{|\boldsymbol{\alpha}|=k+q+3} \zeta(\alpha_0, \ldots, \alpha_k, \alpha_{k+1}+1) \binom{\alpha_{k+1}+q}{q+1}$$
$$+ 2 \sum_{|\boldsymbol{\alpha}|=k+q+2} \zeta(1, \alpha_0, \ldots, \alpha_{k-1}, \alpha_k+1) \binom{\alpha_{k+1}+q}{q+1}$$
$$= \sum_{j=0}^{k} \sum_{m=0}^{q} (-1)^j (j+1) \binom{q}{m} \zeta(\{1\}^j, m+2) \zeta(\{1\}^{k-j}, q-m+2).$$

In our next considerations, we employ the identity produced from the expectation of hypergeometric distribution in the theory of probability. For positive integers G, B and n,

$$\sum_{g=0}^{n} \binom{G}{g}\binom{B}{n-g} g = \frac{nG}{G+B}\binom{G+B}{n}.$$

Lemma 4.5.5. *For a pair of nonnegative integers q, b with $0 \le b \le q$, we have*

$$\sum_{g=0}^{b}\sum_{m=0}^{q}\binom{b-g}{m}\binom{q}{q-m}m = q\binom{q+b}{q+1}.$$

Proof. By the expectation of hypergeometric distribution, the inner sum is equal to

$$\frac{q(b-g)}{q+(b-g)}\binom{q+b-g}{q}.$$

Therefore we rewrite the double sum as

$$q\sum_{g=0}^{b-1}\frac{(q+b-g-1)!}{q!(b-g-1)!},$$

which is equal to

$$q\binom{q+b}{q+1}.$$

In the same manner, we obtain the following. $\square$

Lemma 4.5.6. *For a pair of nonnegative integers q, b with $0 \le b \le q$, we have*

$$\sum_{g=0}^{b}\sum_{m=0}^{q}\binom{b-g}{q-m}\binom{q}{m}m = q\binom{q+b}{q}.$$

Lemma 4.5.7. *For a pair of nonnegative integers q, a, b with $a+b = q$, we have*

$$\sum_{g=0}^{b}\sum_{m=0}^{q}\binom{b-g}{m-a}\binom{q}{q-m}m = q\binom{q+b}{b+1}.$$

Proof. Let $m' = m - a$ be a new dummy variable in place of m. Then the inner sum is equal to

$$\sum_{m'=0}^{b}\binom{b-g}{m'}\binom{q}{b-m'}(m'+a),$$

or

$$a\binom{q+b-g}{b} + \frac{b(b-g)}{q+(b-g)}\binom{q+b-g}{b},$$

or

$$a\binom{q+b-g}{b} + \frac{b(q-g)}{q+(b-g)}\binom{q+b-g}{b} - \frac{ab}{q+(b-g)}\binom{q+b-g}{b},$$

or

$$a\binom{q+b-g}{b} + b\binom{q+b-g-1}{b} - a\binom{q+b-g-1}{b-1}.$$

Sum over g from 0 to b yields our assertion. $\qquad\square$

Lemma 4.5.8. *For a pair of nonnegative integers* q, b *with* $0 \le b \le q$, *we have*

$$\sum_{g=0}^{b}\sum_{m=0}^{q}\binom{b-g}{b-m}\binom{q}{m}m = q\binom{q+b}{b} - q\binom{q-1}{b}.$$

Proof. Rewrite the inner sum as

$$b\sum_{m=0}^{b}\binom{b-g}{b-m}\binom{q}{m} - \sum_{m=0}^{b}\binom{b-g}{b-m}\binom{q}{m}(b-m),$$

which is equal to

$$b\binom{q+b-g}{b} - b\binom{q+b-g-1}{b} + a\binom{q+b-g-1}{b-1}.$$

So we obtain our assertion with a similar procedure as the previous lemma.

$\qquad\square$

Based on Lemmas 4.5.5–4.5.8, we obtain another counterpart of Theorem 16.2.

Theorem 4.5.9. *For a pair of nonnegative integers* k, q *with* k *odd, then we have*

$$q\sum_{|\alpha|=k+r+3}\zeta(\alpha_0,\dots,\alpha_k,\alpha_{k+1}+1)\left\{\binom{q+\alpha_{k+1}-2}{q-1} - \binom{q+\alpha_{k+1}-2}{q+1}\right\}$$

$$= \sum_{j=0}^{k}\sum_{m=1}^{q}(-1)^j m\binom{q}{m}\zeta(\{1\}^j, m+2)\zeta(\{1\}^{k-j}, q-m+2).$$

4.6   Exercises

1. For a pair of nonnegative integers q, r, show that

$$\sum_{m+n=r} \binom{q+m}{q} \lambda^m (\lambda+1)^n = \frac{(q+r+1)!}{q!r!} \int_0^1 (\lambda+1-x)^r x^q dx$$

and the coefficient of λ^ℓ $(0 \le \ell \le r)$ of the above polynomial is given by

$$\binom{q+r+1}{\ell}.$$

2. For a pair of nonnegative integers q, r, show that

$$\sum_{m+n=r} \binom{q+n}{q} (\lambda+1)^n = \frac{(q+r+1)!}{q!r!} \int_0^1 (1+\lambda x)^r x^q dx$$

and the coefficient of λ^ℓ $(0 \le \ell \le r)$ of the above polynomial is given by

$$\binom{q+\ell}{\ell} \binom{q+r+1}{r-\ell}.$$

Hint: Note that

$$\frac{1}{(1-x)^{m+1}} = \sum_{a=0}^{\infty} \binom{m+a}{m} x^a$$

and

$$\frac{1}{[1-(\mu+1)x]^{n+1}} = \sum_{b=0}^{\infty} \binom{n+b}{n} (\mu+1)^b x^b$$

so that the given sum is just the coefficient of x^k in the power series expansion at $x = 0$ of the product

$$\frac{1}{(1-x)^{m+1}} \cdot \frac{1}{[1-(\mu+1)x]^{n+1}}.$$

3. Find the coefficient of μ^j $(0 \le j \le k)$ of the polynomial

$$\sum_{a+b=k} \binom{m+a}{m} \binom{n+b}{n} (\mu+1)^b.$$

4. Show that

$$\sum_{\ell=0}^{r} \binom{n+\ell}{\ell} \binom{m+n+r+1}{r-\ell} (-1)^{\ell} = \binom{m+r}{m}$$

for nonnegative integers m, n and r.

CHAPTER 5

Multiple Zeta Values of Height Two

Multiple zeta values of height two appear to be the form $\zeta(\{1\}^a, m + 2, \{1\}^b, n+2)$ which is interesting in its own way. Through shuffle products of a sum of multiple zeta values and a multiple zeta value of height one, we are able to evaluate a sum of multiple zeta values in terms of single zeta values.

The usual way to obtain a shuffle relation from the shuffle product of a sum of multiple zeta values and a multiple zeta value of height one is to begin with some particular integrals with parameters over the product of two simplices of dimension two. We carry out the process in the final section.

5.1 Sums of Multiple Zeta Values of Height Two

Let k and r be nonnegative integers. Then multiple zeta values of depth $k + 2$, weight $k + r + 4$ and height two appear to be the form

$$\zeta(\{1\}^a, m + 2, \{1\}^b, n + 2)$$

with $a+b=k$ and $m+n=r$. In this section, we shall consider some special kind of integrals which will produce particular sums of multiple zeta values of height two such as

$$\sum_{a+b=k}\sum_{m+n=r}\zeta(\{1\}^a, m+2, \{1\}^b, n+2),$$

$$\sum_{a+b=k}(-1)^b\sum_{m+n=r}\zeta(\{1\}^a, m+2, \{1\}^b, n+2),$$

$$\sum_{a+b=k}\sum_{m+n=r}(-1)^n\zeta(\{1\}^a, m+2, \{1\}^b, n+2)$$

and

$$\sum_{a+b=k}(-1)^b\sum_{m+n=r}(-1)^n\zeta(\{1\}^a, m+2, \{1\}^b, n+2)$$

after the shuffle process. Also some linear combinations of the above sums can be evaluated in terms of single zeta values.

Theorem 5.1.1. *For a pair of nonnegative integers k, r, we have*

$$\sum_{a+b=k}\sum_{m+n=r}\zeta(\{1\}^a, m+2, \{1\}^b, n+2)\{(-1)^n + (-1)^{b+n+1} + (-1)^{b+r}\}$$

$$= (-1)^{k+1}\zeta(k+r+4) - \{1 + (-1)^r + (-1)^{k+r}\}\zeta(\{1\}^{k+1}, r+3)$$

$$+ \sum_{j=0}^{k}\sum_{\ell=0}^{r}(-1)^{j+\ell}\zeta(j+r-\ell+2)\zeta(\{1\}^{k-j}, \ell+2).$$

Proof. Consider the integral

$$\frac{1}{k!r!}\iint_{R_1 \times R_2}\left(-\log\frac{1-t_1}{1-t_2} + \log\frac{1}{1-u_1}\right)^k\left(\log\frac{t_2}{t_1} - \log\frac{u_2}{u_1}\right)^r$$

$$\times\frac{dt_1 dt_2}{(1-t_1)t_2}\frac{du_1 du_2}{(1-u_1)u_2}$$

with $R_1 : 0 < t_1 < t_2 < 1$ and $R_2 : 0 < u_1 < u_2 < 1$. Expanding the factors of the integrand by the binomial theorem, the value of the integral is

$$\sum_{j=0}^{k}\sum_{\ell=0}^{r}(-1)^{j+\ell}\zeta(j+r-\ell+2)\zeta(\{1\}^{k-j}, \ell+2).$$

As usual, we decompose $R_1 \times R_2$ into 6 simplices D_j $(j = 1, 2, 3, 4, 5, 6)$ obtaining from all possible interlacings of t_1, t_2 and u_1, u_2. Here are the values of integrations over D_j $(j = 1, 2, 3, 4, 5, 6)$ in terms of multiple zeta values:

(1)
$$\sum_{a+b=k} \sum_{m+n=r} (-1)^n \zeta(\{1\}^a, m+2, \{1\}^b, n+2),$$

(2)
$$\sum_{a+b=k} (-1)^b \sum_{m+n=r} (-1)^m \sum_{|\boldsymbol{\alpha}|=b+n+1} \zeta(\{1\}^a, m+2, \alpha_0, \ldots, \alpha_b + 1),$$

(3)
$$\sum_{a+b=k} (-1)^b \sum_{m+n=r} (-1)^n \zeta(\{1\}^a, m+1, \{1\}^b, n+3),$$

(4)
$$\sum_{a+b=k} (-1)^b \sum_{m+n=r} \sum_{|\boldsymbol{\alpha}|=b+n+2} \zeta(\{1\}^a, m+1, \alpha_0, \ldots, \alpha_b + 1),$$

(5)
$$\sum_{a+b=k} (-1)^b \sum_{m+n=r} (-1)^m \sum_{|\boldsymbol{\alpha}|=b+n+2} \zeta(\{1\}^a, m+1, \alpha_0, \ldots, \alpha_b + 1),$$

and

(6)
$$\sum_{a+b=k} (-1)^b \sum_{m+n=r} (-1)^r \zeta(\{1\}^a, m+1, \{1\}^b, n+3).$$

Note that the sum of (2) and (5) is equal to

$$\sum_{a+b=k} (-1)^b \sum_{|\boldsymbol{\alpha}|=b+r+2} \zeta(\{1\}^{a+1}, \alpha_0, \ldots, \alpha_b + 1)$$
$$+ (-1)^r \sum_{a+b=k} (-1)^b \zeta(\{1\}^a, r+2, \{1\}^b, 2)$$

while (4) can be rewritten as

$$\sum_{a+b=k} (-1)^b \sum_{|\boldsymbol{\alpha}|=b+r+3} \zeta(\{1\}^a, \alpha_0, \alpha_1, \ldots, \alpha_{b+1} + 1)$$
$$- \sum_{a+b=k} (-1)^b \zeta(\{1\}^a, r+2, \{1\}^b, 2).$$

Combine together to yield

$$(-1)^k \zeta(k+r+4) + \zeta(\{1\}^{k+1}, r+3)$$
$$+ \{-1 + (-1)^r\} \sum_{a+b=k} (-1)^b \zeta(\{1\}^a, r+2, \{1\}^b, 2).$$

On the other hand, we rewrite (3) and (6) as

$$- \sum_{a+b=k} (-1)^b \sum_{m+n=r} (-1)^n \zeta(\{1\}^a, m+2, \{1\}^b, n+2)$$
$$+ (-1)^r \frac{1+(-1)^k}{2} \zeta(\{1\}^{k+1}, r+3) + \sum_{a+b=k} (-1)^b \zeta(\{1\}^a, r+2, \{1\}^b, 2)$$

and

$$\sum_{a+b=k} (-1)^b \sum_{m+n=r} (-1)^r \zeta(\{1\}^a, m+2, \{1\}^b, n+2)$$
$$+ (-1)^r \frac{1+(-1)^k}{2} \zeta(\{1\}^{k+1}, r+3)$$
$$- (-1)^r \sum_{a+b=k} (-1)^b \zeta(\{1\}^a, r+2, \{1\}^b, 2).$$

Consequently, the total sum is

$$\sum_{a+b=k} \sum_{m+n=r} \zeta(\{1\}^a, m+2, \{1\}^b, n+2)\{(-1)^n + (-1)^{b+n+1} + (-1)^{b+r}\}$$
$$+ (-1)^k \zeta(k+r+4) + \{1 + (-1)^r + (-1)^{k+r}\} \zeta(\{1\}^{k+1}, r+3),$$

and hence our assertion follows.   □

Remark 5.1.2. If we consider the integral

$$\frac{1}{k!r!} \iint_{R_1 \times R_2} \left(-\log \frac{1-t_1}{1-t_2} + \log \frac{1}{1-u_1}\right)^k \left(\log \frac{t_2}{t_1} - \log \frac{1}{u_2}\right)^r$$
$$\times \frac{dt_1 dt_2}{(1-t_1)t_2} \frac{du_1 du_2}{(1-u_1)u_2}$$

instead, we obtain the same identity after the shuffle process as demonstrated as above.

To produce sums of multiple zeta values of the particular form

$$\sum_{a+b=k} \sum_{m+n=r} \zeta(\{1\}^a, m+2, \{1\}^b, n+2)$$

which is a sum of multiple zeta values of depth $k + 2$, weight $k + r + 4$ and height two, we are going to carry out the shuffle process with the integral

$$\frac{1}{k!r!} \iint_{R_1 \times R_2} \left(-\log \frac{1 - t_1}{1 - t_2} + \log \frac{1}{1 - u_1} \right)^k \left(\log \frac{t_2}{t_1} + \log \frac{1}{u_2} \right)^r$$
$$\times \frac{dt_1 dt_2}{(1 - t_1)t_2} \frac{du_1 du_2}{(1 - u_1)u_2}$$

which has the value

$$\sum_{j=0}^{k} \sum_{\ell=0}^{r} (-1)^j \zeta(j + r - \ell + 2) \zeta(\{1\}^{k-j}, \ell + 2).$$

We need the following proposition to transform from a sum of multiple zeta values into another sum.

Proposition 5.1.3. *For a pair of integers $k, r \geq 0$, we have*

$$\sum_{n+\ell=r} \sum_{|\boldsymbol{\alpha}|=k+n+2} \zeta(\alpha_0, \ldots, \alpha_k, \alpha_{k+1} + \ell + 2)$$
$$= \sum_{m+n=r} \sum_{|\boldsymbol{\alpha}|=k+n+2} \zeta(\{1\}^{m+1}, \alpha_0, \ldots, \alpha_n + 1).$$

Proof. Consider the integral

$$\frac{1}{k!r!} \int_D \left(\log \frac{1 - u_2}{1 - u_3} \right)^k \left(\log \frac{u_4}{u_1} \right)^r \frac{du_1}{1 - u_1} \frac{du_2}{1 - u_2} \frac{du_3}{u_3} \frac{du_4}{u_4}$$

with $D : 0 < u_1 < u_2 < u_3 < u_4 < 1$. Substituting the second factor of the integrand by

$$\sum_{m+n+\ell=r} \frac{r!}{m!n!\ell!} \left(\log \frac{u_2}{u_1} \right)^m \left(\log \frac{u_3}{u_2} \right)^n \left(\log \frac{u_4}{u_3} \right)^{\ell},$$

the value of the integration is

$$\sum_{m+n+\ell=r} \sum_{|\boldsymbol{\alpha}|=k+n+2} \zeta(m + 1, \alpha_0, \alpha_1, \ldots, \alpha_k + \ell + 2).$$

Rewrite it as

$$\sum_{n+\ell=r} \sum_{|\boldsymbol{\alpha}|=k+n+2} \zeta(\alpha_0, \ldots, \alpha_k, \alpha_{k+1} + \ell + 2).$$

On the other hand, under the change of variables:

$$v_1 = 1 - u_4, \quad v_2 = 1 - u_3, \quad v_3 = 1 - u_2, \quad v_4 = 1 - u_1$$

the integral is

$$\frac{1}{k!r!} \int_D \left(\log \frac{1 - v_1}{1 - v_4} \right)^r \left(\log \frac{v_3}{v_2} \right)^k \frac{dv_1}{1 - v_1} \frac{dv_2}{1 - v_2} \frac{dv_3}{v_3} \frac{dv_4}{v_4}.$$

Substituting the first factor of the integrand by

$$\sum_{m+n+\ell=r} \frac{r!}{m!n!\ell!} \left(\log \frac{1 - v_1}{1 - v_2} \right)^m \left(\log \frac{1 - v_2}{1 - v_3} \right)^n \left(\log \frac{1 - v_3}{1 - v_4} \right)^\ell$$

the value of the integral is

$$\sum_{m+n=r} \sum_{|\boldsymbol{\alpha}|=n+k+2} \zeta(\{1\}^{m+1}, \alpha_0, \ldots, \alpha_n + 1).$$

Thus our identity is proved.   □

Here we present the details of shuffle process in order to evaluate all the integrations over D_j $(j = 1, 2, 3, 4, 5, 6)$ in terms of multiple zeta values.

(1) On the simplex $D_1 : 0 < t_1 < t_2 < u_1 < u_2 < 1$, we replace the first factor of the integrand by

$$\sum_{a+b=k} \binom{k}{a} \left(\log \frac{1}{1 - t_1} \right)^a \left(\log \frac{1 - t_2}{1 - u_1} \right)^b$$

and the other factor by its binomial expansion

$$\sum_{m+n=r} \binom{r}{m} \left(\log \frac{t_2}{t_1} \right)^m \left(\log \frac{1}{u_2} \right)^n.$$

Consequently, the value of the integration over D_1 is

$$\sum_{a+b=k} \sum_{m+n=r} \zeta(\{1\}^a, m + 2, \{1\}^b, n + 2).$$

(2) On the simplex $D_2 : 0 < u_1 < u_2 < t_1 < t_2 < 1$, we replace two factors of the integrand by

$$\sum_{a+b=k} (-1)^b \binom{k}{a} \left(\log \frac{1}{1 - u_1} \right)^a \left(\log \frac{1 - t_1}{1 - t_2} \right)^b$$

and

$$\sum_{m+n+\ell=r} 2^n \frac{r!}{m!n!\ell!} \left(\log \frac{t_1}{u_2}\right)^m \left(\log \frac{t_2}{t_1}\right)^n \left(\log \frac{1}{t_2}\right)^\ell$$

respectively. So that the value of the integration over D_2 is

$$\sum_{a+b=k} (-1)^b \sum_{m+n+\ell=r} 2^n \sum_{|\alpha|=n+b+1} \zeta(\{1\}^a, m+2, \alpha_0, \ldots, \alpha_b+\ell+1)$$

which can be rewritten as

$$\sum_{a+b=k} (-1)^b \sum_{|\alpha|=b+r+3} \zeta(\{1\}^a, \alpha_0, \alpha_1, \ldots, \alpha_{b+1}+1)C(b)$$

with

$$C(b) = \begin{cases} 0, & \text{if } \alpha_0 = 1; \\ 2^{r+3-\alpha_0}(1 - 2^{-\alpha_{b+1}}), & \text{if } \alpha_0 \geq 2. \end{cases}$$

(3) There are possible cancelations among terms from the values of integrations over D_3 and D_6 so we evaluate them at the same time. The value of the integration over D_3 is

$$\sum_{a+b=k} (-1)^b \sum_{n+\ell=r} \sum_{|\alpha|=b+n+2} \zeta(\{1\}^a, \alpha_0, \ldots, \alpha_b, \alpha_{b+1}+\ell+2)$$

while the value of the integration over D_6 is

$$\sum_{a+b=k} (-1)^b \sum_{n+\ell=r} \sum_{|\alpha|=b+n+1} \zeta(\{1\}^{a+1}, \alpha_0, \ldots, \alpha_b+\ell+2).$$

After the cancelation, only the leading term of the first sum with $a = 0$ and $b = k$, and the final term of the second sum with $a = k$ and $b = 0$ survive. Therefore the sum is

$$(-1)^k \sum_{n+\ell=r} \sum_{|\alpha|=k+n+2} \zeta(\alpha_0, \ldots, \alpha_k, \alpha_{k+1}+\ell+2) + (r+1)\zeta(\{1\}^{k+1}, r+3).$$

By Proposition 5.1.3, it is equal to

$$(-1)^k \sum_{m+n=r} \sum_{|\alpha|=n+k+2} \zeta(\{1\}^{m+1}, \alpha_0, \ldots, \alpha_n+1) + (r+1)\zeta(\{1\}^{k+1}, r+3).$$

(4) The values of the integrations over D_4 and D_5 are hard to express directly. However if we take the dual under the change of variables $t \mapsto 1-t$

as we have done in Proposition 5.1.3, then the expressions are much easier. Furthermore, there are also possible cancelations. Two values produced from integrations over D_4 and D_5 with duality are

$$\sum_{m+n+\ell=r} 2^n \sum_{a+b+c=k} (-1)^{k+c} \sum_{\substack{|\boldsymbol{\alpha}|=n+a+1 \\ |\boldsymbol{\beta}|=\ell+b+2}} \zeta(\{1\}^m, \alpha_0, \ldots, \alpha_n, \beta_0, \ldots, \beta_\ell+c+1)$$

and

$$\sum_{m+n+\ell=r} 2^n \sum_{a+b+c=k} (-1)^{k+c} \sum_{\substack{|\boldsymbol{\alpha}|=n+a+1 \\ |\boldsymbol{\beta}|=\ell+b+1}} \zeta(\{1\}^m, \alpha_0, \ldots, \alpha_n, \beta_0, \ldots, \beta_\ell+c+2).$$

After cancelation, only the leading term of the first sum with $c = 0$ and the final term of the second sum with $c = k - a$ survive. Therefore the sum is

$$(-1)^k \sum_{m+n+\ell=r} 2^n \sum_{a+b=k} \sum_{\substack{|\boldsymbol{\alpha}|=n+a+1 \\ |\boldsymbol{\beta}|=\ell+b+2}} \zeta(\{1\}^m, \alpha_0, \ldots, \alpha_n, \beta_0, \ldots, \beta_\ell + 1)$$

$$+ (-1)^k \sum_{m+n+\ell=r} 2^n \sum_{a+b=k} (-1)^b \sum_{|\boldsymbol{\alpha}|=n+a+1} \zeta(\{1\}^m, \alpha_0, \ldots, \alpha_n, \{1\}^\ell, b + 3)$$

which we rewrite as

$$(-1)^k \sum_{m+n=r} (2^{n+1} - 1) \sum_{|\boldsymbol{\alpha}|=n+k+3} \zeta(\{1\}^m, \alpha_0, \ldots, \alpha_n, \alpha_{n+1} + 1)$$

$$+ \sum_{m+n+\ell=r} 2^n \sum_{a+b=k} (-1)^{k+b+1} \sum_{|\boldsymbol{\alpha}|=n+a+2} \zeta(\{1\}^m, \alpha_0, \ldots, \alpha_n, \{1\}^\ell, b + 2)$$

$$+ (2^{r+2} - r - 3) \zeta(\{1\}^{r+1}, k + 3).$$

By the change of variable: $t \mapsto 1 - t$, the second term is equal to

$$\sum_{a+b=k} (-1)^{k+b+1} \sum_{m+n+\ell=r} 2^n \sum_{|\boldsymbol{\beta}|=n+b+2} \zeta(\{1\}^a, m+1+\beta_0, \ldots, \beta_b, \beta_{b+1}+\ell+1),$$

or

$$- \sum_{a+b=k} (-1)^b \sum_{|\boldsymbol{\alpha}|=r+b+3} \zeta(\{1\}^a, \alpha_0, \ldots, \alpha_b, \alpha_{b+1} + 1)\widetilde{C}(b)$$

with

$$\widetilde{C}(b) = 2^{r+2} \left(1 - 2^{1-\alpha_0}\right) \left(1 - 2^{-\alpha_{b+1}}\right).$$

Now we are ready to write down the shuffle relation resulted from our particular integral.

Theorem 5.1.4. *For a pair of integers $k, r \geq 0$, we have*

$$\sum_{a+b=k} \sum_{m+n=r} \zeta(\{1\}^a, m+2, \{1\}^b, n+2)$$

$$+ (-1)^k \sum_{m+n=r} 2^{n+1} \sum_{|\alpha|=n+k+3} \zeta(\{1\}^m, \alpha_0, \ldots, \alpha_n, \alpha_{n+1}+1)$$

$$+ \sum_{a+b=k} (-1)^b \sum_{|\alpha|=b+r+3} \zeta(\{1\}^a, \alpha_0, \ldots, \alpha_b, \alpha_{b+1}+1) \left\{ C(b) - (-1)^k \widetilde{C}(b) \right\}$$

$$= - (-1)^k \zeta(k+r+4) - \left\{ 2^{r+2} - 2 + (-1)^k \right\} \zeta(\{1\}^{k+1}, r+3)$$

$$+ \sum_{j=0}^{k} \sum_{\ell=0}^{r} (-1)^j \zeta(j+r-\ell+2) \zeta(\{1\}^{k-j}, \ell+2)$$

with

$$C(b) = \begin{cases} 0, & \text{if } \alpha_0 = 1; \\ 2^{r+3-\alpha_0} \left(1 - 2^{-\alpha_{b+1}}\right), & \text{if } \alpha_0 \geq 2 \end{cases}$$

and

$$\widetilde{C}(b) = 2^{r+2} \left(1 - 2^{1-\alpha_0}\right) \left(1 - 2^{-\alpha_{b+1}}\right).$$

Remark 5.1.5. If we make a change of variable: $t \mapsto 1 - t$, the value of the integration over D_2 is

$$\sum_{m+n+\ell=r} 2^n \sum_{a+b=k} (-1)^a \sum_{|\alpha|=a+n+1} \zeta(\{1\}^m, \alpha_0, \ldots, \alpha_n+1, \{1\}^\ell, b+2)$$

which can be decomposed as the difference

$$\sum_{m+n=r} 2^{n+1} \sum_{a+b=k} (-1)^a \sum_{|\alpha|=a+n+2} \zeta(\{1\}^m, \alpha_0, \ldots, \alpha_n, b+2)$$

$$- \sum_{m+n+\ell=r} 2^n \sum_{a+b=k} (-1)^a \sum_{|\alpha|=a+n+1} \zeta(\{1\}^m, \alpha_0, \ldots, \alpha_n, \{1\}^\ell, b+2).$$

Make another change of variable: $t \mapsto 1 - t$, the above difference is equal to

$$\sum_{a+b=k} (-1)^b \sum_{n+\ell=r} 2^{n+1} \sum_{|\alpha|=n+b+2} \zeta(\{1\}^a, \alpha_0+1, \ldots, \alpha_b, \alpha_{b+1}+\ell+1)$$

$$- \sum_{a+b=k} (-1)^b \sum_{m+n+\ell=r} 2^n \sum_{|\alpha|=n+b+2} \zeta(\{1\}^a, \alpha_0+m+1, \ldots, \alpha_b, \alpha_{b+1}+\ell+1).$$

5.2   Weighted Sums of Multiple Zeta Values of Height Two

A slight change in the second factor of the integrand of (5.1.1) will produce some interesting weighted sums of multiple zeta values of height two. Here we consider the integral

$$\frac{1}{k!r!} \iint_{R_1 \times R_2} \left(-\log \frac{1-t_1}{1-t_2} + \log \frac{1}{1-u_1} \right)^k \left(\log \frac{t_2}{t_1} - \log \frac{1}{u_1} \right)^r$$
$$\times \frac{dt_1 dt_2}{(1-t_1)t_2} \frac{du_1 du_2}{(1-u_1)u_2}$$

with the value

$$\sum_{j=0}^{k} \sum_{\ell=0}^{r} (-1)^{j+\ell} (\ell+1) \zeta(j+r-\ell+2) \zeta(\{1\}^{k-j}, \ell+2).$$

As usual, we have to evaluate the integration over six simplices obtained from all possible interlacings of t_1, t_2 and u_1, u_2. The integrations over D_1 : $0 < t_1 < t_2 < u_1 < u_2 < 1$, D_2 : $0 < u_1 < u_2 < t_1 < t_2 < 1$, D_3 : $0 < t_1 < u_1 < t_2 < u_2 < 1$ and D_6 : $0 < u_1 < t_1 < t_2 < u_2 < 1$ are easily to be evaluated. These values are given by

$$\sum_{a+b=k} \sum_{m+n=r} (-1)^n (n+1) \zeta(\{1\}^a, m+2, \{1\}^b, n+2),$$

$$\sum_{a+b=k} (-1)^b \sum_{m+n=r} (-1)^r (m+1) \zeta(\{1\}^a, m+2, \{1\}^b, n+2),$$

$$\sum_{a+b=k} (-1)^b \sum_{m+n=r} (-1)^n (n+1) \zeta(\{1\}^a, m+1, \{1\}^b, n+3)$$

and

$$\sum_{a+b=k} (-1)^b \sum_{m+n=r} (-1)^r (n+1) \zeta(\{1\}^a, m+1, \{1\}^b, n+3),$$

respectively.

Combining together, the total sum is

$$(r+1) \sum_{a+b=k} \sum_{m+n=r} \zeta(\{1\}^a, m+2, \{1\}^b, n+2)(-1)^{b+r}$$

$$+ \sum_{a+b=k} \sum_{m+n=r} \zeta(\{1\}^a, m+2, \{1\}^b, n+2) \left\{(-1)^n(n+1) - (-1)^{b+n} n\right\}$$

$$+ \frac{1}{2}(r+1)\left(1 + (-1)^k + (-1)^r + (-1)^{k+r}\right) \zeta(\{1\}^{k+1}, r+3).$$

On the other hand, the values of integrations over $D_4 : 0 < t_1 < u_1 < u_2 < t_2 < 1$ and $D_5 : 0 < u_1 < t_1 < u_2 < t_2 < 1$ are

$$\sum_{a+b=k} (-1)^b \sum_{m+n=r} (-1)^n \sum_{|\boldsymbol{\alpha}|=b+2} \zeta(\{1\}^a, m+1, \alpha_0, \ldots, \alpha_b + n + 1)$$

and

$$\sum_{a+b=k} (-1)^b \sum_{m+n=r} (-1)^r \sum_{|\boldsymbol{\alpha}|=b+2} \zeta(\{1\}^a, m+1, \alpha_0, \ldots, \alpha_b + n + 1).$$

The sum is

$$2 \sum_{a+b=k} (-1)^b \sum_{\substack{m=0 \\ m:\text{even}}}^{r} (-1)^r \sum_{|\boldsymbol{\alpha}|=b+2} \zeta(\{1\}^a, m+1, \alpha_0, \ldots, \alpha_b + r - m + 1).$$

Consequently, we obtain the following shuffle relation.

Theorem 5.2.1. *For a pair of integers $k, r \geq 0$, we have*

$$(r+1) \sum_{a+b=k} \sum_{m+n=r} \zeta(\{1\}^a, m+2, \{1\}^b, n+2)(-1)^{b+r}$$

$$+ \sum_{a+b=k} \sum_{m+n=r} \zeta(\{1\}^a, m+2, \{1\}^b, n+2) \left\{(-1)^n(n+1) + (-1)^{b+n} n\right\}$$

$$+ 2 \sum_{a+b=k} \sum_{\substack{m=0 \\ m:\text{even}}}^{r} \sum_{|\boldsymbol{\alpha}|=b+2} \zeta(\{1\}^a, m+1, \alpha_0, \ldots, \alpha_b + r - m + 1)(-1)^{b+r}$$

$$+ \frac{1}{2}(r+1)\left\{1 + (-1)^k + (-1)^r + (-1)^{k+r}\right\} \zeta(\{1\}^{k+1}, r+3)$$

$$= \sum_{j=0}^{k} \sum_{\ell=0}^{r} (-1)^{j+\ell}(\ell+1)\zeta(j+r-\ell+2)\zeta(\{1\}^{k-j}, \ell+2).$$

5.3   The Shuffle Product Formula of a Sum and Others

Let k, r, j, ℓ be integers such that $0 \le j \le k$ and $0 \le \ell \le r$. In order to obtain the shuffle product formula of

$$\zeta(j + r - \ell + 2)\zeta(\{1\}^{k-j}, \ell + 2)$$

we consider integrals of the form

$$\frac{1}{k!r!} \iint_{R_1 \times R_2} \left(\mu \log \frac{1 - t_1}{1 - t_2} + \log \frac{1}{1 - u_1} \right)^k \left(\log \frac{t_2}{t_1} + \lambda \log \frac{1}{u_2} \right)^r$$
$$\times \frac{dt_1 dt_2}{(1 - t_1)t_2} \frac{du_1 du_2}{(1 - u_1)u_2},$$

where R_1 is the simplex $0 < t_1 < t_2 < 1$, R_2 is the simplex $0 < u_1 < u_2 < 1$ and μ, λ are complex numbers. Once we expand two factors of the integral as

$$\sum_{j=0}^{k} \binom{k}{j} \mu^j \left(\log \frac{1 - t_1}{1 - t_2} \right)^j \left(\log \frac{1}{1 - u_1} \right)^{k-j}$$

and

$$\sum_{\ell=0}^{r} \binom{r}{\ell} \lambda^\ell \left(\log \frac{t_2}{t_1} \right)^{r-\ell} \left(\log \frac{1}{u_2} \right)^\ell,$$

respectively, we see immediately that the value of the integrand is

$$\sum_{j=0}^{k} \sum_{\ell=0}^{r} \mu^j \lambda^\ell \zeta(j + r - \ell + 2)\zeta(\{1\}^{k-j}, \ell + 2).$$

However, if we decompose $R_1 \times R_2$ into 6 simplices as follow:

$$D_1 : \ 0 < t_1 < t_2 < u_1 < u_2 < 1,$$
$$D_2 : \ 0 < u_1 < u_2 < t_1 < t_2 < 1,$$
$$D_3 : \ 0 < t_1 < u_1 < t_2 < u_2 < 1,$$
$$D_4 : \ 0 < t_1 < u_1 < u_2 < t_2 < 1,$$
$$D_5 : \ 0 < u_1 < t_1 < u_2 < t_2 < 1 \text{ and}$$
$$D_6 : \ 0 < u_1 < t_1 < t_2 < u_2 < 1$$

and then evaluate the integrals over D_1, D_2, D_3, D_4, D_5 and D_6 one by one, we obtain 6 sums of multiple zeta values as follow:

(5.3.1)
$$\sum_{a+b+c=k} (\mu+1)^b \sum_{n+p=r} \lambda^p \sum_{|\alpha|=b+n+1} \zeta(\{1\}^a, \alpha_0, \ldots, \alpha_b+1, \{1\}^c, p+2),$$

(5.3.2)
$$\sum_{a+b=k} \mu^b \sum_{m+n+p=r} \lambda^{m+p}(\lambda+1)^n \sum_{|\alpha|=b+n+1} \zeta(\{1\}^a, m+2, \alpha_0, \ldots, \alpha_b+p+1),$$

(5.3.3)
$$\sum_{a+b=k} \{(\mu+1)^b - \mu^{b+1}\} \sum_{n+p=r} \lambda^p \sum_{|\alpha|=b+n+2} \zeta(\{1\}^a, \alpha_0, \ldots, \alpha_b, \alpha_{b+1}+p+2),$$

(5.3.4)
$$\sum_{a+b+c=k} \{(\mu+1)^{b+1}\mu^c - \mu^{b+c+1}\} \sum_{m+n+p=r} (\lambda+1)^n \lambda^p$$
$$\times \sum_{\substack{|\alpha|=b+m+2 \\ |\beta|=c+n+1}} \zeta(\{1\}^a, \alpha_0, \ldots, \alpha_b, \alpha_{b+1}+\beta_0, \ldots, \beta_c+p+1),$$

(5.3.5)
$$\sum_{a+b+c=k} \mu^{b+c} \sum_{m+n+p=r} (\lambda+1)^n \lambda^p$$
$$\times \sum_{\substack{|\alpha|=b+m+1 \\ |\beta|=c+n+1}} \zeta(\{1\}^{a+1}, \alpha_0, \ldots, \alpha_b+\beta_0, \ldots, \beta_c+p+1)$$

and

(5.3.6)
$$\sum_{a+b=k} \mu^b \sum_{n+p=r} \lambda^p \sum_{|\alpha|=b+n+1} \zeta(\{1\}^{a+1}, \alpha_0, \ldots, \alpha_b+p+2).$$

The sum of (5.3.3) and (5.3.6) is equal to

(5.3.7)
$$\sum_{a+b=k} (\mu+1)^{b+1} \sum_{n+p=r} \lambda^p \sum_{|\alpha|=b+n+2} \zeta(\{1\}^a, \alpha_0, \ldots, \alpha_{b+1}+p+2)$$
$$-\mu^{k+1} \sum_{n+p=r} \lambda^p \sum_{|\alpha|=k+n+2} \zeta(\alpha_0, \ldots, \alpha_k, \alpha_{k+1}+p+2)$$
$$+\sum_{n+p=r} \lambda^p \zeta(\{1\}^{k+1}, r+3),$$

while the sum of (5.3.4) and (5.3.5) is equal to

(5.3.8)
$$\sum_{a+b+c=k} (\mu+1)^{b+1} \mu^c \sum_{m+n+p=r} (\lambda+1)^n \lambda^p$$
$$\times \sum_{\substack{|\alpha|=b+m+2 \\ |\beta|=c+n+1}} \zeta(\{1\}^a, \alpha_0, \dots, \alpha_b, \alpha_{b+1}+\beta_0, \dots, \beta_c+p+1)$$
$$- \mu^{k+1} \sum_{b+c=k} \sum_{m+n+p=r} (\lambda+1)^n \lambda^p$$
$$\times \sum_{\substack{|\alpha|=b+m+2 \\ |\beta|=c+n+1}} \zeta(\{1\}^a, \alpha_0, \dots, \alpha_b, \alpha_{b+1}+\beta_0, \dots, \beta_c+p+1)$$
$$+ \sum_{a+c=k} \mu^c \sum_{m+n+p=r} (\lambda+1)^n \lambda^p$$
$$\times \sum_{|\beta|=c+n+1} \zeta(\{1\}^{a+1}, m+1+\beta_0, \dots, \beta_c+p+1).$$

Of course, the value of the integral is also equal to the sum of (5.3.1), (5.3.2), (5.3.7) and (5.3.8). Separate the coefficients of $\mu^j \lambda^\ell$ of both sides of the resulted identity we get the following theorem.

Theorem 5.3.1. *For integers k, r, j and ℓ such that $0 \le j \le k$ and $0 \le \ell \le r$, we have*

$$\zeta(j+r-\ell+2)\zeta(\{1\}^{k-j}, \ell+2)$$
$$= \sum_{a+b+c=k} \binom{b}{j} \sum_{|\alpha|=b+r-\ell+1} \zeta(\{1\}^a, \alpha_0, \dots, \alpha_b+1, \{1\}^c, \ell+2)$$
$$+ \sum_{m+n+p=r} \binom{n}{r-\ell} \sum_{|\alpha|=j+n+1} \zeta(\{1\}^{k-j}, m+2, \alpha_0, \dots, \alpha_j+p+1)$$
$$+ \sum_{a+b=k} \binom{b+1}{j} \sum_{|\alpha|=b+r-\ell+2} \zeta(\{1\}^a, \alpha_0, \dots, \alpha_{b+1}+\ell+2)$$
$$+ \delta_{j,0}\zeta(\{1\}^{k+1}, r+3)$$
$$+ \sum_{a+b+c=k} \binom{b+1}{j-c} \sum_{m+n+p=r} \binom{n}{\ell-p}$$
$$\times \sum_{\substack{|\alpha|=b+m+2 \\ |\beta|=c+n+1}} \zeta(\{1\}^a, \alpha_0, \dots, \alpha_b, \alpha_{b+1}+\beta_0, \dots, \beta_c+p+1)$$

$$+ \sum_{m+n+p=r} \binom{n}{\ell-p} \sum_{|\boldsymbol{\beta}|=j+n+1} \zeta(\{1\}^{k-j+1}, m+1+\beta_0, \ldots, \beta_j+p+1).$$

Remark 5.3.2. If we take the dual of each sums of multiple zeta values, we obtain the following shuffle product formula

$$\zeta(j+r-\ell+2)\zeta(\{1\}^{k-j}, \ell+2)$$

$$= \sum_{a+b+c=k} \binom{b}{j} \sum_{|\boldsymbol{\alpha}|=r-\ell+b+1} \zeta(\{1\}^{\ell}, c+2, \alpha_0, \ldots, \alpha_{r-\ell}+a+1)$$

$$+ \sum_{m+n+p=r} \binom{n}{r-\ell} \sum_{|\boldsymbol{\alpha}|=j+n+1} \zeta(\{1\}^p, \alpha_0, \ldots, \alpha_n+1, \{1\}^m, k-j+2)$$

$$+ \sum_{a+b=k} \binom{b+1}{j} \sum_{|\boldsymbol{\alpha}|=r-\ell+b+1} \zeta(\{1\}^{\ell+1}, \alpha_0, \ldots, \alpha_{r-\ell}+a+1)$$

$$+ \sum_{a+b+c=k} \binom{b+1}{j-c} \sum_{m+n+p=r} \binom{n}{\ell-p}$$

$$\times \sum_{\substack{|\boldsymbol{\alpha}|=n+c+1 \\ |\boldsymbol{\beta}|=m+b+2}} \zeta(\{1\}^p, \alpha_0, \ldots, \alpha_n, \beta_0, \ldots, \beta_m+a+1)$$

$$+ \sum_{m+n+p=r} \binom{n}{\ell-p} \sum_{|\boldsymbol{\alpha}|=n+j+1} \zeta(\{1\}^p, \alpha_0, \ldots, \alpha_n, \{1\}^m, k-j+3)$$

$$+ \delta_{j,0}\zeta(\{1\}^{r+1}, k+3).$$

5.4   Exercises

1. Begin with the integral

$$\frac{1}{k!r!} \iint_{R_1 \times R_2} \left(\mu \log \frac{1-t_1}{1-t_2} + \log \frac{1}{1-u_1} \right)^k \left(\log \frac{t_2}{t_1} + \lambda \log \frac{u_2}{u_1} \right)^r$$

$$\times \frac{dt_1 dt_2}{(1-t_1)t_2} \frac{du_1 du_2}{(1-u_1)u_2}$$

which has the value

$$\sum_{j=0}^{k} \sum_{\ell=0}^{r} \mu^j \lambda^{\ell} \zeta(j+r-\ell+2)\zeta(\{1\}^{k-j}, \ell+2)$$

to obtain the shuffle product formula of $\zeta(j+r-\ell+2)$ and $\zeta(\{1\}^{k-j}, \ell+2)$ for integers k, j, r, ℓ with $0 \le j \le k$ and $0 \le \ell \le r$.

2. Compare the result obtained in previous problem with those shuffle relations obtained in section 5.3.

Part III

Applications of Shuffle Relations in Combinatorics

The shuffle product of two multiple zeta values of weight m and n will produce $\binom{m+n}{m}$ multiple zeta values of weight $m + n$. Behind each shuffle relation obtained from shuffle products there is a combinatorial identity obtained from just counting the number of multiple zeta values produced from shuffle products.

Not only we can produce Pascal identity from shuffle products, but we also create several generalizations. The classical Vandermonde's convolutions are extended to the utmost, vector versions of all types of convolutions which are hard to be proved otherwise.

Some particular double or triple integrals over a simplex of dimension two or three represent sums of multiple zeta values. By counting the number of multiple zeta values in two different ways, we also obtain new combinatorial identities. Especially those integrals obtained from shuffle products of two sets of multiple zeta values, the resulting combinatorial identities are very interesting, but again hard to be proved as usual.

CHAPTER 6

Generalizations of Pascal Identity

The binomial coefficient defined by

$$\binom{m}{n} = \frac{m!}{n!(m-n)!}$$

for integers m, n with $0 \leq n \leq m$ comes from the binomial expansion

$$(1+x)^m = \sum_{n=0}^{m} \binom{m}{n} x^n,$$

has the further relation

$$\binom{m+1}{n+1} = \binom{m}{n} + \binom{m}{n+1}$$

known as Pascal identity.

Through shuffle products of multiple zeta values, we are able to express the generating function

$$\sum_{j=0}^{k} \sum_{\ell=0}^{r} \mu^j \lambda^\ell \binom{k+r+4}{j+\ell+2}$$

in terms of simple polynomial functions in μ, λ. As a consequence we produce the identity

$$\binom{k+r+4}{j+\ell+2} = \sum_{p+q=k} \left\{ \binom{p}{j}\binom{q+r+3}{\ell+1} + \binom{p}{k-j}\binom{q+r+3}{r-\ell+1} \right\}$$

which is a generalization of Pascal identity.

6.1 Applications of Shuffle Products in Combinatorics

The shuffle product of two multiple zeta values of weight m and n respectively, will produce $\binom{m+n}{m}$ multiple zeta values of weight $m + n$. If we are able to count the total numbers of multiple zeta values producing from a shuffle product of two multiple zeta values, we obtain an identity among binomial coefficients which might be of interesting in combinatorics. Presently, we demonstrate the general procedure with the well-known Euler decomposition theorem obtained actually from a shuffle process.

For a pair of integer r, ℓ with $0 \le \ell \le r$, Euler decomposition theorem asserted that

$$\zeta(\ell+2)\zeta(r-\ell+2) = \sum_{|\alpha|=r+3} \zeta(\alpha_1, \alpha_2+1) \left\{ \binom{\alpha_2}{\ell+1} + \binom{\alpha_2}{r-\ell+1} \right\}.$$

Such an identity can be obtained via the shuffle product of $\zeta(\ell+2)$ and $\zeta(r-\ell+2)$. Therefore the number of multiple zeta values appeared in the summation of right-hand side of the above identity must be equal to

$$\binom{r+4}{\ell+2}.$$

In other words, we have the identity

$$\sum_{\alpha=1}^{r+2} \left\{ \binom{\alpha}{\ell+1} + \binom{\alpha}{r-\ell+1} \right\} = \binom{r+4}{\ell+2}$$

or with an elementary calculation that

$$\binom{r+3}{\ell+2} + \binom{r+3}{\ell+1} = \binom{r+4}{\ell+2}$$

which is the well-known Pascal identity.

To develop further generalizations of Pascal identity, we begin with a more general shuffle relation producing from the shuffle product of multiple zeta values $\zeta(\{1\}^j, r - \ell + 2)$ and $\zeta(\{1\}^{k-j}, \ell + 2)$ with $0 \le j \le k$ and $0 \le \ell \le r$ as follows:

(6.1.1)
$$
\zeta(\{1\}^j, r - \ell + 2)\zeta(\{1\}^{k-j}, \ell + 2)
$$
$$
= \sum_{p+q=k} \binom{p}{j} \sum_{|\alpha|=q+r+3} \zeta(\{1\}^p, \alpha_0, \ldots, \alpha_q, \alpha_{q+1} + 1)\binom{\alpha_{q+1}}{\ell+1}
$$
$$
+ \sum_{p+q=k} \binom{p}{k-j} \sum_{|\alpha|=q+r+3} \zeta(\{1\}^p, \alpha_0, \ldots, \alpha_q, \alpha_{q+1} + 1)\binom{\alpha_{q+1}}{r-\ell+1}.
$$

During our process of obtaining combinational identities through shuffle relations, we need the following assertions concerning binomial coefficients which might be old or new. Anyway, we give proofs of them and will use them repeatedly.

Proposition 6.1.1. *For nonnegative integers q, r and a complex number λ, we have*

$$
\sum_{h+\ell=r} \binom{q+h}{q} (\lambda+1)^\ell = \frac{(q+r+1)!}{q!r!} \int_0^1 (\lambda+1-\lambda x)^r x^q dx
$$
$$
= \sum_{\ell=0}^{r} \binom{q+r+1}{\ell} \lambda^{r-\ell}.
$$

Proof. Rewrite the sum as

$$
(\lambda+1)^r \sum_{h=0}^{r} \binom{q+h}{q}\left(\frac{1}{\lambda+1}\right)^h
$$

so that it is equal to the coefficient of x^q in the polynomial of

$$
(\lambda+1)^r \sum_{h=0}^{r}\left(x + \frac{1}{\lambda+1}\right)^{q+h},
$$

or

$$
(\lambda+1)^r \left\{\left(x + \frac{1}{\lambda+1}\right)^{q+r+1} - \left(x + \frac{1}{\lambda+1}\right)^q\right\}\left(x - \frac{\lambda}{\lambda+1}\right)^{-1}.
$$

Express $\left(x + \dfrac{1}{\lambda + 1} \right)^{q+r+1}$ as a polynomial in $\left(x - \dfrac{\lambda}{\lambda + 1} \right)$ via its binomial expansion

$$
\left(x + \frac{1}{\lambda + 1} \right)^{q+r+1} = \left(x - \frac{\lambda}{\lambda + 1} + 1 \right)^{q+r+1}
$$

$$
= \sum_{\ell=0}^{q+r+1} \binom{q+r+1}{\ell} \left(x - \frac{\lambda}{\lambda + 1} \right)^{\ell}.
$$

Only those terms with $\ell \geq q + 1$ have contribution to the coefficient of x^q, after dividing by $\left(x - \dfrac{\lambda}{\lambda + 1} \right)$. Therefore, the coefficient of x^q is

$$
(\lambda + 1)^r \sum_{\ell=q+1}^{q+r+1} \binom{q+r+1}{\ell} \binom{\ell - 1}{q} \left(-\frac{\lambda}{\lambda + 1} \right)^{\ell - q - 1},
$$

or

$$
\frac{(q+r+1)!(\lambda + 1)^r}{q!r!} \sum_{\ell=0}^{r} \binom{r}{\ell} \left(-\frac{\lambda}{\lambda + 1} \right)^{\ell} \frac{1}{\ell + q + 1}.
$$

This is precisely equal to the value of the integral

$$
\frac{(q+r+1)!}{q!r!} \int_0^1 (\lambda + 1 - \lambda x)^r \, x^q \, dx.
$$

The second part is trivial, so we omit the proof. $\qquad\qquad\qquad\square$

Proposition 6.1.2. *For a pair of nonnegative integers q, r and complex numbers μ, λ, we have*

$$
\sum_{m+n=q} (\mu + 1)^m \sum_{h+\ell=r} \binom{n + h}{n} (\lambda + 1)^{\ell}
$$

$$
= \frac{(q+r+2)!}{q!r!} \int_0^1 \int_0^y \left[(\mu + 1)(1 - y) + x \right]^q \left[(\lambda + 1) y - \lambda x \right]^r \, dx dy.
$$

Proof. By the previous proposition, the inner sum is equal to

$$
\frac{(n+r+1)!}{n!r!} \int_0^1 (\lambda + 1 - \lambda x)^r \, x^n dx,
$$

or

$$
(r + 1) \binom{n + r + 1}{r + 1} \int_0^1 (\lambda + 1 - \lambda x)^r \, x^n \, dx.
$$

Next we have to find the sum of

$$\sum_{m+n=q} (\mu+1)^m \binom{n+r+1}{r+1} x^n$$

or

$$(\mu+1)^q \sum_{n=0}^{q} \binom{n+r+1}{r+1} \left(\frac{x}{\mu+1}\right)^n.$$

Again by the previous proposition, the above sum is equal to

$$\frac{(q+r+2)!}{q!(r+1)!} \int_0^1 [(\mu+1)(1-y)+yx]^q \, y^{r+1} \, dy.$$

Therefore the whole sum is equal to

$$\frac{(q+r+2)!}{q!r!} \int_0^1 \int_0^1 [(\mu+1)(1-y)+yx]^q \, [(\lambda+1)-\lambda x]^r \, y^{r+1} \, dx dy.$$

A simple change of variable yields our assertion.   □

Now we are ready to prove a generalization of Pascal identity.

Theorem 6.1.3. *For integers* k, r, j, ℓ *with* $0 \le j \le k$ *and* $0 \le \ell \le r$, *we have*

$$\sum_{p+q=k} \binom{p}{j}\binom{q+r+3}{r-\ell+1} + \sum_{p+q=k} \binom{p}{k-j}\binom{q+r+3}{\ell+1} = \binom{k+r+4}{j+r-\ell+2}.$$

Proof. Fix $\alpha_{q+1} = \alpha$ with $1 \le \alpha \le r+2$, the number of positive integral solutions to the equation

$$\alpha_0 + \alpha_1 + \cdots + \alpha_q = q+r+3-\alpha$$

is

$$\binom{q+r+2-\alpha}{q}.$$

Therefore, the number of multiple zeta values in the first summation of (6.1.1) is

$$\sum_{p+q=k} \binom{p}{j} \sum_{\alpha=1}^{r+2} \binom{q+r+2-\alpha}{q}\binom{\alpha}{\ell+1}.$$

The inner summation can be simplified further. By Proposition 6.1.1, we have

$$\sum_{\alpha=0}^{r+2} \binom{q+r+2-\alpha}{q} (\lambda+1)^{\alpha} = \sum_{g=0}^{r+2} \binom{q+r+3}{g} \lambda^{r+2-g}.$$

Comparing the coefficient of $\lambda^{\ell+1}$ on the both sides, we get that

$$\sum_{\alpha=0}^{r+2} \binom{q+r+2-\alpha}{q} \binom{\alpha}{\ell+1} = \binom{q+r+3}{r-\ell+1}.$$

On the other hand, the number of multiple zeta values in the second summation of (6.1.1) is equal to

$$\sum_{p+q=k} \binom{p}{k-j} \sum_{\alpha=1}^{r+2} \binom{q+r+2-\alpha}{q} \binom{\alpha}{r-\ell+1},$$

which is equal to

$$\sum_{p+q=k} \binom{p}{k-j} \binom{q+r+3}{\ell+1}.$$

Consequently, our assertion follows.   □

For our further convenience, we let

$$(6.1.2) \qquad I_1(m,n;\lambda) = \frac{(m+n+1)!}{m!n!} \int_0^1 (\lambda+1-\lambda x)^n x^m \, dx$$

and

$$(6.1.3)$$
$$I_2(q,r;\mu,\lambda)$$
$$= \frac{(q+r+2)!}{q!r!} \int_0^1 \int_0^y [(\mu+1)(1-y)+x]^q [(\lambda+1)y - \lambda x]^r \, dx dy$$

which represent polynomials

$$\sum_{a+b=n} \binom{m+a}{m} (\lambda+1)^b$$

and

$$\sum_{m+n=q} (\mu+1)^m \sum_{a+b=r} \binom{n+a}{n} (\lambda+1)^b,$$

respectively. The generating function for the binomial coefficients $\binom{m}{n}$, $(n = 0, 1, \ldots, m)$, is just the polynomial $(1 + x)^m$. In the following, we are going to find a generating function of binomial coefficients

$$\binom{k + r + 4}{j + r - \ell + 2}, \quad 0 \le j \le k, \quad 0 \le \ell \le r$$

in terms of $I_2(k, r + 2; \mu, \lambda)$ and $\mu^k \lambda^{r+2} I_2(k, r + 2; \mu^{-1}, \lambda^{-1})$.

Theorem 6.1.4. *For a pair of nonnegative integers k and r, we have*

$$\sum_{j=0}^{k} \sum_{\ell=0}^{r} \binom{k + r + 4}{j + r - \ell + 2} \mu^j \lambda^{\ell+1}$$

$$= I_2\left(k, r + 2; \mu, \lambda\right) + \mu^k \lambda^{r+2} I_2\left(k, r + 2; \frac{1}{\mu}, \frac{1}{\lambda}\right)$$

$$- \sum_{p+q=k} (\mu + 1)^p \left\{ \lambda^{r+2} + \binom{q + r + 3}{r + 2} \right\}$$

$$- \sum_{p+q=k} (\mu + 1)^p \mu^q \left\{ 1 + \binom{q + r + 3}{r + 2} \lambda^{r+2} \right\}$$

with

$$I_2\left(k, r + 2; \mu, \lambda\right) = \sum_{p+q=k} (\mu + 1)^p \sum_{a+b=r+2} \binom{q + a}{q} (\lambda + 1)^b$$

$$= \frac{(k + r + 4)!}{k!(r + 2)!} \int_0^1 \int_0^y \left[(\mu + 1)(1 - y) + x\right]^k$$

$$\times \left[(\lambda + 1) y - \lambda x\right]^{r+2} dx dy.$$

Proof. It is a direct calculation by Propositions 6.1.1 and 6.1.2. Indeed,

$$\sum_{j=0}^{k} \sum_{\ell=0}^{r} \sum_{p+q=k} \binom{p}{j} \binom{q + r + 3}{r - \ell + 1} \mu^j \lambda^{\ell+1}$$

$$= \sum_{p+q=k} (\mu + 1)^p \sum_{\ell=0}^{r} \binom{q + r + 3}{r - \ell + 1} \lambda^{\ell+1}$$

$$= \sum_{p+q=k} (\mu + 1)^p \left\{ \sum_{g=0}^{r+2} \binom{q + r + 3}{g} \lambda^{r+2-g} - \lambda^{r+2} - \binom{q + r + 3}{r + 2} \right\}$$

$$= \sum_{p+q=k} (\mu + 1)^p \sum_{\alpha=0}^{r+2} \binom{q + r + 2 - \alpha}{q} (\lambda + 1)^\alpha$$

$$- \sum_{p+q=k} (\mu+1)^p \left\{ \lambda^{r+2} + \binom{q+r+3}{r+2} \right\}$$

$$= I_2(k, r+2; \mu, \lambda) - \sum_{p+q=k} (\mu+1)^p \left\{ \lambda^{r+2} + \binom{q+r+3}{r+2} \right\}.$$

On the other hand, we have

$$\sum_{j=0}^{k} \sum_{\ell=0}^{r} \sum_{p+q=k} \binom{p}{k-j} \binom{q+r+3}{\ell+1} \mu^j \lambda^{\ell+1}$$

$$= \sum_{j=0}^{k} \sum_{\ell=0}^{r} \sum_{p+q=k} \binom{p}{j} \binom{q+r+3}{r-\ell+1} \mu^{k-j} \lambda^{r+2-(\ell+1)}$$

$$= \mu^k \lambda^{r+2} I_2 \left(k, r+2; \frac{1}{\mu}, \frac{1}{\lambda} \right) - \sum_{p+q=k} (\mu+1)^p \mu^q \left\{ 1 + \binom{q+r+3}{r+2} \lambda^{r+2} \right\}.$$

This leads to our assertion.   □

Remark 6.1.5. Theorem 6.1.4 comes from also the shuffle process of the following integral

$$\frac{1}{k!r!} \iint_{R_1 \times R_2} \left(\mu \log \frac{1}{1-t_1} + \log \frac{1}{u_1} \right)^k \left(\log \frac{1}{t_2} + \lambda \log \frac{1}{u_2} \right)^r$$

$$\times \frac{dt_1 dt_2}{(1-t_1)t_2} \frac{du_1 du_2}{(1-u_1)u_2},$$

which is equal to

$$\sum_{j=0}^{k} \sum_{\ell=0}^{r} \mu^j \lambda^\ell \zeta \left(\{1\}^j, r-\ell+2 \right) \zeta \left(\{1\}^{k-j}, \ell+2 \right).$$

Each double weighted sum formula we produced before through shuffle process will provide a combinatorial identity.

Theorem 6.1.6. *For a pair of integers $k, r \geq 0$ with $k+r$ even, we have*

$$\frac{1}{2} \sum_{j=0}^{k} \sum_{\ell=0}^{r} (-1)^{j+\ell} \binom{k+r+4}{j+r-\ell+2} = \binom{k+r+2}{k+1} + (-1)^k.$$

Proof. The above combinatorial identity comes from the shuffle relation

$$\sum_{|\alpha|=k+r+3} \zeta(\alpha_0,\ldots,\alpha_k,\alpha_{k+1}+1) + (-1)^k \zeta(\{1\}^{k+1},r+3)$$

$$= \frac{1}{2}\sum_{j=0}^{k}\sum_{\ell=0}^{r}(-1)^{j+\ell}\zeta(\{1\}^j,r-\ell+2)\zeta(\{1\}^{k-j},\ell+2).$$

Note that the number of positive integral solutions to the equation

$$\alpha_0 + \alpha_1 + \cdots + \alpha_{k+1} = k+r+3$$

is

$$\binom{k+r+2}{k+1}.$$

So our assertion follows from our previous consideration. □

Theorem 6.1.7. *For a pair of integers $k,r \geq 0$ with k even, we have*

$$\frac{1}{2}\sum_{j=0}^{k}\sum_{\ell=0}^{r}(-1)^j\binom{k+r+4}{j+r-\ell+2} = \sum_{p=1}^{r+1}\binom{k+r+3}{p}.$$

Proof. The above identity is based on the shuffle relation

$$\sum_{|\alpha|=k+r+3} 2^{\alpha_{k+1}}\zeta(\alpha_0,\ldots,\alpha_k,\alpha_{k+1}+1)$$

$$= \sum_{|\alpha|=k+r+3} \zeta(\alpha_0,\ldots,\alpha_k,\alpha_{k+1}+1) + \zeta(\{1\}^{k+1},r+3)$$

$$+ \frac{1}{2}\sum_{j=0}^{k}\sum_{\ell=0}^{r}(-1)^j\,\zeta(\{1\}^j,r-\ell+2)\zeta(\{1\}^{k-j},\ell+2).$$

The number of multiple zeta values on the left side of the above is

$$\sum_{\alpha=1}^{r+2}\binom{k+r+2-\alpha}{k}2^\alpha \quad \text{and} \quad 2\sum_{\beta=0}^{r+1}\binom{k+\beta}{k}2^{r+1-\beta}.$$

By Proposition 6.1.1, the value is equal to

$$2\sum_{p=0}^{r+1}\binom{k+r+2}{p}.$$

On the other hand, the number of multiple zeta values in the sum

$$\sum_{|\boldsymbol{\alpha}|=k+r+3} \zeta(\alpha_0,\ldots,\alpha_k,\alpha_{k+1}+1)$$

is $\binom{k+r+2}{k+1}$. Consequently, our assertion follows from Pascal identity.   □

Theorem 6.1.8. *For a pair of nonnegative integers k and r with k even, we have*

$$\frac{1}{2}\sum_{j=0}^{k}\sum_{\ell=0}^{r}(-1)^j\binom{r}{\ell}\binom{k+r+4}{j+r-\ell+2}=\binom{k+2r+3}{r+1}.$$

Proof. We begin with the shuffle relation

$$\sum_{|\boldsymbol{\alpha}|=k+r+3} \zeta(\alpha_0,\ldots,\alpha_k,\alpha_{k+1}+1)\binom{\alpha_{k+1}+r}{r+1}$$

$$=\frac{1}{2}\sum_{j=0}^{k}\sum_{\ell=0}^{r}(-1)^j\binom{r}{\ell}\zeta(\{1\}^j,r-\ell+2)\zeta(\{1\}^{k-j},\ell+2).$$

The number of multiple zeta values on the left is

$$\sum_{\alpha=1}^{r+2}\binom{k+r+2-\alpha}{k}\binom{\alpha+r}{r+1}$$

or

$$\sum_{\alpha=0}^{r+1}\binom{k+\alpha}{\alpha}\binom{2r+2-\alpha}{r+1}.$$

According to Proposition 6.1.1, we have

$$\sum_{\alpha=0}^{2r+2}\binom{k+\alpha}{k}(\lambda+1)^{2r+2-\alpha}=\sum_{\ell=0}^{2r+2}\binom{k+2r+3}{\ell}\lambda^{2r+2-\ell}.$$

Comparing the coefficients of λ^{r+1} on the both sides leads to the identity

$$\sum_{\alpha=0}^{2r+2}\binom{k+\alpha}{k}\binom{2r+2-\alpha}{r+1}=\binom{k+2r+3}{r+1}.$$

When $\alpha \geq r+2$, the binomial coefficient

$$\binom{2r+2-\alpha}{r+1}$$

vanishes by convention. Therefore, the upper limit of α can be changed from $2r+2$ to $r+1$, and hence our assertion follows.   □

Theorem 6.1.9. *For a pair of integers $k, r \geq 0$ with r even, we have*

$$\sum_{j=0}^{k} \sum_{\substack{\ell=0 \\ \ell:even}}^{r} \binom{k+r+4}{j+r-\ell+2} = \sum_{p+q=k} 2^{p+1} \sum_{\ell=0}^{r+1} \binom{q+r+2}{\ell}.$$

Proof. It is based on the shuffle relation

$$\sum_{p+q=k} 2^p \sum_{|\boldsymbol{\alpha}|=q+r+3} \zeta(\{1\}^p, \alpha_0, \ldots, \alpha_q, \alpha_{q+1}+1)2^{\alpha_{q+1}}$$

$$= \sum_{j=0}^{k} \sum_{\substack{\ell=0 \\ \ell:even}}^{r} \zeta(\{1\}^j, r-\ell+2)\zeta(\{1\}^{k-j}, \ell+2).$$

The number of multiple zeta values on the left is

$$\sum_{p+q=k} 2^p \sum_{\alpha=1}^{r+2} \binom{q+r+2-\alpha}{q}2^\alpha,$$

or

$$\sum_{p+q=k} 2^p \sum_{\alpha=0}^{r+1} \binom{q+\alpha}{\alpha}2^{r+2-\alpha}.$$

By Proposition 6.1.1, the inner sum is equal to

$$2\sum_{\ell=0}^{r+1} \binom{q+r+2}{\ell},$$

and then follows our assertion. □

Theorem 4.3.1 is another generalization of Euler decomposition theorem producing from a shuffle process. By counting the number of multiple zeta values, we get the following.

Theorem 6.1.10. *For integers q, r, m, ℓ with $0 \leq m \leq q$, $0 \leq \ell \leq r$, we* *have*

$$\sum_{n=0}^{r}\left\{\binom{q-m+n+1}{n-\ell}+\binom{m+n+1}{n+\ell-r}\right\}$$

$$+\sum_{a+b=q}\sum_{g=0}^{b}\left\{\binom{r+g+1}{r-\ell}\binom{b-g}{m-a}+\binom{r+g+1}{\ell}\binom{b-g}{q-m}\right.$$

$$+\binom{r+g+1}{\ell}\binom{b-g}{b-m}+\binom{r+g+1}{r-\ell}\binom{b-g}{m}\right\}$$

$$=\binom{q+r+4}{m+r-\ell+2}.$$

Remark 6.1.11. If we restrict $q = m = 0$ in Theorem 6.1.10, we get that

$$\sum_{n=0}^{r}\left\{\binom{n+1}{n-\ell}+\binom{n+1}{n+\ell-r}\right\}+2\binom{r+1}{r-\ell}+2\binom{r+1}{\ell}=\binom{r+4}{\ell+2}$$

or with an elementary calculation as

$$\binom{r+2}{\ell+2}+\binom{r+2}{r-\ell+1}+2\binom{r+1}{r-\ell}+2\binom{r+1}{\ell}=\binom{r+4}{\ell+2}$$

which is equivalent to the Pascal identity. On the other hand, if we restrict $r = \ell = 0$, we get that

$$2+\sum_{a+b=q}\sum_{g=0}^{b}\left\{\binom{b-g}{m-a}+\binom{b-g}{q-m}+\binom{b-g}{b-m}+\binom{b-g}{m}\right\}=\binom{q+4}{m+2}.$$

Such an identity can be proved directly. For example, we have

$$\sum_{a+b=q}\sum_{g=0}^{b}\binom{b-g}{m}=\sum_{b=0}^{q}\sum_{g=m}^{b}\binom{g}{m}=\sum_{b=0}^{q}\binom{b+1}{m+1}=\binom{q+2}{m+2}$$

and

$$\sum_{a+b=q}\sum_{g=0}^{b}\binom{b-g}{b-m}=\sum_{b=m}^{q}\sum_{g=0}^{b}\binom{g}{b-m}=\sum_{b=m}^{q}\binom{b+1}{b-m+1}$$

$$=\sum_{b=m}^{q}\binom{b+1}{m}=\binom{q+2}{m+1}-1.$$

Therefore, the identity is equivalent to

$$\binom{q+2}{m+2} + \binom{q+2}{m} + 2\binom{q+2}{m+1} = \binom{q+4}{m+2}$$

which is equivalent to Pascal identity.

A generating function for binomial coefficients

$$\binom{q+r+4}{m+r-\ell+2}, \quad 0 \le m \le q, \quad 0 \le \ell \le r$$

can be expressed in terms of integrals I_1 and I_2 given by (6.1.2) and (6.1.3).

Theorem 6.1.12. *For a pair of integers* $q, r \ge 0$, *we have*

$$\sum_{m=0}^{q} \sum_{\ell=0}^{r} \mu^m \lambda^\ell \binom{q+r+4}{m+r-\ell+2}$$

$$= \sum_{m+n=q} \sum_{\ell=0}^{r} \left\{ \mu^m I_1(n, \ell; \lambda) + \mu^n \lambda^r I_1\left(n, \ell; \frac{1}{\lambda}\right) \right\}$$

$$+ \sum_{m+n=q} (\mu^m + 1) \left\{ I_2(n, r; \mu, \lambda) + \mu^n \lambda^r I_2\left(n, r; \frac{1}{\mu}, \frac{1}{\lambda}\right) \right\},$$

where I_1 *and* I_2 *are shown in (6.1.2) and (6.1.3).*

Theorem 6.1.13. *For a pair of nonnegative integers* q, r *with* $q + r$ *is even, we have*

$$\frac{1}{2} \sum_{m=0}^{q} \sum_{\ell=0}^{r} (-1)^{m+\ell} \binom{q+r+4}{m+r-\ell+2}$$

$$= \binom{q+r+1}{r+1} + \sum_{n=0}^{q} (-1)^{q-n} \left\{ \binom{n+r+1}{r} + \binom{n+r}{r} \right\}.$$

Theorem 6.1.14. *For a pair of nonnegative integers* q, r *with* r *is even, we have*

$$\frac{1}{2} \sum_{m=0}^{q} \sum_{\ell=0}^{r} (-1)^\ell \binom{q+r+4}{m+r-\ell+2} = 2 \sum_{m=0}^{q} \binom{q+r+2}{m+1} - \binom{q+r+2}{r+1} + 1.$$

Theorem 6.1.15. *For a pair of integers* $q, r \geq 0$ *with* q *is even, we have*

$$\frac{1}{2} \sum_{m=0}^{q} \sum_{\ell=0}^{r} (-1)^m \binom{q+r+4}{m+r-\ell+2}$$

$$= \sum_{n=0}^{q} (-1)^n \sum_{\ell=0}^{r} \left\{ \binom{n+r+1}{\ell} + \binom{n+r+2}{\ell+1} \right\} + \sum_{\ell=0}^{r} \binom{q+r+2}{\ell+1}$$

$$- \sum_{n=0}^{q} (-1)^n \binom{n+r+2}{r+1} + \frac{1}{2} (1 + (-1)^q) - 2^{r+1} + 1.$$

Theorem 6.1.16. *For a pair of integers* $q, r \geq 0$, *we have*

$$\frac{1}{2} \sum_{m=0}^{q} \sum_{\ell=0}^{r} \binom{q+r+4}{m+r-\ell+2}$$

$$= \sum_{n=0}^{q} \sum_{a+b=n} 2^{a+1} \sum_{\ell=0}^{r} \binom{r+b+1}{\ell}$$

$$+ \sum_{\ell=0}^{r} \binom{q+r+3}{\ell+2} - \binom{q+r+3}{q+1} + q + r + 5.$$

Remark 6.1.17. The shuffle relations corresponding to identities shown from Theorem 6.1.13 to Theorem 6.1.16 are as follows:

$$\sum_{m+n=q} (-1)^m \sum_{h+\ell=r} \zeta(m+h+2, n+\ell+2) \binom{n+\ell}{\ell}$$

$$+ \sum_{m+n=q} \{1 + (-1)^m\} \zeta(m+1, n+r+3) \binom{n+r}{n}$$

$$= \frac{1}{2} \sum_{m=0}^{q} \sum_{\ell=0}^{r} (-1)^{m+\ell} \zeta(m+r-\ell+2) \zeta(q-m+\ell+2),$$

$$\sum_{m+n=q} \sum_{h+\ell=r} \zeta(m+h+2, n+\ell+2) \binom{n+\ell}{n}$$

$$+ 2 \sum_{m+n=q} \zeta(m+1, n+r+3) \sum_{a=0}^{n} \binom{n+r+1}{a}$$

$$= \frac{1}{2} \sum_{m=0}^{q} \sum_{\ell=0}^{r} (-1)^{\ell} \zeta(m+r-\ell+2) \zeta(q-m+\ell+2),$$

$$\sum_{m+n=q} (-1)^m \sum_{h+\ell=r} \zeta(m+h+2, n+\ell+2) \sum_{p=0}^{\ell} \binom{n+\ell+1}{p}$$

$$+ \sum_{m+n=q} \{1+(-1)^n\} \zeta(m+1, n+r+3) \sum_{\ell=0}^{r} \binom{n+r+1}{\ell}$$

$$= \frac{1}{2} \sum_{m=0}^{q} \sum_{n=0}^{r} (-1)^m \zeta(m+r-\ell+2) \zeta(q-m+\ell+2),$$

and

$$\sum_{m+n=q} \sum_{h+\ell=r} \zeta(m+h+2, n+\ell+2) \sum_{p=0}^{\ell} \binom{n+\ell+1}{p}$$

$$+ 2 \sum_{m+n=q} \zeta(m+1, n+r+2) \sum_{a+b=n} 2^a \sum_{\ell=0}^{r} \binom{r+b+1}{\ell}$$

$$= \frac{1}{2} \sum_{m=0}^{q} \sum_{\ell=0}^{r} \zeta(m+r-\ell+2) \zeta(q-m+\ell+2).$$

6.2 Hypergeometric Distribution

According to the hypergeometric distribution from theory of probability, the probability of getting g good elements and b bad elements from a population of size $N = G + B$ containing G good and B bad elements is

$$\binom{G}{g} \binom{B}{b} \Big/ \binom{G+B}{g+b},$$

and hence

$$\sum_{g=0}^{n} \binom{G}{g} \binom{B}{n-g} = \binom{G+B}{n}.$$

On the other hand, from the polynomial identity

$$\sum_{h+n=r} \binom{g+h}{g} (\lambda+1)^n = \sum_{\ell=0}^{r} \binom{g+r+1}{\ell} \lambda^{r-\ell},$$

by comparing the coefficients of λ^ℓ, we conclude that

$$\sum_{h+n=r} \binom{g+h}{g} \binom{n}{\ell} = \binom{g+r+1}{r-\ell} = \binom{g+r+1}{g+\ell+1}.$$

Here we first write down an identity of similar kind:

$$\sum_{a+b=k} \sum_{m+n=r} \binom{a+m}{a}\binom{b+n+1}{b+1} = \sum_{\alpha=1}^{r+1}\binom{k+\alpha}{k}\alpha.$$

How to prove such formula is the main purpose of this section. We give a proof and look for possible extensions.

Theorem 6.2.1. *For a pair of nonnegative integers k, r, we have*

$$\sum_{a+b=k} \sum_{m+n=r} \binom{a+m}{a}\binom{b+n+1}{b+1} = \frac{(k+r+2)!}{(k+2)\cdot k!r!}.$$

Proof. Consider the sum of multiple zeta values given by

$$\sum_{a+b=k} \sum_{m+n=r} \sum_{\substack{|\alpha|=m+a+1 \\ |\beta|=n+b+2}} \zeta(\alpha_0,\ldots,\alpha_a+\beta_0,\ldots,\beta_b,\beta_{b+1}+1).$$

Note that the number of multiple zeta values in the above sum is

$$\sum_{a+b=k} \sum_{m+n=r} \binom{a+m}{a}\binom{b+n+1}{b+1}$$

since the numbers of positive integral solutions to linear equations

$$\alpha_0 + \cdots + \alpha_a = a+m+1 \quad \text{and} \quad \beta_0 + \cdots + \beta_b + \beta_{b+1} = b+n+2$$

are

$$\binom{a+m}{a} \quad \text{and} \quad \binom{b+n+1}{b+1},$$

respectively. On the other hand, the sum can be rewritten as

$$\sum_{|\alpha|=k+r+3} \sum_{j=0}^{k}(\alpha_j - 1)\zeta(\alpha_0,\ldots,\alpha_k,\alpha_{k+1}+1).$$

Under the condition $|\alpha| = k+r+3$, the inner sum

$$\sum_{j=0}^{k}(\alpha_j - 1) = \alpha_0 + \cdots + \alpha_k - (k+1)$$

$$= k+r+3 - \alpha_{k+1} - (k+1)$$

$$= r+2 - \alpha_{k+1}.$$

Therefore, the sum is equal to

$$\sum_{|\alpha|=k+r+3} \zeta(\alpha_0,\ldots,\alpha_k,\alpha_{k+1}+1)(r+2-\alpha_{k+1}).$$

The number of multiple zeta values in the above sum is

$$\sum_{\alpha=1}^{r+2} \binom{k+r+2-\alpha}{k}(r+2-\alpha),$$

or

$$\sum_{\beta=1}^{r+1} \beta\binom{k+\beta}{k},$$

which is equal to

$$\frac{(k+r+2)!}{(k+2)\cdot k!r!}.$$

Thus our assertion is proved.   □

Similar considerations lead to different combinatorial identities. For example, from the identity

$$\sum_{a+b+c=k} 2^a \sum_{m+n=r} \sum_{\substack{|\alpha|=b+m+1 \\ |\beta|=c+n+2}} \zeta(\{1\}^a,\alpha_0,\ldots,\alpha_b+\beta_0,\ldots,\beta_c,\beta_{c+1}+1)$$

$$= \sum_{p+q=k} 2^p \sum_{|\alpha|=q+r+3} \zeta(\{1\}^p,\alpha_0,\ldots,\alpha_q,\alpha_{q+1}+1)(r+2-\alpha_{q+1}),$$

we get the combinatorial identity

$$\sum_{a+b+c=k} 2^a \sum_{m+n=r} \binom{b+m}{b}\binom{c+n+1}{c+1} = \sum_{j=0}^{k} \sum_{\alpha=1}^{r+1} \binom{k+\alpha+1}{j}\alpha,$$

which is difficult to be proved otherwise.

Two combinatorial identities obtained above are just special examples of applications of shuffle relations producing from shuffle products of multiple zeta values. Here we need two more general shuffle relations we already proved before. Recall that the usual notation for sums of multiple zeta values

$$E(\mu,\lambda) = \sum_{p+q=k} \mu^p \sum_{|\alpha|=q+r+3} \zeta(\{1\}^p,\alpha_0,\ldots,\alpha_q,\alpha_{q+1}+1)\lambda^{\alpha_{q+1}},$$

so that

$$E(0, \lambda) = \sum_{|\boldsymbol{\alpha}|=k+r+3} \zeta(\alpha_0, \ldots, \alpha_k, \alpha_{k+1} + 1)\lambda^{\alpha_{k+1}}$$

and

$$E(2, \lambda) = \sum_{p+q=k} 2^p \sum_{|\boldsymbol{\alpha}|=q+r+3} \zeta(\{1\}^p, \alpha_0, \ldots, \alpha_q, \alpha_{q+1} + 1)\lambda^{\alpha_{q+1}}.$$

Theorem 6.2.2. *For a pair of nonnegative integers k, r and complex numbers μ, λ, ν with $\lambda \neq \mu$, $\mu\lambda\nu \neq 0$, we have*

(6.2.1)

$$\sum_{a+b=k} \sum_{m+n+p=r} \mu^m \lambda^n \nu^p \sum_{\substack{|\boldsymbol{\alpha}|=a+m+1 \\ |\boldsymbol{\beta}|=b+n+1}} \zeta(\alpha_0, \ldots, \alpha_a + \beta_0, \ldots, \beta_b, p + 2)$$

$$= \frac{1}{(\lambda - \mu)\nu} \left\{ \lambda^{r+2} E\left(0, \frac{\nu}{\lambda}\right) - \mu^{r+2} E\left(0, \frac{\nu}{\mu}\right) \right\}.$$

Theorem 6.2.3. *For a pair of nonnegative integers k, r and complex numbers μ, λ, ν with $\lambda \neq \mu$, $\mu\lambda\nu \neq 0$, we have*

(6.2.2)

$$\sum_{a+b+c=k} 2^a \sum_{m+n+p=r} \mu^m \lambda^n \nu^p$$

$$\times \sum_{\substack{|\boldsymbol{\alpha}|=b+m+1 \\ |\boldsymbol{\beta}|=c+n+1}} \zeta(\{1\}^a, \alpha_0, \ldots, \alpha_b + \beta_0, \ldots, \beta_c, p + 2)$$

$$= \frac{1}{(\lambda - \mu)\nu} \left\{ \lambda^{r+2} E\left(2, \frac{\nu}{\lambda}\right) - \mu^{r+2} E\left(2, \frac{\nu}{\mu}\right) \right\}.$$

Remark 6.2.4. The sums of multiple zeta values of (6.2.1) and (6.2.2) come from the values of integrals

$$\frac{1}{k!r!} \iint_{0<t_1<t_2<u_1<u_2<1} \left(\log \frac{1-t_1}{1-u_1} \right)^k \left(\mu \log \frac{t_2}{t_1} + \lambda \log \frac{u_1}{t_2} + \nu \log \frac{u_2}{u_1} \right)^r$$

$$\times \frac{dt_1 dt_2}{(1-t_1)t_2} \frac{du_1 du_2}{(1-u_1)u_2}$$

and

$$\frac{1}{k!r!} \iint_{0<t_1<t_2<u_1<u_2<1} \left(\log \frac{1}{1-t_1} + \log \frac{1}{1-u_1} \right)^k$$

$$\times \left(\mu \log \frac{t_2}{t_1} + \lambda \log \frac{u_1}{t_2} + \nu \log \frac{u_2}{u_1} \right)^r \frac{dt_1 dt_2}{(1-t_1)t_2} \frac{du_1 du_2}{(1-u_1)u_2},$$

respectively.

Simply by counting the numbers of multiple zeta values on the both sides of (6.2.1) and (6.2.2), we obtain the following combinatorial identities.

Theorem 6.2.5. *For a pair of nonnegative integers k, r and complex numbers μ, λ, ν with $\lambda \neq \mu$, $\mu\lambda\nu \neq 0$, we have*

$$\sum_{a+b=k} \sum_{m+n+p=r} \mu^m \lambda^n \nu^p \binom{a+m}{a}\binom{b+n}{b}$$
$$= \sum_{\beta=1}^{r+1} \binom{k+\beta}{k}\left(\frac{\lambda^\beta - \mu^\beta}{\lambda - \mu}\right)\nu^{r+1-\beta}.$$

Theorem 6.2.6. *For a pair of nonnegative integers k, r and complex numbers μ, λ, ν with $\lambda \neq \mu$, $\mu\lambda\nu \neq 0$, we have*

(6.2.3)

$$\sum_{a+b+c=k} 2^a \sum_{m+n+p=r} \mu^m \lambda^n \nu^p \binom{b+m}{b}\binom{c+n}{c}$$
$$= \sum_{j=0}^{k}\sum_{\beta=1}^{r+1} \binom{k+\beta+1}{j}\left(\frac{\lambda^\beta - \mu^\beta}{\lambda - \mu}\right)\nu^{r+1-\beta}.$$

Remark 6.2.7. We use the identity

$$\sum_{p+q=k} 2^p \binom{q+\beta}{\beta} = \sum_{j=0}^{k} \binom{k+\beta+1}{j}$$

in order to get (6.2.3).

To produce more combinatorial identities of similar kind, we are going to produce new shuffle relations from two sums of multiple zeta values. Consider integral of the form

$$\frac{1}{k!r!} \iint_{R_1 \times R_2} \left(\mu \log \frac{1-t_1}{1-t_2} + \log \frac{1-u_1}{1-u_2}\right)^k \left(\log \frac{t_2}{t_1} + \lambda \log \frac{u_2}{u_1}\right)^r$$
$$\times \frac{dt_1 dt_2}{(1-t_1)t_2}\frac{du_1 du_2}{(1-u_1)u_2}.$$

The value of the above integral is

$$\sum_{j=0}^{k}\sum_{\ell=0}^{r} \mu^j \lambda^\ell\, \zeta(j+r-\ell+2)\zeta(k-j+\ell+2).$$

Through the usual shuffle process, i.e., decompose $R_1 \times R_2$ into 6 simplices and the evaluate the integral over each simplex one by one in terms of multiple zeta values. This leads to the identity

(6.2.4)

$$\sum_{j=0}^{k}\sum_{\ell=0}^{r}\mu^j\lambda^\ell\zeta(j+r-\ell+2)\zeta(k-j+\ell+2)$$

$$=\sum_{j=0}^{k}\sum_{\ell=0}^{r}\sum_{\substack{|\boldsymbol{\alpha}|=j+r-\ell+1\\|\boldsymbol{\beta}|=k-j+\ell+1}}\zeta(\alpha_0,\ldots,\alpha_j+1,\beta_0,\ldots,\beta_{k-j}+1)\left\{\mu^j\lambda^\ell+\mu^{k-j}\lambda^{r-\ell}\right\}$$

$$+\sum_{a+b+c=k}\sum_{m+n+p=r}\sum_{\substack{|\boldsymbol{\alpha}|=a+m+1\\|\boldsymbol{\beta}|=b+n+1\\|\boldsymbol{\gamma}|=c+p+1}}\zeta(\alpha_0,\ldots,\alpha_a+\beta_0,\ldots,\beta_b+\gamma_0,\ldots,\gamma_c+1)$$

$$\times\left\{\mu^a(\mu+1)^b(\lambda+1)^n\lambda^p+\mu^c(\mu+1)^b(\lambda+1)^n\lambda^m\right.$$
$$\left.+(\mu+1)^b\lambda^{m+p}(\lambda+1)^n+\mu^{a+c}(\mu+1)^b(\lambda+1)^n\right\}.$$

Note that the number of multiple zeta values of weight $j+r-\ell+2$ in the sum

$$\sum_{|\boldsymbol{\alpha}|=j+r-\ell+1}\zeta(\alpha_0,\ldots,\alpha_j+1)$$

is $\binom{j+r-\ell}{j}$. By counting the number of multiple zeta values on the right-hand side of (6.2.4), we obtain the following.

Theorem 6.2.8. *For a pair of nonnegative integers k,r and complex numbers μ, λ, we have*

(6.2.5)

$$\sum_{a+b+c=k}\sum_{m+n+p=r}\binom{a+m}{a}\binom{b+n}{b}\binom{c+p}{c}P(\mu,\lambda)$$

$$=\sum_{j=0}^{k}\sum_{\ell=0}^{r}\mu^j\lambda^\ell\left\{\binom{k+r+4}{j+r-\ell+2}-2\right\}\binom{j+r-\ell}{j}\binom{k-j+\ell}{\ell}$$

with

$$P(\mu,\lambda)=\mu^a(\mu+1)^b(\lambda+1)^n\lambda^p+u^c(\mu+1)^b(\lambda+1)^n\lambda^m$$
$$+(\mu+1)^b\lambda^{m+p}(\lambda+1)^n+\mu^{a+c}(\mu+1)^b(\lambda+1)^n.$$

If we separate the coefficients of $\mu^j \lambda^\ell$ on the both sides of (6.2.5), we obtain the following.

Theorem 6.2.9. *For a pair of nonnegative integers k, r and integers j, ℓ with $0 \leq j \leq k$, $0 \leq \ell \leq r$, we have*

$$\sum_{a+b+c=k} \sum_{m+n+p=r} \binom{a+m}{a}\binom{b+n}{b}\binom{c+p}{c}$$
$$\times \left\{ \binom{b}{j-a}\binom{n}{\ell-p} + \binom{b}{j-c}\binom{n}{\ell-m} \right.$$
$$\left. + \binom{b}{j}\binom{n}{r-\ell} + \binom{b}{k-j}\binom{n}{\ell} \right\}$$
$$= \left\{ \binom{k+r+4}{j+r-\ell+2} - 2 \right\} \binom{j+r-\ell}{j}\binom{k-j+\ell}{\ell}.$$

Three particular cases corresponding $\mu = \lambda = 1$, $\mu = -1$, $\lambda = 1$ and $\mu = \lambda = -1$ of Theorem 6.2.8 worth mention here.

Theorem 6.2.10. *For a pair of nonnegative integers k, r, we have*

$$\sum_{a+b\leq k} \sum_{|\boldsymbol{\alpha}|=k+r+3} 2^{\alpha_{a+1}+\alpha_{a+2}+\cdots+\alpha_{a+b}} \left(2^{\alpha_{a+b+1}} - 2 \right)$$

$$(\boldsymbol{\alpha} = (\alpha_0, \alpha_1, \ldots, \alpha_k, \alpha_{k+1}))$$

$$= \frac{1}{2} \sum_{j=0}^{k} \sum_{\ell=0}^{r} \left\{ \binom{k+r+4}{j+r-\ell+2} - 2 \right\} \binom{j+r-\ell}{j}\binom{k-j+\ell}{\ell}.$$

Proof. When $\mu = \lambda = 1$, the shuffle relation in (6.2.5) is

$$4 \sum_{a+b+c=k} 2^b \sum_{m+n+p=r} 2^n$$

$$\times \sum_{\substack{|\boldsymbol{\alpha}|=a+m+1 \\ |\boldsymbol{\beta}|=b+n+1 \\ |\boldsymbol{\gamma}|=c+p+1}} \zeta(\alpha_0, \ldots, \alpha_a, \beta_0, \ldots, \beta_b + \gamma_0, \ldots, \gamma_c + 1)$$

$$+ 2 \sum_{j=0}^{k} \sum_{\ell=0}^{r} \sum_{\substack{|\boldsymbol{\alpha}|=j+r-\ell+1 \\ |\boldsymbol{\beta}|=k-j+\ell+1}} \zeta(\alpha_0, \ldots, \alpha_j + 1, \beta_0, \ldots, \beta_{k-j+1} + 1)$$

$$= \sum_{j=0}^{k} \sum_{\ell=0}^{r} \zeta(j+r-\ell+2)\zeta(k-j+\ell+2).$$

The first sum of multiple zeta values can be rewritten as

$$2 \sum_{a+b \leq k} \sum_{|\alpha|=k+r+3} \zeta(\alpha_0, \ldots, \alpha_k, \alpha_{k+1} + 1) 2^{\alpha_{a+1} + \cdots + \alpha_{a+b}} (2^{\alpha_{a+b+1}} - 2).$$

By counting the number of multiple zeta values and note that the shuffle process is between $\binom{j+r-\ell+2}{j}$ multiple zeta values of weight $j+r-\ell+2$ and $\binom{k-j+\ell+2}{\ell}$ multiple zeta value of weight $k-j+\ell+2$, we get our assertion. □

Theorem 6.2.11. *For a pair of nonnegative integers k, r with k even, we have*

$$\sum_{\substack{j=0 \\ j:\text{even}}}^{k} \sum_{|\alpha|=k+r+3} (2^{\alpha_{j+1}} - 2) \qquad (\boldsymbol{\alpha} = (\alpha_0, \alpha_1, \ldots, \alpha_k, \alpha_{k+1}))$$

$$= \frac{1}{2} \sum_{j=0}^{k} \sum_{\ell=0}^{r} (-1)^j \left\{ \binom{k+r+4}{j+r-\ell+2} - 2 \right\} \binom{j+r-\ell}{j} \binom{k-j+\ell}{\ell}.$$

Theorem 6.2.12. *For a pair of nonnegative integers k, r with $k+r$ even, we have*

$$\sum_{a+b=k} \sum_{m+n=r} \binom{a+m}{a} \binom{b+n}{b} \{1 + (-1)^{a+n}\}$$

$$= \frac{1}{2} \sum_{j=0}^{k} \sum_{\ell=0}^{r} (-1)^{j+\ell} \left\{ \binom{k+r+4}{j+r-\ell+2} - 2 \right\} \binom{j+r-\ell}{j} \binom{k-j+\ell}{\ell}.$$

Some alternating sums of multiple zeta values are equal to particular single sums of multiple zeta values through shuffle process. Here are two examples.

Theorem 6.2.13. *For a pair of even integers $k, r \geq 0$, we have*

$$\sum_{j=0}^{k} \sum_{\ell=0}^{r} (-1)^\ell \sum_{\substack{|\alpha|=j+r-\ell+1 \\ |\beta|=k-j+\ell+1}} \zeta(\alpha_0, \alpha_1, \ldots, \alpha_j + \beta_j, \beta_{j+1}, \ldots, \beta_k, 2)$$

$$= \sum_{|\alpha|=k+r+2} \zeta(\alpha_0, \alpha_1, \ldots, \alpha_k, 2).$$

Theorem 6.2.14. *For a pair of even integers $k, r \geq 0$, we have*

$$\sum_{j=0}^{k}\sum_{\ell=0}^{r}(-1)^{\ell}\sum_{\substack{|\boldsymbol{\alpha}|=j+r-\ell+1 \\ |\boldsymbol{\beta}|=k-j+\ell+2}}\zeta(\alpha_0,\alpha_1,\ldots,\alpha_j+\beta_j,\beta_{j+1},\ldots,\beta_k,\beta_{k+1}+1)$$

$$+ \sum_{\substack{0\leq\ell\leq r \\ \ell:\text{even}}}\sum_{|\boldsymbol{\alpha}|=k+\ell+1}\zeta(\alpha_0,\alpha_1,\ldots,\alpha_k,r-\ell+3)$$

$$= \zeta(k+r+4).$$

Indeed, Theorem 6.2.14 is just a special case of Theorem 6.2.5 with $\mu = 1$, $\lambda = \nu = -1$. Of course, these shuffle relation also have combinatorial implication. Here are the results.

Theorem 6.2.15. *For a pair of even integers $k, r \geq 0$, we have*

$$\sum_{j=0}^{k}\sum_{\ell=0}^{r}(-1)^{\ell}\binom{j+r-\ell}{j}\binom{k-j+\ell}{\ell} = \binom{k+r+1}{k}.$$

Theorem 6.2.16. *For a pair of integers $k, r \geq 0$ with r even, we have*

$$\sum_{j=0}^{k}\sum_{\ell=0}^{r}(-1)^{\ell}\binom{j+r-\ell}{j}\binom{k-j+\ell+1}{k-j+1} = \frac{1}{2}\sum_{\beta=0}^{r+1}\binom{k+\beta}{k}\left(1+(-1)^{\beta+1}\right).$$

6.3 The Generating Function of Three Variables

Instead of evaluating the generating function

$$G_3(\xi,\mu,\lambda) = \sum_{j=0}^{k}\sum_{m=0}^{q}\sum_{\ell=0}^{r}C(k+q+r+4,j+m+\ell+2)\xi^j\mu^m\lambda^\ell,$$

we evaluate another generating function with a slight change

$$G_3'(\xi,\mu,\lambda) = \sum_{j=0}^{k}\sum_{m=0}^{q}\sum_{\ell=0}^{r}C(k+q+r+4,j+m+r-\ell+2)\xi^j\mu^m\lambda^\ell,$$

which is more convenient to deal with. These two polynomials are relative by the relation

$$G_3(\xi,\mu,\lambda) = \lambda^r G_3'(\xi,\mu,\lambda^{-1}).$$

For nonnegative integers k, q, r and complex numbers ξ, μ, λ, let

$$F_{\xi,\mu,\lambda}(t_1, t_2, u_1, u_2) = \left(\xi \log \frac{1}{1 - t_1} + \log \frac{1}{1 - u_1} \right)^k$$
$$\times \left(\mu \log \frac{t_2}{t_1} + \log \frac{u_2}{u_1} \right)^q \left(\log \frac{1}{t_2} + \lambda \log \frac{1}{u_2} \right)^r.$$

In order to evaluate $G_3'(\xi, \mu, \lambda)$, we consider a particular integral with $F_{\xi,\mu,\lambda}(t_1, t_2, u_1, u_2)$ as the integral and its region of integration is a product of two simplices of dimension two,

$$(6.3.1) \qquad \frac{1}{k!q!r!} \iint_{R_1 \times R_2} F_{\xi,\mu,\lambda}(t_1, t_2, u_1, u_2) \frac{dt_1 dt_2}{(1 - t_1)t_2} \frac{du_1 du_2}{(1 - u_1)u_2}$$

where $R_1 : 0 < t_1 < t_2 < 1$ and $R_2 : 0 < u_1 < u_2 < 1$.

For nonnegative integers n, ℓ, p, we can express the multiple zeta value $\zeta(\{1\}^n, \ell + p + 2)$ as Drinfeld integral

$$\int_{E_{n+\ell+p+2}} \prod_{i=1}^{n+1} \frac{dt_i}{1 - t_i} \prod_{j=n+1}^{n+\ell+1} \frac{dt_j}{t_j} \prod_{k=n+\ell+2}^{n+\ell+p+2} \frac{dt_k}{t_k}.$$

Fix t_{n+1} and $t_{n+\ell+2}$ and integrate with respect to the remaining variables $t_1, t_2, \ldots, t_n; t_{n+1}, t_{n+2}, \ldots, t_{n+\ell+1}; t_{n+\ell+3}, \ldots, t_{m+\ell+p+2}$ we get that

$$\zeta(\{1\}^n, \ell+p+2) = \frac{1}{n!\ell!p!} \int_{E_2} \left(\log \frac{1}{1 - t_1} \right)^n \left(\log \frac{t_2}{t_1} \right)^\ell \left(\log \frac{1}{t_2} \right)^p \frac{dt_1 dt_2}{(1 - t_1)t_2}.$$

It follows that the value of the integral $(6.3.1)$ is

$$\sum_{j=0}^{k} \sum_{m=0}^{q} \sum_{\ell=0}^{r} \xi^j \mu^m \lambda^\ell \zeta(\{1\}^j, m + r - \ell + 2)\zeta(\{1\}^{k-j}, q - m + \ell + 2).$$

As the number of multiple zeta values producing from the shuffle product of $\zeta(\{1\}^j, m + r - \ell + 2)$ and $\zeta(\{1\}^{k-j}, q - m + \ell + 2)$ is

$$C(k + q + r + 4, j + m + r - \ell + 2).$$

Therefore, after a shuffle process with the integral (4.4) and then count the total number of multiple zeta values, we are able to express $G_3'(\xi, \mu, \lambda)$ in a different way.

Presently, the shuffle process is to decompose $R_1 \times R_2$ into 6 four-dimensional simplices obtaining from all possible interlacings of t_1, t_2 and u_1, u_2 as follow

$$D_1 : \quad 0 < t_1 < t_2 < u_1 < u_2 < 1,$$
$$D_2 : \quad 0 < u_1 < u_2 < t_1 < t_2 < 1,$$
$$D_3 : \quad 0 < t_1 < u_1 < t_2 < u_2 < 1,$$
$$D_4 : \quad 0 < t_1 < u_1 < u_2 < t_2 < 1,$$
$$D_5 : \quad 0 < u_1 < t_1 < u_2 < t_2 < 1 \text{ and}$$
$$D_6 : \quad 0 < u_1 < t_1 < t_2 < u_2 < 1,$$

and then evaluate the integrals over D_j $(j = 1, 2, 3, 4, 5, 6)$ one by one in terms of multiple zeta values. Then count the number of multiple zeta values resulted from integration over D_j $(j = 1, 2, 3, 4, 5, 6)$. Here is the details of calculation.

(1) On the simplex $D_1 : 0 < t_1 < t_2 < u_1 < u_2 < 1$, we change the three factors of $F_{\xi,\mu,\lambda}$ into

$$\left\{ (\xi + 1) \log \frac{1}{1 - t_1} + \log \frac{1 - t_1}{1 - t_2} + \log \frac{1 - t_2}{1 - u_1} \right\}^k, \quad \left(\mu \log \frac{t_2}{t_1} + \log \frac{u_2}{u_1} \right)^q$$

and

$$\left\{ \log \frac{u_1}{t_2} + \log \frac{u_2}{u_1} + (\lambda + 1) \log \frac{1}{u_2} \right\}^r$$

and replace them by their multinomial expansions

$$\sum_{a+b+c=k} \frac{k!}{a!b!c!} (\xi + 1)^a \left(\log \frac{1}{1 - t_1} \right)^a \left(\log \frac{1 - t_1}{1 - t_2} \right)^b \left(\log \frac{1 - t_2}{1 - u_1} \right)^c,$$

$$\sum_{m+n=q} \binom{q}{m} \mu^m \left(\log \frac{t_2}{t_1} \right)^m \left(\log \frac{u_2}{u_1} \right)^n$$

and

$$\sum_{d+e+f=r} \frac{r!}{d!e!f!} (\lambda + 1)^f \left(\log \frac{u_1}{t_2} \right)^d \left(\log \frac{u_2}{u_1} \right)^e \left(\log \frac{1}{u_2} \right)^f.$$

Therefore the values of the integral over D_1, in terms of multiple zeta values, is

(6.3.2)

$$\sum_{a+b+c=k} (\xi+1)^a \sum_{m+n=q} \mu^m \sum_{d+e+f=r} \binom{n+e}{n} (\lambda+1)^f$$

$$\times \sum_{\substack{|\boldsymbol{\alpha}|=b+m+1 \\ |\boldsymbol{\beta}|=c+d+1}} \zeta(\{1\}^a, \alpha_0, \alpha_1, \ldots, \alpha_b + \beta_0, \beta_1, \ldots, \beta_c, n+e+f+2).$$

For fixed a, n, e, f, the number of multiple zeta values in the fourth summation of (6.3.2) is

$$\binom{b+m}{n}\binom{c+d}{c}$$

which is the number of positive integral solutions to the linear equations

$$\alpha_0 + \alpha_1 + \cdots + \alpha_b = b+m+1, \quad \beta_0 + \beta_1 + \cdots + \beta_c = c+d+1.$$

Therefore, the contribution to $G_3'(\xi, \mu, \lambda)$ from the integration over D_1 is

$$\sum_{a+b+c=k} \sum_{m+n=q} \mu^m \sum_{d+e+f=r} \binom{c+d}{c}$$

$$\times \left\{ (\xi+1)^a \binom{b+m}{m} \right\} \left\{ \binom{n+e}{n} (\lambda+1)^f \right\}$$

which will be denoted by $H_1(\xi, \mu, \lambda)$. Note that

$$\sum_{a+b=p} (\xi+1)^a \binom{b+m}{m} = I_1(m, p; \xi)$$

and

$$\sum_{e+f=g} \binom{n+e}{n} (\lambda+1)^f = I_1(n, g; \lambda).$$

So we express $H_1(\xi, \mu, \lambda)$ as

$$\sum_{p+c=k} \sum_{m+n=q} \sum_{d+g=r} \mu^m \binom{c+d}{c} I_1(m, p; \xi) I_1(n, g; \lambda).$$

(2) On the simplex $D_2 : 0 < u_1 < u_2 < t_1 < t_2 < 1$, we just exchange the roles of t_1, t_2 and u_1, u_2, so that the contribution to $G_3'(\xi, \mu, \lambda)$ from the integration over D_2 is

$$H_2(\xi, \mu, \lambda) = \xi^k \mu^q \lambda^r H_1(\xi^{-1}, \mu^{-1}, \lambda^{-1})$$

or more explicit as

$$H_2(\xi, \mu, \lambda) = \sum_{p+c=k} \sum_{m+n=q} \sum_{d+g=r} \mu^n \binom{c+d}{c} \xi^k I_1(m, p; \xi^{-1}) \lambda^r I_1(n, g; \lambda^{-1}).$$

(3) On the simplex $D_3 : 0 < t_1 < u_1 < t_2 < u_2 < 1$, we first change the factors of integral $F_{\xi,\mu,\lambda}(t_1, t_2, u_1, u_2)$ as

$$\left\{ (\xi+1) \log \frac{1}{1-t_1} + \log \frac{1-t_1}{1-u_1} \right\}^k, \quad \left(\mu \log \frac{u_1}{t_1} + (\mu+1) \log \frac{t_2}{u_1} + \log \frac{u_2}{t_2} \right)^q$$

and

$$\left\{ \log \frac{u_2}{t_2} + (\lambda+1) \log \frac{1}{u_2} \right\}^r$$

and then substitute them by their multinomial expansions,

$$\sum_{a+b=k} \binom{k}{a} (\xi+1)^a \left(\log \frac{1}{1-t_1} \right)^a \left(\log \frac{1-t_1}{1-u_1} \right)^b,$$

$$\sum_{m+n+p=q} \frac{q!}{m!n!p!} \mu^m (\mu+1)^n \left(\log \frac{u_1}{t_1} \right)^m \left(\log \frac{t_2}{u_1} \right)^n \left(\log \frac{u_2}{t_2} \right)^p$$

and

$$\sum_{d+e=r} \binom{r}{d} (\lambda+1)^e \left(\log \frac{u_2}{t_2} \right)^d \left(\log \frac{1}{u_2} \right)^e,$$

respectively. So that the value of the integral over D_2 is

$$\sum_{a+b=k} (\xi+1)^a \sum_{m+n+p=q} \mu^m (\mu+1)^n \sum_{d+e=r} \binom{p+d}{p} (\lambda+1)^e$$

$$\times \sum_{|\alpha|=b+m+1} \zeta(\{1\}^a, \alpha_0, \alpha_1, \ldots, \alpha_b, n+p+r+2).$$

It follows that the contribution to $G_3'(\xi, \mu, \lambda)$ from the integration over D_3 is

$$H_3(\xi, \mu, \lambda) = \sum_{a+b=k} \sum_{m+n+p=q} \mu^m \sum_{d+e=r} \left\{ (\xi+1)^a \binom{b+m}{m} \right\}$$

$$\times \left\{ (\mu+1)^n \binom{p+d}{p} (\lambda+1)^e \right\}$$

or

$$\sum_{m+n=q} \mu^m I_1(m, k; \xi) I_2(n, r; \lambda).$$

(4) The evaluation of the integrals over D_4, D_5 and D_6 are similar to the evaluation of the integral over D_3, so we write down the contribution directly.

$$H_4(\xi, \mu, \lambda) = \sum_{m+n=q} \mu^q I_1(k, m; \xi) \cdot \lambda^r I_2(n, r; \mu^{-1}, \lambda^{-1}),$$

$$H_5(\xi, \mu, \lambda) = \sum_{m+n=q} \mu^n \xi^k I_1(k, m; \xi^{-1}) \cdot \lambda^r I_2(n, r; \mu^{-1}, \lambda^{-1}),$$

$$H_6(\xi, \mu, \lambda) = \sum_{m+n=q} \xi^k I_1(k, m; \xi^{-1}) I_2(n, r; \mu, \lambda).$$

In the final, sum $H_j(\xi, \mu, \lambda)$ $(j = 1, 2, 3, 4, 5, 6)$ together to yield the generating function $G_3'(\xi, \mu, \lambda)$. Under the relation

$$G_3(\xi, \mu, \lambda) = \lambda^r G_3'(\xi, \mu, \lambda^{-1})$$

we finally obtain $G_3(\xi, \mu, \lambda)$.

Theorem 6.3.1. *For nonnegative integers k, q, r, we have*

$$G_3(\xi, \mu, \lambda) = \sum_{j=0}^{k} \sum_{m=0}^{q} \sum_{\ell=0}^{r} C(k+q+r+4, j+m+\ell+2) \xi^j \mu^m \lambda^\ell$$

$$= \sum_{p+c=q} \sum_{m+n=q} \sum_{d+g=r} \binom{c+d}{c} \left\{ \mu^m I_1(m, p; \xi) \cdot \lambda^r I_1(n, g; \lambda^{-1}) \right.$$

$$\left. + \mu^n \xi^k I_1(m, p; \xi^{-1}) \cdot I_1(n, g; \lambda) \right\}$$

$$+ \sum_{m+n=q} \left\{ \mu^m I_1(m, k; \xi) + \xi^k I_1(m, k; \xi^{-1}) \right\}$$

$$\times \left\{ \lambda^r I_2(n, r; \mu, \lambda^{-1}) + \mu^n I_2(n, r; \mu^{-1}, \lambda) \right\}.$$

Based on the above theorem, we separate the coefficient of $\xi^j \mu^m \lambda^\ell$ and obtain the following theorem.

Theorem 6.3.2. *For integers k, q, r, j, m, ℓ such that $0 \leq j \leq k$, $0 \leq m \leq q$ and $0 \leq \ell \leq r$, we have*

$$C(k+q+r+4, j+m+\ell+2)$$

$$= \sum_{a+b=k} \sum_{c+d=r} \left\{ \binom{m+a+1}{a-j} \binom{q-m+d+1}{d+\ell-r} \right.$$

$$\left. + \binom{q-m+a+1}{a+j-k} \binom{m+d+1}{d-\ell} \right\}$$

$$+ \sum_{a+b+c=q} \binom{a+k+1}{k-j} \left\{ \binom{b}{m-a} \binom{c+r+1}{\ell} + \binom{b}{q-m} \binom{c+r+1}{r-\ell} \right\}$$

$$+ \sum_{a+b+c=q} \binom{a+k+1}{j} \left\{ \binom{b}{m-c} \binom{c+r+1}{r-\ell} + \binom{b}{m} \binom{c+r+1}{\ell} \right\}.$$

Some interesting consequences are obtained simply by restricting $k = 0$ or $q = 0$ in the above theorem.

Corollary 6.3.3. *For integers q, r, m, ℓ with $0 \leq m \leq q$ and $0 \leq \ell \leq r$, we have*

$$C(q+r+4, m+\ell+2)$$

$$= \binom{q-m+r+2}{\ell} + \binom{m+r+2}{r-\ell}$$

$$+ \sum_{a+b+c=q} \left\{ \binom{b}{m-a} \binom{c+r+1}{\ell} + \binom{b}{q-m} \binom{c+r+1}{r-\ell} \right.$$

$$\left. + \binom{b}{m-c} \binom{c+r+1}{r-\ell} + \binom{b}{m} \binom{c+r+1}{r-\ell} \right\}.$$

Corollary 6.3.4. *For integers k, r, j, ℓ with $0 \leq j \leq k$ and $0 \leq \ell \leq r$, we have*

$$C(k+r+4, j+\ell+2)$$

$$= \sum_{a+b=k} \sum_{c+d=r} \binom{b+c}{b} \left\{ \binom{a+1}{j+1} \binom{d+1}{r-\ell+1} + \binom{a+1}{k-j+1} \binom{d+1}{\ell+1} \right\}$$

$$+ \binom{k+2}{j+1} \binom{r+2}{\ell+1}.$$

6.4   Exercises

1. Show that for integers k, j, r, ℓ with $0 \le j \le k$ and $0 \le \ell \le r$

$$\binom{k+r+4}{j+\ell+2}$$
$$= \sum_{m+p=r+2} \left\{ \binom{m+k+1}{j}\binom{p}{\ell+1} + \binom{m+k+1}{k-j}\binom{p}{r-\ell+1} \right\}.$$

2. For a pair of nonnegative integers k, r, prove that

$$\sum_{a+b=k} \sum_{m+n=r} \binom{a+m}{a}\binom{b+n}{b}(-1)^{b+n} = \frac{1}{2}\left\{1 + (-1)^{k+r}\right\}\binom{k+r}{k}.$$

3. Suppose that k, m, n, p, j are integers with $0 \le j \le k$ and $m, n, p \ge 0$. Show that

$$\sum_{a+b+c=k} \binom{a+n}{a}\binom{b+m}{b}\binom{c+p}{c}\binom{b}{j}$$
$$= \binom{m+j}{j}\binom{m+n+p+k+2}{k-j}.$$

4. Suppose that k, j, r, ℓ are integers with $0 \le j \le k$ and $0 \le \ell \le r$. Prove that

$$\sum_{a+b+c=k} \sum_{m+n+p=r} \binom{a+m}{a}\binom{b+n}{b}\binom{c+p}{c}\binom{b}{j}\binom{n}{r-\ell}$$
$$= \binom{j+r-\ell}{j}\binom{k+r-\ell+2}{k-j}\binom{k+r+2}{\ell}.$$

CHAPTER 7

Combinatorial Identities of Convolution Type

Certain sums of products of binomial coefficients represents the number of multiple zeta values in a sum of multiple zeta values. Therefore shuffle relations among multiple zeta values then lead to combinatorial identities. In particular, the classical Vandermonde's convolution

$$\sum_{m+n=r} \binom{a+m}{a} \binom{b+n}{b} = \binom{a+b+r+1}{r}$$

can be extended to more general form as

$$\sum_{|\boldsymbol{\beta}|=r} \binom{\alpha_1+\beta_1}{\beta_1} \binom{\alpha_2+\beta_2}{\beta_2} \cdots \binom{\alpha_n+\beta_n}{\beta_n} = \binom{|\boldsymbol{\alpha}|+r+n-1}{r}.$$

7.1 Some Particular Combinatorial Identities

The particular sum of products of binomial coefficients

$$\sum_{a+b=k} \sum_{m+n=r} \binom{a+m}{a} \binom{b+n}{b}$$

173

is used to denote the number of multiple zeta values in the sum

$$\text{(7.1.1)} \qquad \sum_{a+b=k} \sum_{m+n=r} \sum_{\substack{|\boldsymbol{\alpha}|=a+m+1 \\ |\boldsymbol{\beta}|=b+n+1}} \zeta(\alpha_0, \alpha_1, \ldots, \alpha_a + \beta_0, \beta_1, \ldots, \beta_b + 1).$$

The above sum can be rewritten as

$$\sum_{|\boldsymbol{\alpha}|=k+r+2} \sum_{j=0}^{k} (\alpha_j - 1)\zeta(\alpha_0, \alpha_1, \ldots, \alpha_k + 1)$$

or simply

$$(r+1) \sum_{|\boldsymbol{\alpha}|=k+r+2} \zeta(\alpha_0, \alpha_1, \ldots, \alpha_k + 1).$$

Consequently, we conclude that

$$\sum_{a+b=k} \sum_{m+n=r} \binom{a+m}{a}\binom{b+n}{b} = (r+1)\binom{k+r+1}{k} = \frac{(k+r+1)!}{k!r!}.$$

Sums of multiple zeta values as shown in (7.1.1) or their similars appear in some special shuffle relations. Therefore, we can employ to evaluate some alternating sums of binomial coefficients.

Theorem 7.1.1. *For a pair of nonnegative integers k, r with $k+r$ is even, then we have*

$$\sum_{j=0}^{k}\sum_{\ell=0}^{r}(-1)^{j+\ell}(r-\ell+1)(\ell+1)C(k+r+4, j+r-\ell+2) = (2k+2r+6)C(k+r, k).$$

Proof. Consider the particular integral

$$\frac{1}{k!r!}\iint_{R_1\times R_2}\left(\log\frac{1-t_1}{1-u_1}\right)^k\left(\log\frac{u_1}{t_1}\right)^r\frac{dt_1dt_2}{(1-t_1)t_2}\frac{du_1du_2}{(1-u_1)u_2}$$

which has the value

$$\sum_{j=0}^{k}\sum_{\ell=0}^{r}(-1)^{j+\ell}(r-\ell+1)(\ell+1)\zeta(\{1\}^j, r-\ell+2)\zeta(\{1\}^{k-j}, \ell+2).$$

Therefore, after the shuffle process, the number of multiple zeta values appears must be

$$\text{(7.1.2)} \qquad \sum_{j=0}^{k}\sum_{\ell=0}^{r}(-1)^{j+\ell}(r-\ell+1)(\ell+1)C(k+r+4, j+r-\ell+2).$$

On the other hand, the number of multiple zeta values appear in the value of integrations over D_1, D_2, D_3, D_4, D_5 and D_6 are given by

(1) $\displaystyle\sum_{a+b=k}\sum_{m+n=r} \binom{a+m}{a}\binom{b+n}{b} = \frac{(k+r+1)!}{k!r!}$,

(2) $\displaystyle(-1)^{k+r}\sum_{a+b=k}\sum_{m+n=r} \binom{a+m}{a}\binom{b+n}{b} = (-1)^{k+r}\frac{(k+r+1)!}{k!r!}$,

(3) $C(k+r,k)$,

(4) $C(k+r,k)$,

(5) $(-1)^{k+r}C(k+r,k)$ and

(6) $(-1)^{k+r}C(k+r,k)$.

When $k+r$ is even, the sum of the above is $(2k+2r+6)C(k+r,k)$. Of course, this number is equal to the number shown in (7.1.2). □

The following sum of triple products of binomial coefficients

$$(7.1.3) \qquad \sum_{a+b+c=k}\sum_{m+n+p=r} \binom{a+m}{a}\binom{b+n}{b}\binom{c+p}{c}$$

is used to denoted the number of multiple zeta values in the sum
(7.1.4)
$$\sum_{a+b+c=k}\sum_{m+n+p=r}\sum_{\substack{|\alpha|=a+m+1\\|\beta|=b+n+1\\|\gamma|=c+p+1}} \zeta(\alpha_0,\alpha_1,\ldots,\alpha_a,\beta_0,\beta_1,\ldots,\beta_b+\gamma_0,\gamma_1,\ldots,\gamma_c+1)$$

since the number of positive integral solutions to the linear equations:

$$\alpha_0 + \alpha_1 + \ldots + \alpha_a = a + m + 1,$$
$$\beta_0 + \beta_1 + \ldots + \beta_b = b + n + 1,$$
$$\gamma_0 + \gamma_1 + \ldots + \gamma_c = c + p + 1$$

are

$$\binom{a+m}{a}, \quad \binom{b+n}{b}, \quad \binom{c+p}{c}$$

respectively. Sums of multiple zeta values as shown in (7.1.4) appear in shuffle relations obtaining from a particular integral. This leads to the evaluation of (7.1.3).

Theorem 7.1.2. *For a pair of nonnegative integers k and r, we have*

$$\sum_{a+b+c=k}\sum_{m+n+p=r}\binom{a+m}{a}\binom{b+n}{b}\binom{c+p}{c}$$

$$=\frac{1}{2}\sum_{j=0}^{k}\sum_{\ell=0}^{r}(-1)^{j+\ell}(r-\ell+1)(k-j+1)C(k+r+4,j+r-\ell+2)$$

$$-\frac{1}{2}\left\{2k+2r+3+(-1)^{k+r}\right\}C(k+r,r).$$

Proof. Consider the particular integrals

$$\frac{1}{k!r!}\iint_{R_1\times R_2}\left(\log\frac{1-t_1}{1-u_2}\right)^k\left(\log\frac{u_2}{t_1}\right)^r\frac{dt_1dt_2}{(1-t_1)t_2}\frac{du_1du_2}{(1-u_1)u_2}$$

which have the value

$$\sum_{j=0}^{k}\sum_{\ell=0}^{r}(-1)^{j+\ell}(r-\ell+1)(k-j+1)\zeta(\{1\}^j,r-\ell+2)\zeta(\{1\}^{k-j},\ell+2).$$

Consequently, after a shuffle process, the number of multiple zeta values produced is equal to

$$(7.1.5)\quad\sum_{j=0}^{k}\sum_{\ell=0}^{r}(-1)^{j+\ell}(r-\ell+1)(k-j+1)C(k+r+4,j+r-\ell+2).$$

On the other hand, the number of multiple zeta values appear in the values of integrations over D_1, D_2, D_3, D_4, D_5 and D_6 are

(1) $\displaystyle\sum_{a+b+c=k}\sum_{m+n+p=r}\binom{a+m}{a}\binom{b+n}{b}\binom{c+p}{c},$

(2) $(-1)^{k+r}C(k+r,r),$

(3) $\displaystyle\sum_{a+b+c=k}\sum_{m+n+p=r}\binom{a+m}{a}\binom{b+n}{b}\binom{c+p}{c},$

(4) $\displaystyle\sum_{a+b=k}\sum_{m+n=r}\binom{a+m}{a}\binom{b+n}{b}=\frac{(k+r+1)!}{k!r!},$

(5) $C(k+r,k)$ and

(6) $\displaystyle\sum_{a+b=k}\sum_{m+n=r}\binom{a+m}{a}\binom{b+n}{b}=\frac{(k+r+1)!}{k!r!}.$

Our assertion then follows from

$$(1) + (2) + (3) + (4) + (5) + (6) = (7.1.5).$$

$\square$

Remark 7.1.3. My Ph.D student Tung Yang Li proved that

$$\sum_{a+b+c=k} \sum_{m+n+p=r} \binom{a+m}{a}\binom{b+n}{b}\binom{c+p}{c} = \frac{(k+r+2)!}{2!k!r!}$$

or more general that

$$\sum_{|\boldsymbol{\alpha}|=k} \sum_{|\boldsymbol{\beta}|=r} \binom{\alpha_1+\beta_1}{\alpha_1}\binom{\alpha_2+\beta_2}{\alpha_2}\cdots\binom{\alpha_n+\beta_n}{\alpha_n} = \frac{(k+r+n-1)!}{(n-1)!k!r!}.$$

He employed the fact that

$$\frac{1}{(1-x)^{m+1}} = \sum_{n=0}^{\infty} \binom{m+n}{n} x^n$$

and hence that

$$\sum_{|\boldsymbol{\beta}|=r} \binom{\alpha_1+\beta_1}{\alpha_1}\binom{\alpha_2+\beta_2}{\alpha_2}\cdots\binom{\alpha_n+\beta_n}{\alpha_n}$$

is just the coefficient of x^r in the product

$$\prod_{j=1}^{n} \frac{1}{(1-x)^{\alpha_j+1}}.$$

Consequently, the value is equal to

$$\binom{k+r+n-1}{r}.$$

Sum over all α with $|\boldsymbol{\alpha}| = k$ then yield our result.

7.2 A Generating Function for Products

The product

$$\binom{j+\ell}{j}\binom{k+r+4}{j+\ell+2}$$

is the number of multiple zeta values produced from the shuffle product of the multiple zeta value $\zeta(\{1\}^{k-j}, r - \ell + 2)$ and the sum of multiple zeta values

$$\sum_{|\alpha|=j+\ell+1} \zeta(\alpha_0, \alpha_1, \ldots, \alpha_j + 1).$$

In order to evaluate the generating function

(7.2.1)
$$\sum_{j=0}^{k} \sum_{\ell=0}^{r} \binom{j+\ell}{j} \binom{k+r+4}{j+\ell+2} \mu^j \lambda^\ell,$$

we consider the integral

(7.2.2)
$$\frac{1}{k!r!} \iint_{R_1 \times R_2} \left(\mu \log \frac{1-t_1}{1-t_2} + \log \frac{1}{1-u_1} \right)^k \left(\lambda \log \frac{t_2}{t_1} + \log \frac{1}{u_2} \right)^r$$
$$\times \frac{dt_1 dt_2}{(1-t_1)t_2} \frac{du_1 du_2}{(1-u_1)u_2}$$

which has the value

$$\sum_{j=0}^{k} \sum_{\ell=0}^{r} \mu^j \lambda^\ell \zeta(j + \ell + 2)\zeta(\{1\}^{k-j}, r - \ell + 2).$$

We need the following propositions which are simple consequences of Propositions 6.1.1 and 6.1.2.

Proposition 7.2.1. *Suppose that q, r are nonnegative integers and μ is a complex number. Then we have*

$$\sum_{m=0}^{r} \binom{q+m}{q} (\mu+1)^m = \frac{(q+r+1)!}{q!r!} \int_0^1 x^q (1+\mu x)^r dx$$
$$= \sum_{\ell=0}^{r} \frac{(q+r+1)!}{q!\ell!(r-\ell)!} \cdot \frac{\mu^\ell}{q+\ell+1}.$$

Proposition 7.2.2. *Suppose that k, m are nonnegative integers and μ is a complex numbers. Then we have*

$$\sum_{a+b+c=k} \binom{b+m}{b} (\mu+1)^b = \frac{(k+m+2)!}{k!m!} \int_0^1 \int_0^y (1+\mu x)^k x^m dx dy.$$

Here are two immediate consequences of the above propositions.

Corollary 7.2.3. *Suppose that q, r, ℓ are integers with $0 \le \ell \le r$ and $q \ge 0$. Then*

$$\sum_{m=0}^{r} \binom{q+m}{q}\binom{m}{\ell} = \binom{q+\ell}{\ell}\binom{q+r+1}{r-\ell}.$$

Corollary 7.2.4. *Suppose that k, m, j are integers with $0 \le j \le k$ and $m \ge 0$. Then we have*

$$\sum_{a+b+c=k} \binom{b+m}{b}\binom{b}{j} = \binom{m+j}{j}\binom{k+m+2}{k-j}.$$

For our convenience, we let

$$I_3(q,r;\mu) = \frac{(q+r+1)!}{q!r!} \int_0^1 (1+\mu x)^r x^q dx$$

and

$$I_4(m,k;\mu) = \frac{(k+m+2)!}{k!m!} \int_0^1 \int_0^y (1+\mu x)^k x^m dx dy.$$

First we outline our general procedure. The shuffle process with the integral (7.2.1) is equivalent to decompose $R_1 \times R_2$ into 6 simplices of dimension four and then evaluate the integral over each simplex in terms of multiple zeta values. By counting the number of multiple zeta values, we can express the generating function (7.2.2) as a sum of 6 polynomials in μ and λ.

Here are the details of evaluations.

(1) On the simplex $D_1 : 0 < t_1 < t_2 < u_1 < u_2 < 1$, we substitute the factors

$$\left(\mu \log \frac{1-t_1}{1-t_2} + \log \frac{1}{1-u_1}\right)^k \quad \text{and} \quad \left(\lambda \log \frac{t_2}{t_1} + \log \frac{1}{u_2}\right)^r$$

by their multinomial expansions

$$\sum_{a+b+c=k} \frac{k!}{a!b!c!} (\mu+1)^b \left(\log \frac{1}{1-t_1}\right)^a \left(\log \frac{1-t_1}{1-t_2}\right)^b \left(\log \frac{1-t_2}{1-u_1}\right)^c$$

and

$$\sum_{m+n=r} \binom{r}{m} \lambda^m \left(\log \frac{t_2}{t_1}\right)^m \left(\log \frac{1}{u_2}\right)^n.$$

So that the value of the integration over D_1 is

$$\sum_{a+b+c=k} (\mu+1)^b \sum_{m+n=r} \lambda^m$$

$$\times \sum_{|\alpha|=b+m+1} \zeta(\{1\}^a, \alpha_0, \alpha_1, \ldots, \alpha_b+1, \{1\}^c, n+2).$$

The number of multiple zeta values in the inner sum is

$$\binom{b+m}{b}.$$

So the contribution to the generating function is

$$\sum_{a+b+c=k} (\mu+1)^b \sum_{m+n=r} \lambda^m \binom{b+m}{b}$$

or

$$(7.2.3) \qquad \sum_{m+n=r} \lambda^m I_4(m, k; \mu).$$

(2) On the simplex $D_2 : 0 < u_1 < u_2 < t_1 < t_2 < 1$, we substitute two factors in the integrand by

$$\sum_{a+b=k} \binom{k}{a} \mu^b \left(\log \frac{1}{1-u_1}\right)^a \left(\log \frac{1-t_1}{1-t_2}\right)^b$$

and

$$\sum_{m+n+p=r} \frac{r!}{m!n!p!} (\lambda+1)^n \left(\log \frac{t_1}{u_2}\right)^m \left(\log \frac{t_2}{t_1}\right)^n \left(\log \frac{1}{t_2}\right)^p,$$

so that the value of the integration over D_2 is

$$\sum_{a+b=k} \mu^b \sum_{m+n+p=r} (\lambda+1)^n \sum_{|\alpha|=b+n+1} \zeta(\{1\}^a, m+2, \alpha_0, \alpha_1, \ldots, \alpha_b+p+1).$$

Therefore the contribution to the generating function is

$$\sum_{a+b=k} \mu^b \sum_{m+n+p=r} (\lambda+1)^n \binom{b+n}{b}$$

or

$$(7.2.4) \qquad \sum_{a+b=k} \mu^b I_4(b, r; \lambda).$$

(3) The value of the integration over D_3 is

$$\sum_{a+b+c=k} (\mu+1)^b \mu^c \sum_{m+n+p=r} \lambda^{m+n}$$

$$\times \sum_{\substack{|\alpha|=b+m+1 \\ |\beta|=c+n+1}} \zeta(\{1\}^a, \alpha_0, \alpha_1, \ldots, \alpha_b, \beta_0, \beta_1, \ldots, \beta_c + p + 2)$$

and the contribution to the generating function is

$$\sum_{a+b+c=k} (\mu+1)^b \mu^c \sum_{m+n+p=r} \lambda^{m+n} \binom{b+m}{b}\binom{c+n}{c}.$$

Note that

$$\sum_{m+n=g} \binom{b+m}{b}\binom{c+n}{c} = \binom{b+c+g+1}{b+c+1} = \binom{b+c+g+1}{g}$$

and hence the sum is equal to

$$\sum_{a+b=k} \{(\mu+1)^{b+1} - \mu^{b+1}\} \sum_{m+p=r} \lambda^m \binom{b+m+1}{m}$$

or

(7.2.5) $$\sum_{m+p=r} \lambda^m \{I_3(m, k+1; \mu) - I_3(m, k+1; \mu-1)\}.$$

(4) The value of the integration over D_4 is

$$\sum_{a+b+c=k} \{(\mu+1)^{b+1} - \mu^{b+1}\} \mu^c \sum_{m+n+p=r} \lambda^m (\lambda+1)^n$$

$$\times \sum_{\substack{|\alpha|=b+m+2 \\ |\beta|=c+n+1}} \zeta(\{1\}^a, \alpha_0, \alpha_1, \ldots, \alpha_{b+1} + \beta_0, \beta_1, \ldots, \beta_c + p + 1)$$

while the value of the integration over D_5 is

$$\sum_{a+b+c=k} \mu^{b+c} \sum_{m+n+p=r} \lambda^m (\lambda+1)^n$$

$$\times \sum_{\substack{|\alpha|=b+m+1 \\ |\beta|=c+n+1}} \zeta(\{1\}^{a+1}, \alpha_0, \alpha_1, \ldots, \alpha_b + \beta_0, \beta_1, \ldots, \beta_c + p + 1).$$

Add together to yield that the total is

$$\sum_{a+b+c=k} (\mu+1)^{b+1}\mu^c \sum_{m+n+p=r} \lambda^m(\lambda+1)^n$$

$$\times \sum_{\substack{|\boldsymbol{\alpha}|=b+m+2 \\ |\boldsymbol{\beta}|=c+n+1}} \zeta(\{1\}^a, \alpha_0, \alpha_1, \ldots, \alpha_{b+1}+\beta_0, \beta_1, \ldots, \beta_c+p+1)$$

$$+ \sum_{a+c=k} \mu^c \sum_{m+n+p=r} \lambda^m(\lambda+1)^n$$

$$\times \sum_{|\boldsymbol{\beta}|=c+n+1} \zeta(\{1\}^{a+1}, m+1+\beta_0, \beta_1, \ldots, \beta_c+p+1)$$

$$-\mu^{k+1} \sum_{b+c=k} \sum_{m+n+p=r} \lambda^m(\lambda+1)^n$$

$$\times \sum_{\substack{|\boldsymbol{\alpha}|=b+m+2 \\ |\boldsymbol{\beta}|=c+n+1}} \zeta(\alpha_0, \alpha_1, \ldots, \alpha_{b+1}+\beta_0, \beta_1, \ldots, \beta_c+p+1)$$

and the contribution to the generation function is

(7.2.6)
$$\sum_{a+b=k} \sum_{m+n=r} \mu^b I_3(m, a+1; \mu) \cdot \lambda^m I_3(b, n; \lambda)$$

$$-\mu^{k+1} \sum_{b+c=k} \sum_{m+n=r} \binom{b+m+1}{m} \cdot \lambda^m I_3(c, n; \lambda).$$

(5) The contribution from the integration over D_6 to the generating function is

$$\sum_{a+b=k} \mu^b \sum_{m+n=r} \lambda^m \binom{b+m}{b}$$

or

(7.2.7)
$$\sum_{m+n=r} \lambda^m I_3(m, k; \mu-1).$$

Our assertion then follows from the fact

$$(7.2.1) = (7.2.3) + (7.2.4) + (7.2.5) + (7.2.6) + (7.2.7).$$

Theorem 7.2.5. *Suppose that* k, r *are nonnegative integers. Then we have*

$$\sum_{j=0}^{k}\sum_{\ell=0}^{r}\binom{j+\ell}{j}\binom{k+r+4}{j+\ell+2}\mu^{j}\lambda^{\ell}$$

$$= \sum_{m+n=r}\lambda^{m}\{I_{4}(m,k;\mu) + I_{3}(m,k+1;\mu)\} + \sum_{a+b=k}\mu^{b}I_{4}(b,r;\lambda)$$

(7.2.8)
$$+ \sum_{a+b=k}\sum_{m+n=r}\mu^{b}I_{3}(m,a+1;u)\cdot\lambda^{m}I_{3}(b,n;\lambda)$$

$$- \mu^{k+1}\sum_{m+n=r}\lambda^{m}\binom{k+m+1}{m}$$

$$- \mu^{k+1}\sum_{b+c=k}\sum_{m+n=r}\binom{b+m+1}{m}\lambda^{m}I_{3}(c,n;\lambda)$$

with

$$I_{3}(m,r;\mu) = \sum_{\ell=0}^{r}\binom{m+\ell}{m}(\mu+1)^{\ell} = \frac{(m+r+1)!}{m!r!}\int_{0}^{1}(1+\mu x)^{r}x^{m}dx$$

and

$$I_{4}(q,k;\mu) = \frac{(k+q+2)!}{k!q!}\int_{0}^{1}\int_{0}^{y}(1+\mu x)^{k}x^{q}dxdy.$$

After separating the coefficients of $\mu^{j}\lambda^{\ell}$ from (7.2.8) we obtain the following.

Theorem 7.2.6. *For integers* k, r, j, ℓ *with* $0 \leq j \leq k$ *and* $0 \leq \ell \leq r$ *we have*

$$\binom{j+\ell}{j}\binom{k+r+4}{j+\ell+2}$$
$$= \binom{j+\ell}{j}\left\{\binom{k+\ell+2}{k-j} + \binom{k+\ell+1}{k-j} + \binom{r+j+2}{r-\ell}\right\}$$
$$+ \sum_{a+b=k}\sum_{m+n=r}\binom{m+a+2}{k-j+1}\binom{n+b+1}{r-\ell}\binom{\ell-m+b}{b}\binom{j-b+m}{m}.$$

Remark 7.2.7. The restriction $k = j = 0$ gives the identity

$$\binom{r+4}{\ell+2} = \binom{r+2}{\ell+2} + (\ell+3) + \sum_{\substack{m+n=r\\\ell\geq m}}\frac{(m+2)(n+1)!}{(r-\ell)!(\ell-m+1)!}$$

which is quite different from the classical Pascal identity

$$\binom{r+4}{\ell+2} = \binom{r+3}{\ell+1} + \binom{r+3}{\ell+2}.$$

On the other hand, the restriction $r = \ell = 0$ give the identity

$$\binom{k+3}{j+1} = 1 + \sum_{\substack{a+b=k \\ b \leq j}} \frac{(a+2)!}{(k-j+1)!(j-b+1)!}.$$

7.3   A Combinatorial Identity of Convolution Type

As an immediate application of the combinatorial identity

$$\sum_{|\boldsymbol{\beta}|=r} \binom{\alpha_1+\beta_1}{\alpha_1}\binom{\alpha_2+\beta_2}{\alpha_2}\cdots\binom{\alpha_n+\beta_n}{\alpha_n}$$
$$= \binom{|\boldsymbol{\alpha}|+r+n-1}{|\boldsymbol{\alpha}|+n-1} = \binom{|\boldsymbol{\alpha}|+r+n-1}{r},$$

we prove another identity of the similar type through the shuffle product of a sum of multiple zeta values and a multiple zeta value of the form $\zeta(\{1\}^m, n+2)$.

Theorem 7.3.1. *For nonnegative integers p, q, r, ℓ, m, n, we have*

$$\sum_{a+b=q}\sum_{c+d=r}\binom{a+c+m+p+1}{m}\binom{a+c}{a}\binom{b+d}{b}\binom{b+d+n+\ell+2}{n+1}$$
$$= \binom{q+r}{q}\sum_{\alpha=0}^{q+r}\binom{\alpha+m+p+1}{m}\binom{\alpha^*+n+\ell+2}{n+1}$$

with $\alpha^* = q+r-\alpha$.

Such a combinatorial identity comes from the shuffle product of the sum

$$(*) \qquad\qquad \sum_{|\boldsymbol{\alpha}|=q+r+1} \zeta(\{1\}^p, \alpha_0, \alpha_1, \ldots, \alpha_q + \ell + 1)$$

which consists of $\binom{q+r}{q}$ multiple zeta values of weight $w_1 = p+q+r+\ell+2$ and the multiple zeta value $\zeta(\{1\}^m, n+2)$ of weight $w_2 = m+n+2$. After the shuffle product, it will produce totally

$$\binom{q+r}{q}\binom{w_1+w_2}{w_1}$$

multiple zeta values of weight $w_1 + w_2$. Also according to the identity mentioned in the beginning, we have

(7.3.1) $$\binom{w_1+w_2}{w_1} = \sum_{\alpha=0}^{w_1} \binom{\alpha+m}{m}\binom{\alpha^*+n+1}{n+1}$$

with $\alpha^* = w_1 - \alpha = p + q + r + \ell + 2 - \alpha$.

Now we proceed with the shuffle process. First we express the sum $(*)$ and the multiple zeta value $\zeta(\{1\}^m, n+2)$ by the double integral

$$\frac{1}{p!q!r!\ell!} \iint_{E_2} \left(\log\frac{1}{1-t_1}\right)^p \left(\log\frac{1-t_1}{1-t_2}\right)^q \left(\log\frac{t_2}{t_1}\right)^r \left(\log\frac{1}{t_2}\right)^\ell \frac{dt_1 dt_2}{(1-t_1)t_2}$$

and the integral

$$\frac{1}{m!(n+1)!} \int_0^1 \left(\log\frac{1}{1-u}\right)^m \left(\log\frac{1}{u}\right)^{n+1} \frac{du}{1-u},$$

respectively. We decompose $E_2 \times [0,1]$ into three nonoverlapping simplices of dimension three:

$$D_1 : \ 0 < t_1 < t_2 < u < 1,$$
$$D_2 : \ 0 < t_1 < u < t_2 < 1 \text{ and}$$
$$D_3 : \ 0 < u < t_1 < t_2 < 1.$$

On the first simplex $D_1 : \ 0 < t_1 < t_2 < u < 1$, we substitute the factor

$$\left(\log\frac{1}{1-u}\right)^m = \left(\log\frac{1}{1-t_1} + \log\frac{1-t_1}{1-t_2} + \log\frac{1-t_2}{1-u}\right)^m$$

by its multinomial expansion

$$\sum_{|\boldsymbol{m}|=m} \frac{m!}{m_1!m_2!m_3!} \left(\log\frac{1}{1-t_1}\right)^{m_1} \left(\log\frac{1-t_1}{1-t_2}\right)^{m_2} \left(\log\frac{1-t_2}{1-u}\right)^{m_3}$$

and the other factor $\left(\log\frac{1}{t_2}\right)$ by

$$\sum_{|\boldsymbol{\ell}|=\ell} \frac{\ell!}{\ell_1!\ell_2!} \left(\log\frac{u}{t_2}\right)^{\ell_1} \left(\log\frac{1}{u}\right)^{\ell_2}$$

so that the value of integration over D_1, in terms of multiple zeta value, is

$$
\sum_{|\boldsymbol{m}|=m} \sum_{|\boldsymbol{\ell}|=\ell} \sum_{\substack{|\boldsymbol{\alpha}|=q+r+m_2+1 \\ |\boldsymbol{\beta}|=m_3+\ell_1+1}} \binom{p+m_1}{p}\binom{q+m_2}{q}\binom{\ell_2+n+1}{n+1}
$$

$$
\times \zeta(\{1\}^{p+m_1}, \alpha_0, \alpha_1, \ldots, \alpha_{m_2+q}+\beta_0, \beta_1, \ldots, \beta_{m_3}, \ell_2+n+2).
$$

In particular, the number of multiple zeta values in the above sum is

$$
\sum_{|\boldsymbol{\ell}|=\ell} \sum_{|\boldsymbol{m}|=m} \binom{m_1+p}{p}\binom{m_2+q+r}{q+r}\binom{m_3+\ell_1}{\ell_1}\binom{\ell_2+n+1}{n+1}\binom{q+r}{q}.
$$

The inner sum

$$
\sum_{|\boldsymbol{m}|=m} \binom{m_1+p}{p}\binom{m_2+q+r}{q+r}\binom{m_3+\ell_1}{\ell_1}
$$

is equal to

$$
\binom{m+p+q+r+\ell_1+2}{m},
$$

and therefore the number of multiple zeta values is

$$
\binom{q+r}{q} \sum_{|\boldsymbol{\ell}|=\ell} \binom{m+p+q+r+\ell_1+2}{m}\binom{\ell_2+n+1}{n+1},
$$

or

$$
(7.3.2) \qquad \binom{q+r}{q} \sum_{\alpha=p+q+r+2}^{w_1} \binom{\alpha+m}{m}\binom{\alpha^*+n+1}{n+1}.
$$

In the same way, we obtain that the number of multiple zeta values appeared in the value of integration over $D_3 : 0 < u < t_1 < t_2 < 1$ is

$$
(7.3.3) \qquad \binom{q+r}{q} \sum_{\alpha=0}^{p} \binom{\alpha+m}{m}\binom{\alpha^*+n+1}{n+1} \qquad (\alpha+\alpha^*=w_1).
$$

On the simplex $D_2 : 0 < t_1 < u < t_2 < 1$, the factors

$$
\left(\log \frac{1-t_1}{1-t_2}\right)^q \quad \text{and} \quad \left(\log \frac{t_2}{t_1}\right)^r
$$

are substituted by

$$
\sum_{a+b=q} \frac{q!}{a!b!} \left(\log \frac{1-t_1}{1-u}\right)^a \left(\log \frac{1-u}{1-t_2}\right)^b
$$

and

$$\sum_{c+d=r} \frac{r!}{c!d!} \left(\log \frac{u}{t_1}\right)^c \left(\log \frac{t_2}{u}\right)^d$$

while the factors

$$\left(\log \frac{1}{1-u}\right)^m \quad \text{and} \quad \left(\log \frac{1}{u}\right)^{n+1}$$

are substituted by

$$\sum_{|\boldsymbol{m}|=m} \frac{m!}{m_1!m_2!} \left(\log \frac{1}{1-t_1}\right)^{m_1} \left(\log \frac{1-t_1}{1-u}\right)^{m_2}$$

and

$$\sum_{|\boldsymbol{n}|=n+1} \frac{(n+1)!}{n_1!n_2!} \left(\log \frac{t_2}{u}\right)^{n_1} \left(\log \frac{1}{t_2}\right)^{n_2}.$$

The number of multiple zeta values appeared in the value of integration over D_2 is

$$\sum_{a+b=q} \sum_{c+d=r} \binom{a+c}{a}\binom{b+d}{b} \left\{ \sum_{|\boldsymbol{m}|=m} \binom{m_1+p}{p}\binom{m_2+a+c}{a+c} \right\}$$

$$\times \left\{ \sum_{|\boldsymbol{n}|=n+1} \binom{n_1+b+d}{b+d}\binom{n_2+\ell}{\ell} \right\}.$$

The sums of two inner summations are

$$\binom{a+c+m+p+1}{m}$$

and

$$\binom{b+d+n+\ell+2}{n+1}$$

respectively. Consequently, the number of multiple zeta values is
(7.3.4)
$$\sum_{a+b=q} \sum_{c+d=r} \binom{a+c+m+p+1}{m}\binom{a+c}{a}\binom{b+d}{b}\binom{b+d+n+\ell+2}{n+1}.$$

Our assertion then follows from

$$(7.3.4) = (7.3.1) - (7.3.2) - (7.3.3).$$

7.4 Another Generating Function of Three Variables

For nonnegative integers m, n, ℓ, we have

$$
\frac{1}{m!n!\ell!} \int_{E_2} \left(\log \frac{1}{t_1} \right)^m \left(\log \frac{t_2}{t_1} \right)^n \left(\log \frac{1}{t_2} \right)^\ell \frac{dt_1 dt_2}{(1-t_1)t_2}
$$

$$
= \sum_{a+b=m} \frac{1}{a!b!n!\ell!} \int_{E_2} \left(\log \frac{t_2}{t_1} \right)^{n+a} \left(\log \frac{1}{t_2} \right)^{\ell+b} \frac{dt_1 dt_2}{(1-t_1)t_2}
$$

$$
= \sum_{a+b=m} \binom{n+a}{n} \binom{\ell+b}{\ell} \zeta(m+n+\ell+2)
$$

$$
= \binom{m+n+\ell+1}{m} \zeta(m+n+\ell+2).
$$

It follows that for nonnegative integers k, q, r and complex numbers ξ, μ, λ we have

$$
\frac{1}{k!q!r!} \iint_{R_1 \times R_2} \left(\log \frac{1}{t_1} + \xi \log \frac{1}{u_1} \right)^k \left(\mu \log \frac{t_2}{t_1} + \log \frac{u_2}{u_1} \right)^q
$$

$$
\times \left(\log \frac{1}{t_2} + \lambda \log \frac{1}{u_2} \right)^r \frac{dt_1 dt_2 du_1 du_2}{(1-t_1)t_2(1-u_1)u_2}
$$

$$
= \sum_{j=0}^{k} \sum_{m=0}^{q} \sum_{\ell=0}^{r} \xi^j \mu^m \lambda^\ell \binom{k-j+m+r-\ell+1}{k-j} \binom{j+q-m+\ell+1}{j}
$$

$$
\times \zeta(k-j+m+r-\ell+2)\zeta(j+q-m+\ell+2),
$$

where $R_1 : 0 < t_1 < t_2 < 1$ and $R_2 : 0 < u_1 < u_2 < 1$.

Now we are going to evaluate the generating function

$$
G(\xi, \mu, \lambda) = \sum_{j=0}^{k} \sum_{m=0}^{q} \sum_{\ell=0}^{r} \xi^j \mu^m \lambda^\ell \binom{k+q+r+4}{k-j+m+r-\ell+2}
$$

$$
\times \binom{k-j+m+r-\ell+1}{k-j} \binom{j+q-m+\ell+1}{j}
$$

in terms of some simple polynomials we have already used before through a shuffle product process. Recall that

$$I_1(m, r; \lambda) = \sum_{a+b=r} (\lambda + 1)^a \binom{b+m}{m}$$

$$= \frac{(m+r+1)!}{m!r!} \int_0^1 x^m (\lambda + 1 - \lambda x)^r dx,$$

$$I_2(k, r; \mu, \lambda)$$

$$= \sum_{a+b=k} (\mu + 1)^a \sum_{m+n=r} (\lambda + 1)^n \binom{b+m}{b}$$

$$= \frac{(k+r+2)!}{k!r!} \int_0^1 \int_0^y [(\mu + 1)(1 - y) + x]^k [(\lambda + 1)y - \lambda x]^r dx dy,$$

and

$$I_3(m, k; \xi) = \sum_{a+b=k} (\xi + 1)^b \binom{m+b}{m}$$

$$= \frac{(m+k+1)!}{m!k!} \int_0^1 x^m (1 + \xi x)^k dx.$$

Also note that the coefficients of λ^ℓ, $\mu^m \lambda^\ell$ and ξ^j in $I_1(m, r; \lambda)$, $I_2(k, r; \mu, \lambda)$ and $I_3(m, k; \xi)$ are

$$\binom{m+r+1}{r-\ell}, \quad \sum_{a+b=k} \binom{a}{j} \binom{b+r+1}{r-\ell} \quad \text{and} \quad \binom{m+j}{j} \binom{m+k+1}{k-j}$$

respectively.

We need a new polynomial in one variable.

Proposition 7.4.1. *For nonnegative integers m, n, k and a complex number ξ, let*

$$P(m, n, k; \xi) = \sum_{a+b=k} (\xi + 1)^b \binom{m+a}{m} \binom{n+b}{n}.$$

Then for $0 \leq j \leq k$, the coefficient of ξ^j of $P(m, n, k; \xi)$ is

$$\binom{n+j}{j} \binom{m+n+k+1}{k-j}.$$

Proof. As

$$\frac{1}{(1-x)^{m+1}} = \sum_{a=0}^{\infty} \binom{m+a}{m} x^a$$

and

$$\frac{1}{[1-(\xi+1)x]^{n+1}} = \sum_{b=0}^{\infty} (\xi+1)^b \binom{n+b}{n} x^b$$

so that $P(m, n, k; \xi)$ is just the coefficient of x^k of the product

$$\frac{1}{(1-x)^{m+1}} \cdot \frac{1}{[1-(\xi+1)x]^{n+1}}.$$

The coefficient of ξ^j of the above product is

$$\frac{1}{j!} \left(\frac{\partial}{\partial \xi}\right)^j \left\{ \frac{1}{(1-x)^{m+1}} \cdot \frac{1}{[1-(\xi+1)x]^{n+1}} \right\} \bigg|_{\xi=0}$$

$$= \binom{n+j}{j} \frac{x^j}{(1-x)^{m+n+j+2}}.$$

Note that the coefficient of ξ^j in $P(m, n, k; \xi)$ is equal to the coefficient of x^k in

$$\binom{n+j}{j} \frac{x^j}{(1-x)^{m+n+j+2}}$$

which is

$$\binom{n+j}{j} \binom{m+n+k+1}{k-j}.$$

$\square$

As a replacement of the shuffle product process, we decompose $R_1 \times R_2$ into 6 simplices of dimension four as follows

$$D_1 : \ 0 < t_1 < t_2 < u_1 < u_2 < 1,$$
$$D_2 : \ 0 < u_1 < u_2 < t_1 < t_2 < 1,$$
$$D_3 : \ 0 < t_1 < u_1 < t_2 < u_2 < 1,$$
$$D_4 : \ 0 < t_1 < u_1 < u_2 < t_2 < 1,$$
$$D_5 : \ 0 < u_1 < t_1 < u_2 < t_2 < 1 \text{ and}$$
$$D_6 : \ 0 < u_1 < t_1 < t_2 < u_2 < 1.$$

Then we evaluate the integration over each simplex in terms of multiple zeta values one by one. By counting the number of multiple zeta values appeared in the values of integrations over these simplices, we obtain the value of $G(\xi, \mu, \lambda)$ in terms of I_1, I_2, I_3, P as shown above.

Theorem 7.4.2. *With notations as above, we have*

$$G(\xi, \mu, \lambda) = \sum_{m+n=q} \sum_{e+f=r} \left\{ \mu^m P(m+e+1, n+f+1, k; \xi) I_1(n, f; \lambda) \right.$$

$$\left. + \xi^k \mu^n \lambda^r P(m+e+1, n+f+1, k; \xi^{-1}) I_1(n, f; \lambda^{-1}) \right\}$$

$$+ \sum_{m+n=q} \left\{ \mu^m P(m, n+r+2, k; \xi) + \xi^k P(m, n+r+2, k; \xi^{-1}) \right\}$$

$$\times \left\{ I_2(n, r; \mu, \lambda) + \mu^n \lambda^r I_2(n, r; \mu^{-1}, \lambda^{-1}) \right\}.$$

Proof. On the first simplex $D_1 : \ 0 < t_1 < t_2 < u_1 < u_2 < 1$, the three factors of the integrand are substituted by

$$\sum_{a+b+c+d=k} \frac{k!}{a!b!c!d!} (\xi+1)^{c+d} \left(\log \frac{t_2}{t_1} \right)^a \left(\log \frac{u_1}{t_2} \right)^b \left(\log \frac{u_2}{u_1} \right)^c \left(\log \frac{1}{u_2} \right)^d,$$

$$\sum_{m+n=q} \frac{q!}{m!n!} \mu^m \left(\log \frac{t_2}{t_1} \right)^m \left(\log \frac{u_2}{u_1} \right)^n$$

and

$$\sum_{e+f+g=r} \frac{r!}{e!f!g!} (\lambda+1)^g \left(\log \frac{u_1}{t_2} \right)^e \left(\log \frac{u_2}{u_1} \right)^f \left(\log \frac{1}{u_2} \right)^g.$$

The partial generating function for the number of multiple zeta values appeared is

$$H_1(\xi, \mu, \lambda) = \sum_{a+b+c+d=k} (\xi+1)^{c+d} \sum_{m+n=q} \mu^m \sum_{e+f+g=r} (\lambda+1)^g$$

$$\times \binom{m+a}{m} \binom{b+e}{e} \binom{c+n+f}{n+f} \binom{d+g}{d} \binom{n+f}{n}.$$

Let $\alpha = a + b$ and $\beta = c + d$ be new dummy variables in place of a, b, c, d.

As

$$\sum_{a+b} \binom{m+a}{m}\binom{b+e}{e} = \binom{\alpha+m+e+1}{\alpha}$$

and

$$\sum_{c+d=\beta} \binom{c+n+f}{n+f}\binom{d+g}{g} = \binom{\beta+n+f+g+1}{\beta}.$$

So $H_1(\xi, \mu, \lambda)$ can be rewritten as

$$\sum_{a+b=k} (\xi+1)^b \sum_{m+n=q} \mu^m \sum_{e+f+g=r} (\lambda+1)^g$$
$$\times \binom{a+m+e+1}{a}\binom{b+n+f+g+1}{b}\binom{n+f}{n},$$

or

$$\sum_{a+b=k} (\xi+1)^b \sum_{m+n=q} \mu^m \sum_{e+f=r} \binom{a+m+e+1}{a}\binom{b+n+f+1}{b} I_1(n, f; \lambda).$$

In terms of our new notations, it is equal to

$$\sum_{m+n=q} \sum_{e+f=r} \mu^m P(m+e+1, n+f+1, k; \xi) I_1(n, f; \lambda).$$

On the domain $D_2 : \; 0 < u_1 < u_2 < t_1 < t_2 < 1$, we exchange the roles of t_1, t_2 and u_1, u_2, so the partial generating function is simply given by

$$\xi^k \mu^q \lambda^r H_1(\xi^{-1}, \mu^{-1}, \lambda^{-1}),$$

or

$$\xi^k \lambda^r \sum_{m+n=q} \sum_{e+f=r} \mu^n P(m+e+1, n+f+1, k; \xi^{-1}) I_1(n, f; \lambda^{-1}).$$

On the domain $D_3 : \; 0 < t_1 < u_1 < t_2 < u_2 < 1$, the three factors of the integrand are substituted by

$$\sum_{a+b+c+d=k} \frac{k!}{a!b!c!d!}(\xi+1)^{b+c+d}\left(\log\frac{u_1}{t_1}\right)^a\left(\log\frac{t_2}{u_1}\right)^b\left(\log\frac{u_2}{t_2}\right)^c\left(\log\frac{1}{u_2}\right)^d,$$

$$\sum_{m+n+p=q} \frac{q!}{m!n!p!}\mu^m(\mu+1)^n\left(\log\frac{u_1}{t_1}\right)^m\left(\log\frac{t_2}{u_1}\right)^n\left(\log\frac{u_2}{t_2}\right)^p$$

and

$$\sum_{e+f=r} \frac{r!}{e!f!}(\lambda+1)^f\left(\log\frac{u_2}{t_2}\right)^e\left(\log\frac{1}{u_2}\right)^f,$$

and the partial generating function for the number of multiple zeta values appeared is given by

$$H_3(\xi, \mu, \lambda) = \sum_{a+b+c+d=k} (\xi + 1)^{b+c+d} \sum_{m+n+p=q} \mu^m (\mu + 1)^n \sum_{e+f=r} (\lambda + 1)^f$$
$$\times \binom{a+m}{m}\binom{b+n}{n}\binom{c+p+e}{p+e}\binom{d+f}{f}\binom{p+e}{p}.$$

Let $b + c + d = \beta$, then

$$\sum_{b+c+d=\beta} \binom{b+n}{n}\binom{c+p+e}{p+e}\binom{d+f}{f} = \binom{\beta+n+p+e+f+2}{\beta}.$$

Consequently, we rewrite $H_3(\xi, \mu, \lambda)$ as

$$\sum_{a+b=k} (\xi+1)^b \sum_{m+n+p=q} \mu^m(\mu+1)^n \sum_{e+f=r} (\lambda+1)^f$$
$$\times \binom{a+m}{m}\binom{b+n+p+r+2}{b}\binom{p+e}{p},$$

or

$$H_3(\xi, \mu, \lambda) = \sum_{m+n=q} \mu^m P(m, n+r+2, k; \xi) I_2(n, r; \mu, \lambda).$$

In the same manner, we obtain the partial generating functions of number of the multiple zeta values appeared in the values of integrations over D_4, D_5 and D_6 as

$$H_4(\xi, \mu, \lambda) = \mu^q \lambda^r \sum_{m+n=q} P(m, n+r+2, k; \xi) I_2(n, r; \mu^{-1}, \lambda^{-1}),$$

$$H_5(\xi, \mu, \lambda) = \xi^k \lambda^r \sum_{m+n=q} \mu^n P(m, n+r+2, k; \xi^{-1}) I_2(n, r; \mu^{-1}, \lambda^{-1}),$$

and

$$H_6(\xi, \mu, \lambda) = \xi^k \sum_{m+n=q} P(m, n+r+2, k; \xi^{-1}) I_2(n, r; \mu, \lambda).$$

Our assertion then follows from

$$G(\xi, \mu, \lambda) = \sum_{j=1}^{6} H_j(\xi, \mu, \lambda).$$

$\square$

For our convenience, for $0 \le j \le k$, $0 \le m \le q$ and $0 \le \ell \le r$, we let

$$\bar{j} = k - j, \quad \bar{m} = q - m \text{ and } \bar{\ell} = r - \ell.$$

After separating the coefficient of $\xi^j \mu^m \lambda^\ell$ in the generating function $G(\xi, \mu, \lambda)$, we obtain the following theorem.

Theorem 7.4.3. *For integers k, q, r and j, m, ℓ with $0 \le j \le k$, $0 \le m \le q$ and $0 \le \ell \le r$ and $\bar{j} = k - j$, $\bar{m} = q - m$, $\bar{\ell} = r - \ell$, we have*

$$\binom{k+q+r+4}{\bar{j}+m+\bar{\ell}+2}\binom{\bar{j}+m+\bar{\ell}+1}{\bar{j}}\binom{j+\bar{m}+\ell+1}{j}$$

$$= \sum_{e+f=r} \left\{ \binom{k+q+r+3}{j}\binom{\bar{m}+f+j+1}{j}\binom{\bar{m}+f+1}{f-\ell} \right.$$

$$\left. + \binom{k+q+r+3}{j}\binom{m+f+\bar{j}+1}{\bar{j}}\binom{m+f+1}{\ell} \right\}$$

$$+ \sum_{a+b+c=q} \left\{ \binom{k+q+r+3}{\bar{j}}\binom{b+c+r+j+2}{j}\binom{b}{m-a}\binom{c+r+1}{r-\ell} \right.$$

$$+ \binom{k+q+r+3}{\bar{j}}\binom{b+c+r+j+2}{j}\binom{b}{\bar{m}}\binom{c+r+1}{\ell}$$

$$+ \binom{k+q+r+3}{j}\binom{b+c+r+\bar{j}+2}{\bar{j}}\binom{b}{m-c}\binom{c+r+1}{\ell}$$

$$\left. + \binom{k+q+r+3}{j}\binom{b+c+r+\bar{j}+2}{\bar{j}}\binom{b}{m}\binom{c+r+1}{r-\ell} \right\}.$$

7.5   Exercises

1. For nonnegative integers m, n, k_1, k_2 and variables ξ_1, ξ_2, let

$$P(m, n, k_1, k_2, \xi_1, \xi_2)$$

$$= \sum_{a_1+b_1=k_1} \sum_{a_2+b_2=k_2} (\xi_1 + 1)^{b_1} (\xi_2 + 1)^{b_2} \binom{m + a_1 + a_2}{m, a_1, a_2} \binom{n + b_1 + b_2}{n, b_1, b_2}.$$

For $0 \le j_1 \le k_1$ and $0 \le j_2 \le k_2$, find the coefficient of $\xi_1^{j_1} \xi_2^{j_2}$ of P.

2. Suppose that k, p, q, r, ℓ, m, n are nonnegative integers. Prove that

$$\sum_{a+b=p}\sum_{c+d=q}\sum_{e+f=r} \binom{a+c+e+m+k+1}{m}\binom{a+c+e}{a,c,e}$$
$$\times \binom{b+d+f}{b,d,f}\binom{b+d+f+n+\ell+2}{n+1}$$
$$= \binom{p+q+r}{p,q,r}\sum_{\alpha=0}^{p+q+r}\binom{\alpha+m+k+1}{m}\binom{\alpha^*+n+\ell+2}{n+1}$$

with $\alpha^* = p + q + r - \alpha$.

3. Find the numbers of multiple zeta values represented by the following integrals:

(a) $\dfrac{1}{m!n!}\displaystyle\iint_{E_2}\left(\log\frac{1-t_1}{1-t_2}\right)^m\left(\log\frac{t_2}{t_1}\right)^n\frac{dt_1dt_2}{(1-t_1)t_2}$

(b) $\dfrac{1}{p!q!r!}\displaystyle\iint_{E_2}\left(\log\frac{1}{1-t_1}\right)^p\left(\log\frac{1}{1-t_2}\right)^q\left(\log\frac{t_2}{t_1}\right)^r\frac{dt_1dt_2}{(1-t_1)t_2}$

(c) $\dfrac{1}{p!q!m!n!}\displaystyle\iint_{0<t_1<u<t_2<1}\left(\log\frac{1-t_1}{1-t_2}\right)^p\left(\log\frac{t_2}{t_1}\right)^q$
$$\times\left(\log\frac{1}{1-u}\right)^m\left(\log\frac{1}{u}\right)^n\frac{dt_1}{1-t_1}\frac{du}{u}\frac{dt_2}{t_2}.$$

CHAPTER 8

Vector Versions of Some Combinatorial Identities

Once we can interpret a combinatorial identity from the counting of numbers of multiple zeta values in certain sum of multiple zeta values, we can always produce vector version of the identity. The well-known Vandermonde's convolution

$$\sum_{a+b=p} \binom{k+a}{k} \binom{q+b}{q} = \binom{k+p+q+1}{p}$$

now becomes the identity

$$\sum_{\boldsymbol{\alpha}+\boldsymbol{\beta}=\boldsymbol{p}} \binom{k+\alpha_1+\alpha_2+\cdots+\alpha_n}{k,\alpha_1,\alpha_2,\ldots,\alpha_n} \binom{q+\beta_1+\beta_2+\cdots+\beta_n}{q,\beta_1,\beta_2,\ldots,\beta_n}$$

$$= \binom{k+q+1+p_1+p_2+\cdots+p_n}{k+q+1,p_1,p_2,\ldots,p_n}.$$

8.1   The Shuffle Product of Two Sums

For a pair of nonnegative integers p and q, the integral

$$\frac{1}{p!q!} \int_{0<t_1<t_2<1} \left(\log \frac{1}{1-t_2}\right)^p \left(\log \frac{1}{t_1}\right)^q \frac{dt_1 dt_2}{(1-t_1)t_2}$$

can be rewritten as

$$\sum_{a+b=p} \sum_{c+d=q} \frac{1}{a!b!c!d!} \int_{0<t_1<t_2<1} \left(\log \frac{1}{1-t_1}\right)^a \left(\log \frac{1-t_1}{1-t_2}\right)^b$$

$$\times \left(\log \frac{t_2}{t_1}\right)^c \left(\log \frac{1}{t_2}\right)^d \frac{dt_1 dt_2}{(1-t_1)t_2},$$

and hence its value, in terms of multiple zeta values, is

$$(8.1.1) \quad \sum_{a+b=p} \sum_{c+d=q} \sum_{|\alpha|=b+c+1} \zeta(\{1\}^a, \alpha_0, \alpha_1, \ldots, \alpha_b + d + 1) \equiv N(p,q).$$

The number of multiple zeta values in the above sum is

$$\sum_{a+b=p} \sum_{c+d=q} \binom{b+c}{b}$$

or by an elementary calculation, is

$$\binom{p+q+2}{p+1} - 1.$$

In order to investigate the combinatorial consequences from the shuffle products of two sums of multiple zeta values as shown in (8.1.1), we consider the integral

$$(8.1.2) \quad \frac{1}{k!r!} \iint_{R_1 \times R_2} \left(\mu \log \frac{1}{1-t_2} + \log \frac{1}{1-u_2}\right)^k \left(\log \frac{1}{t_1} + \lambda \log \frac{1}{u_1}\right)^r$$

$$\times \frac{dt_1 dt_2}{(1-t_1)t_2} \frac{du_1 du_2}{(1-u_1)u_2}$$

where $R_1 : 0 < t_1 < t_2 < 1$ and $R_2 : 0 < u_1 < u_2 < 1$.

The value of integral (8.1.2) is given by

$$\sum_{j=0}^{k} \sum_{\ell=0}^{r} \mu^j \lambda^\ell N(j, r-\ell) N(k-j, \ell)$$

and, after the shuffle product process, the generating function for the number of multiple zeta values produced is

(8.1.3)
$$\sum_{j=0}^{k} \sum_{\ell=0}^{r} \mu^j \lambda^\ell \left\{ \binom{j+r-\ell+2}{j+1} - 1 \right\}$$
$$\times \left\{ \binom{k-j+\ell+2}{\ell+1} - 1 \right\} \binom{k+r+4}{j+r-\ell+2}$$
$$\equiv G(k,r;\mu,\lambda).$$

Recall that for integers $m, n, r \geq 0$, we let

(8.1.4)
$$P(m,n,r;\lambda) = \sum_{a+b=r} \binom{m+a}{m} \binom{n+b}{n} (\lambda+1)^b.$$

Also the coefficient of λ^ℓ $(0 \leq \ell \leq r)$ of $P(m,n,r;\lambda)$ is

$$\binom{n+\ell}{\ell} \binom{r+m+n+1}{r-\ell}.$$

Presently we are going to express $G(k,r;\mu,\lambda)$ in terms of P's through the shuffle products of multiple zeta values.

Theorem 8.1.1. *Notations as shown in* (8.1.3) *and* (8.1.4),

(8.1.5)
$$G(k,r;\mu,\lambda)$$
$$= \sum_{|\alpha|=k} (\mu+1)^{\alpha_1+\alpha_2} P(\alpha_2+\alpha_3+1, \alpha_4+1, r; \lambda)$$
$$+ \sum_{|\alpha|=k} (\mu+1)^{\alpha_1+\alpha_2} \mu^{\alpha_3+\alpha_4} \lambda^r P(\alpha_2+\alpha_3+1, \alpha_4+1, r; \lambda^{-1})$$
$$+ \sum_{|\alpha|=k} \left\{ (\mu+1)^{\alpha_1+\alpha_2+\alpha_3} + (\mu+1)^{\alpha_1+\alpha_2+\alpha_3} \mu^{\alpha_4} \right\}$$
$$\times \left\{ P(\alpha_2, \alpha_3+\alpha_4+2, r; \lambda) + \lambda^r P(\alpha_2, \alpha_3+\alpha_4+2, r; \lambda^{-1}) \right\}.$$

Proof. The shuffle product process is equivalent to decompose $R_1 \times R_2$ into

6 nonoverlapping simplices of dimension four as follows,

$$D_1 : 0 < t_1 < t_2 < u_1 < u_2 < 1,$$
$$D_2 : 0 < u_1 < u_2 < t_1 < t_2 < 1,$$
$$D_3 : 0 < t_1 < u_1 < t_2 < u_2 < 1,$$
$$D_4 : 0 < t_1 < u_1 < u_2 < t_2 < 1,$$
$$D_5 : 0 < u_1 < t_1 < u_2 < t_2 < 1 \text{ and}$$
$$D_6 : 0 < u_1 < t_1 < t_2 < u_2 < 1,$$

and then evaluate the integral (8.1.2) over each simplex one by one in terms of multiple zeta values. Counting the number of multiple zeta values, we get 6 polynomial functions of μ and λ as shown on the right hand side of (8.1.5). Here are the details of the shuffle product process.

On the simplex D_1: $0 < t_1 < t_2 < u_1 < u_2 < 1$, we replace the factors

$$\left(\mu \log \frac{1}{1-t_2} + \log \frac{1}{1-u_2} \right)^k \quad \text{and} \quad \left(\log \frac{1}{t_1} + \lambda \log \frac{1}{u_1} \right)^r$$

by

$$\left\{ (\mu+1) \log \frac{1}{1-t_1} + (\mu+1) \log \frac{1-t_1}{1-t_2} + \log \frac{1-t_2}{1-u_1} + \log \frac{1-u_1}{1-u_2} \right\}^k$$

and

$$\left\{ \log \frac{t_2}{t_1} + \log \frac{u_1}{t_2} + (\lambda+1) \log \frac{u_2}{u_1} + (\lambda+1) \log \frac{1}{u_2} \right\}^r,$$

and then expand into multinomial series expansions

$$\sum_{|\alpha|=k} \frac{k!}{\alpha_1! \alpha_2! \alpha_3! \alpha_4!} (\mu+1)^{\alpha_1+\alpha_2} \left(\log \frac{1}{1-t_1} \right)^{\alpha_1} \left(\log \frac{1-t_1}{1-t_2} \right)^{\alpha_2}$$
$$\times \left(\log \frac{1-t_2}{1-u_1} \right)^{\alpha_3} \left(\log \frac{1-u_1}{1-u_2} \right)^{\alpha_4}$$

and

$$\sum_{|\beta|=r} \frac{r!}{\beta_1! \beta_2! \beta_3! \beta_4!} (\lambda+1)^{\beta_3+\beta_4} \left(\log \frac{t_2}{t_1} \right)^{\beta_1} \left(\log \frac{u_1}{t_2} \right)^{\beta_2}$$
$$\times \left(\log \frac{u_2}{u_1} \right)^{\beta_3} \left(\log \frac{1}{u_2} \right)^{\beta_4},$$

respectively. The partial generating function for the number of multiple zeta values produced from the shuffle product is
(8.1.6)
$$\sum_{|\alpha|=k} (\mu+1)^{\alpha_1+\alpha_2} \sum_{|\beta|=r} (\lambda+1)^{\beta_3+\beta_4} \binom{\alpha_2+\beta_1}{\alpha_2}\binom{\alpha_3+\beta_2}{\alpha_3}\binom{\alpha_4+\beta_3}{\alpha_4}.$$

Note that
$$\sum_{\beta_1+\beta_2=c} \binom{\alpha_2+\beta_1}{\alpha_2}\binom{\alpha_3+\beta_2}{\alpha_3} = \binom{c+\alpha_2+\alpha_3+1}{c}$$

and
$$\sum_{\beta_3+\beta_4=d} \binom{\alpha_4+\beta_3}{\alpha_4} = \binom{d+\alpha_4+1}{d}.$$

Consequently, the inner sum of (8.1.6) is $P(\alpha_2+\alpha_3+1,\alpha_4+1,r;\lambda)$ and the partial generating function is

(8.1.7) $$\sum_{|\alpha|=k} (\mu+1)^{\alpha_1+\alpha_2} P(\alpha_2+\alpha_3+1,\alpha_4+1,r;\lambda).$$

In just the same manner, the partial generating function from the integration over D_2 is
$$\sum_{|\alpha|=k} (\mu+1)^{\alpha_1+\alpha_2}\mu^{\alpha_3+\alpha_4} \sum_{|\beta|=r} \lambda^{\beta_1+\beta_2}(\lambda+1)^{\beta_3+\beta_4}$$
$$\times \binom{\alpha_2+\beta_1}{\alpha_2}\binom{\alpha_3+\beta_2}{\alpha_3}\binom{\alpha_4+\beta_3}{\alpha_4},$$

or

(8.1.8) $$\lambda^r \sum_{|\alpha|=k} (\mu+1)^{\alpha_1+\alpha_2}\mu^{\alpha_3+\alpha_4} P(\alpha_2+\alpha_3+1,\alpha_4+1,r;\lambda^{-1}).$$

On the simplex $D_3: \ 0 < t_1 < u_1 < t_2 < u_2 < 1$, two factors of the integrand are replaced by
$$\left\{(\mu+1)\log\frac{1}{1-t_1} + (\mu+1)\log\frac{1-t_1}{1-u_1} + (\mu+1)\log\frac{1-u_1}{1-t_2} + \log\frac{1-t_2}{1-u_2}\right\}^k$$

and
$$\left\{\log\frac{u_1}{t_1} + (\lambda+1)\log\frac{t_2}{u_1} + (\lambda+1)\log\frac{u_2}{t_2} + (\lambda+1)\log\frac{1}{u_2}\right\}^r,$$

and then expand into multinomial series expansions. The partial generating
function from the integration over D_3 is

$$\sum_{|\alpha|=k} (\mu+1)^{\alpha_1+\alpha_2+\alpha_3} \sum_{|\beta|=r} (\lambda+1)^{\beta_2+\beta_3+\beta_4} \binom{\alpha_2+\beta_1}{\alpha_2} \binom{\alpha_3+\beta_2}{\alpha_3} \binom{\alpha_4+\beta_3}{\alpha_4},$$

or

$$(8.1.9) \qquad \sum_{|\alpha|=k} (\mu+1)^{\alpha_1+\alpha_2+\alpha_3} P(\alpha_2, \alpha_3+\alpha_4+2, r; \lambda).$$

In the same way, we obtain all the partial generating functions from the
integrations over D_4, D_5 and D_6 as follows:

$$(8.1.10) \qquad \sum_{|\alpha|=k} (\mu+1)^{\alpha_1+\alpha_2+\alpha_3} \mu^{\alpha_4} P(\alpha_2, \alpha_3+\alpha_4+2, r; \lambda),$$

$$(8.1.11) \qquad \lambda^r \sum_{|\alpha|=k} (\mu+1)^{\alpha_1+\alpha_2+\alpha_3} \mu^{\alpha_4} P(\alpha_2, \alpha_3+\alpha_4+2, r; \lambda^{-1})$$

and

$$(8.1.12) \qquad \lambda^r \sum_{|\alpha|=k} (\mu+1)^{\alpha_1+\alpha_2+\alpha_3} P(\alpha_2, \alpha_3+\alpha_4+2, r; \lambda^{-1}).$$

Our assertion then follows from the fact that

$$G(k, r; \mu, \lambda) = (8.1.7) + (8.1.8) + (8.1.9) + (8.1.10) + (8.1.11) + (8.1.12).$$

$$\square$$

The coefficient of λ^ℓ in $P(m, n, r; \lambda)$ is

$$\binom{n+\ell}{\ell} \binom{r+m+n+1}{r-\ell}$$

while the coefficient of λ^ℓ in

$$\lambda^r P(m, n, r; \lambda^{-1}) = \sum_{a+b=r} \binom{m+a}{m} \binom{n+b}{n} \lambda^a (\lambda+1)^b$$

is

$$\binom{n+r-\ell}{r-\ell} \binom{r+m+n+1}{\ell}.$$

With these simple facts, we are able to separate the coefficient of $\mu^j \lambda^\ell$ in
$G(k, r; \mu, \lambda)$ as given in (8.1.5).

Theorem 8.1.2. *For integers k, r, j and ℓ with $0 \leq j \leq k$ and $0 \leq \ell \leq r$, we have*

$$\left\{\binom{j+r-\ell+2}{j+1} - 1\right\}\left\{\binom{k-j+\ell+2}{\ell+1} - 1\right\}\binom{k+r+4}{j+r-\ell+2}$$
$$= \sum_{a+b=k}\{H_{a,b}(j,\ell) + H_{a,b}(k-j,r-\ell)$$
$$+ F_{a,b}(j,\ell) + F_{a,b}(k-j,\ell) + F_{a,b}(j,r-\ell) + F_{a,b}(k-j,r-\ell)\}$$

with

$$H_{a,b}(j,\ell) = \binom{a}{j}\left\{\binom{b+\ell+2}{\ell+1} - 1\right\}\left\{\binom{k+r+4}{r-\ell+1} - \binom{b+r+3}{r-\ell+1}\right\}$$

and

$$F_{a,b}(j,\ell) = \left\{\binom{k+1}{j+1} - \binom{a}{j+1}\right\}\binom{b+\ell+2}{\ell}$$
$$\times \left\{\binom{k+r+4}{r-\ell+1} - \binom{b+r+3}{r-\ell+1}\right\}.$$

Proof. The coefficient of $\mu^j \lambda^\ell$ of the sum

$$\sum_{|\alpha|=k}(\mu+1)^{\alpha_1+\alpha_2}P(\alpha_2+\alpha_3+1,\alpha_4+1,r;\lambda)$$

is given by

$$\sum_{|\alpha|=k}\binom{\alpha_1+\alpha_2}{j}\binom{\alpha_4+\ell+1}{\ell}\binom{\alpha_2+\alpha_3+\alpha_4+r+3}{r-\ell}.$$

With $\alpha_1 + \alpha_2 = a$, $\alpha_3 + \alpha_4 = b$ as new dummy variables, the above sum is equal to

$$\sum_{a+b=k}\binom{a}{j}\sum_{\alpha_4=0}^{b}\binom{\alpha_4+\ell+1}{\ell}\sum_{\alpha_2=0}^{a}\binom{\alpha_2+b+r+3}{r-\ell}$$

or

$$\sum_{a+b=k}\binom{a}{j}\left\{\binom{b+\ell+2}{\ell+1} - 1\right\}\left\{\binom{k+r+4}{r-\ell+1} - \binom{b+r+3}{r-\ell+1}\right\}$$

which is

$$\sum_{a+b=k}H_{a,b}(j,\ell).$$

Also the coefficient of $\mu^j \lambda^\ell$ of the polynomial

$$\sum_{|\boldsymbol{\alpha}|=k} (\mu+1)^{\alpha_1+\alpha_2} \mu^{\alpha_3+\alpha_4} \lambda^r P(\alpha_2 + \alpha_3 + 1, \alpha_4 + 1, r; \lambda^{-1})$$

is given by

$$\sum_{|\boldsymbol{\alpha}|=k} \binom{\alpha_1 + \alpha_2}{k-j} \binom{\alpha_4 + r - \ell + 1}{r - \ell} \binom{\alpha_2 + \alpha_3 + \alpha_4 + r + 3}{\ell}$$

which is

$$\sum_{a+b=k} H_{a,b}(k-j, r-\ell).$$

On the other hand, the coefficient of $\mu^j \lambda^\ell$ in

$$\sum_{|\boldsymbol{\alpha}|=k} (\mu+1)^{\alpha_1+\alpha_2+\alpha_3} P(\alpha_2, \alpha_3 + \alpha_4 + 2, r; \lambda)$$

is

$$\sum_{|\boldsymbol{\alpha}|=k} \binom{\alpha_1 + \alpha_2 + \alpha_3}{j} \binom{\alpha_3 + \alpha_4 + \ell + 2}{\ell} \binom{r + \alpha_2 + \alpha_3 + \alpha_4 + 3}{r - \ell}.$$

Again with $\alpha_1 + \alpha_2 = a$, $\alpha_3 + \alpha_4 = b$ as new dummy variables, the sum is equal to

$$\sum_{a+b=k} \binom{b + \ell + 2}{\ell} \sum_{\alpha_3=0}^{b} \binom{a + \alpha_3}{j} \sum_{\alpha_2=0}^{a} \binom{r + \alpha_2 + b + 3}{r - \ell}$$

or

$$\sum_{a+b=k} \binom{b + \ell + 2}{\ell} \left\{ \binom{k+1}{j+1} - \binom{a}{j+1} \right\} \left\{ \binom{k+r+4}{r-\ell+1} - \binom{b+r+3}{r-\ell+1} \right\}$$

which is

$$\sum_{a+b=k} F_{a,b}(j, \ell).$$

The coefficient of $\mu^j \lambda^\ell$ of the remaining three polynomials appear in $G(k, r; \mu, \lambda)$ are

$$\sum_{a+b=k} F_{a,b}(j, r-\ell), \quad \sum_{a+b=k} F_{a,b}(k-j, r-\ell), \quad \text{and} \quad \sum_{a+b=k} F_{a,b}(k-j, \ell).$$

Therefore, our assertion follows.   □

8.2 More Combinatorial Identities of Convolution Type

There is a classical identity appeared in page 169 of [19], given as

$$(8.2.1) \qquad \sum_{0 \leq k \leq \ell} \binom{\ell - k}{m}\binom{q + k}{n} = \binom{\ell + q + 1}{m + n + 1}$$

with integers $\ell, m \geq 0$ and $n \geq q \geq 0$. It can be restated as

$$\sum_{a+b=q} \binom{a + m}{m}\binom{b + n}{n} = \binom{q + m + n + 1}{m + n + 1}$$

with integers $q, m, n \geq 0$. Such an identity is known as a Vandermonde's convolution, because Alexandre Vandermonde wrote a significant paper about it in the later 1700s. However it also appeared in a book by Chu Shih-Chich in China as early as 1303.

The proof is quite simple nowadays. Differentiating both sides of the well-known geometric series expansion

$$\frac{1}{1 - x} = \sum_{k=0}^{\infty} x^k, \quad |x| < 1$$

m-times, we get

$$\frac{1}{(1 - x)^{m+1}} = \sum_{k=0}^{\infty} \binom{m + k}{m} x^k, \quad |x| < 1.$$

In light of the above power series expansion, not only we can prove (8.2.1), but also we can prove a much more general case.

Proposition 8.2.1. *For nonnegative integers* $q, \beta_1, \beta_2, \ldots, \beta_n$ *with* $r = \beta_1 + \beta_2 + \cdots + \beta_n$, *we have*

$$\sum_{|\alpha|=q} \binom{\alpha_1 + \beta_1}{\beta_1}\binom{\alpha_2 + \beta_2}{\beta_2} \cdots \binom{\alpha_n + \beta_n}{\beta_n} = \binom{q + r + n - 1}{r + n - 1}.$$

Proof. The sum of products of n binomial coefficients is just the coefficient of x^q in the product

$$\frac{1}{(1 - x)^{\beta_1+1}} \times \frac{1}{(1 - x)^{\beta_2+1}} \times \cdots \times \frac{1}{(1 - x)^{\beta_n+1}},$$

since

$$\frac{1}{(1-x)^{\beta_j+1}} = \sum_{\alpha_j=0}^{\infty} \binom{\alpha_j + \beta_j}{\beta_j} x^{\alpha_j}, \qquad j = 1, 2, \ldots, n.$$

Therefore, it is the coefficient of x^q of the function

$$\frac{1}{(1-x)^{r+n}},$$

when it is expanded in power series. Indeed it is

$$\binom{q+r+n-1}{r+n-1}.$$

$\square$

Remark 8.2.2. If we sum also over all $\beta_1, \beta_2, \ldots, \beta_n$ with $\beta_1+\beta_2+\cdots+\beta_n = r$, we get

$$\sum_{|\alpha|=q} \sum_{|\beta|=r} \binom{\alpha_1 + \beta_1}{\alpha_1} \binom{\alpha_2 + \beta_2}{\alpha_2} \cdots \binom{\alpha_n + \beta_n}{\alpha_n} = \binom{q+r+n-1}{q, r, n-1}.$$

Vandermonde's convolution (8.2.1) as well as the general forms are useful in counting the numbers of multiple zeta values represented by some particular integrals over simplices. Also they have extensions in vector form as we shall see as below which are difficult to be proved otherwise.

For nonnegative integers k, p, q, we consider the integral

$$(8.2.2) \quad I_1 = \frac{1}{k!p!q!} \int_{E_2} \left(\log \frac{1}{1-t_1}\right)^k \left(\log \frac{1}{1-t_2}\right)^p \left(\log \frac{t_2}{t_1}\right)^q \frac{dt_1}{1-t_1} \frac{dt_2}{t_2}.$$

If we substitute the factor

$$\left(\log \frac{1}{1-t_2}\right)^p = \left(\log \frac{1}{1-t_1} + \log \frac{1-t_1}{1-t_2}\right)^p$$

by its binomial expansion

$$\sum_{a+b=p} \frac{p!}{a!b!} \left(\log \frac{1}{1-t_1}\right)^a \left(\log \frac{1-t_1}{1-t_2}\right)^b,$$

we see immediately that the multiple zeta values represented by I_1 is the sum

$$\sum_{a+b=p} \binom{k+a}{k} \sum_{|\alpha|=b+q+1} \zeta(\{1\}^{k+a}, \alpha_0, \ldots, \alpha_b + 1)$$

and the number of multiple zeta values in the sum is

$$\sum_{a+b=p} \binom{k+a}{k}\binom{q+b}{q}.$$

By Vandermonde's convolution, the number is

$$\binom{k+p+q+1}{p}.$$

On the other hand, if we begin with the integral

$$I_2 = \frac{1}{k! p_1! p_2! \ldots p_n! q!} \int_{E_2} \left(\log \frac{1}{1-t_1}\right)^k \left(\log \frac{1}{1-t_2}\right)^{p_1+p_2+\ldots+p_n}$$
$$\times \left(\log \frac{t_2}{t_1}\right)^q \frac{dt_1}{1-t_1} \frac{dt_2}{t_2}$$

and as before, substitute the factors $\left(\log \frac{1}{1-t_2}\right)^{p_j}$ by

$$\sum_{\alpha_j+\beta_j=p_j} \frac{p_j!}{\alpha_j! \beta_j!} \left(\log \frac{1}{1-t_1}\right)^{\alpha_j} \left(\log \frac{1-t_1}{1-t_2}\right)^{\beta_j}, \quad j=1,2,\ldots,n.$$

Then the number of multiple zeta values represented by I_2 is

$$\sum_{p=\alpha+\beta} \binom{k+\alpha_1+\alpha_2+\cdots+\alpha_n}{k,\alpha_1,\alpha_2,\ldots,\alpha_n}\binom{\beta_1+\beta_2+\cdots+\beta_n+q}{\beta_1,\beta_2,\ldots,\beta_n,q}$$

with $\boldsymbol{p} = (p_1, p_2, \ldots, p_n)$, $\boldsymbol{\alpha} = (\alpha_1, \alpha_2, \ldots, \alpha_n)$ and $\boldsymbol{\beta} = (\beta_1, \beta_2, \ldots, \beta_n)$.

It is obvious that

$$I_2 = \binom{p_1+p_2+\cdots+p_n}{p_1, p_2, \ldots, p_n} I_1.$$

This leads to the following identity.

Theorem 8.2.3. *For nonnegative integers* $k, p_1, p_2, \ldots, p_n, q$ *with* $\boldsymbol{p} = (p_1, p_2, \ldots, p_n)$, *then we have*

$$\sum_{\alpha+\beta=p} \binom{k+\alpha_1+\alpha_2+\cdots+\alpha_n}{k,\alpha_1,\alpha_2,\ldots,\alpha_n}\binom{\beta_1+\beta_2+\cdots+\beta_n+q}{\beta_1,\beta_2,\ldots,\beta_n,q}$$
$$= \binom{p_1+p_2+\cdots+p_n}{p_1, p_2, \ldots, p_n}\binom{k+|\boldsymbol{p}|+q+1}{|\boldsymbol{p}|}.$$

For our convenience, we let $M(\boldsymbol{g})$ be the multinomial coefficient

(8.2.3)
$$\binom{g_1 + g_2 + \cdots + g_n}{g_1, g_2, \ldots, g_n}$$

when $\boldsymbol{g} = (g_1, g_2, \ldots, g_n)$.

In our next consideration, we focus on integrals which arose from shuffle products of two sets of multiple zeta values. For integers k, r, j, ℓ with $0 \le j \le k$ and $0 \le \ell \le r$, consider the integral

$$I_3 = \frac{1}{j!(k-j)!(\ell+1)!(r-\ell+1)!} \iint_{0<u<t<1} \left(\log \frac{1}{1-u}\right)^{k-j} \left(\log \frac{1}{u}\right)^{\ell+1}$$
$$\times \left(\log \frac{1}{1-t}\right)^{j} \left(\log \frac{1}{t}\right)^{r-\ell+1} \frac{du}{1-u}\frac{dt}{1-t}.$$

Substituting the factor

$$\left(\log \frac{1}{u}\right)^{\ell+1} \quad \text{and} \quad \left(\log \frac{1}{1-t}\right)^{j}$$

by their binomial expansions

$$\sum_{c+d=\ell+1} \frac{(\ell+1)!}{c!d!} \left(\log \frac{t}{u}\right)^{c} \left(\log \frac{1}{t}\right)^{d} \quad \text{and}$$

$$\sum_{a+b=j} \frac{j!}{a!b!} \left(\log \frac{1}{1-u}\right)^{a} \left(\log \frac{1-u}{1-t}\right)^{b},$$

we see that the number of multiple zeta values represented by the integral I_3 is

$$\sum_{a+b=j} \binom{k-j+a}{k-j} \sum_{c+d=\ell+1} \binom{c+b}{b}\binom{d+r-\ell+1}{r-\ell+1}.$$

Applying Vandermonde's convolution on the inner sum, the number becomes

$$\sum_{a+b=j} \binom{k-j+a}{k-j}\binom{b+r+3}{\ell+1},$$

or simply

$$\sum_{a+b=k} \binom{a}{k-j}\binom{b+r+3}{\ell+1}.$$

Now we introduce the vectors

$$\boldsymbol{j} = (j_1, j_2, \ldots, j_m) \quad \text{with} \quad j_1 + j_2 + \cdots + j_m = j$$

and

$$\boldsymbol{\ell} = (\ell_1, \ell_2, \ldots, \ell_n) \quad \text{with} \quad \ell_1 + \ell_2 + \cdots + \ell_n = \ell + 1$$

and also employ notations used in vectors

$$|\boldsymbol{j}| = j_1 + j_2 + \cdots + j_m, \quad |\boldsymbol{\ell}| = \ell_1 + \ell_2 + \cdots + \ell_n,$$

$$\boldsymbol{j}! = j_1! j_2! \cdots j_m! \quad \text{and} \quad \boldsymbol{\ell}! = \ell_1! \ell_2! \cdots \ell_n!.$$

Then we consider the integral

$$I_4 = \frac{1}{\boldsymbol{j}!(k-j)!\boldsymbol{\ell}!(r-\ell+1)!} \iint_{0<u<t<1} \left(\log\frac{1}{1-u}\right)^{k-j} \left(\log\frac{1}{u}\right)^{|\boldsymbol{\ell}|}$$
$$\times \left(\log\frac{1}{1-t}\right)^{|\boldsymbol{j}|} \left(\log\frac{1}{t}\right)^{r-\ell+1} \frac{du}{1-u}\frac{dt}{1-t}.$$

The number of multiple zeta values represented by I_4 is

$$\sum_{\boldsymbol{\alpha}+\boldsymbol{\beta}=\boldsymbol{j}} \sum_{\boldsymbol{\gamma}+\boldsymbol{\delta}=\boldsymbol{\ell}} \binom{|\boldsymbol{\alpha}|+k-j}{\alpha_1, \alpha_2, \ldots, \alpha_m, k-j} \binom{|\boldsymbol{\beta}|+|\boldsymbol{\gamma}|}{\beta_1, \beta_2, \ldots, \beta_m, \gamma_1, \gamma_2, \ldots, \gamma_n}$$
$$\times \binom{|\boldsymbol{\delta}|+r-\ell+1}{\delta_1, \delta_2, \ldots, \delta_n, r-\ell+1}$$

or

$$\sum_{\boldsymbol{\alpha}+\boldsymbol{\beta}=\boldsymbol{j}} \sum_{\boldsymbol{\gamma}+\boldsymbol{\delta}=\boldsymbol{\ell}} \binom{|\boldsymbol{\alpha}|+k-j}{|\boldsymbol{\alpha}|} \binom{|\boldsymbol{\beta}|+|\boldsymbol{\gamma}|}{|\boldsymbol{\beta}|} \binom{|\boldsymbol{\delta}|+r-\ell+1}{|\boldsymbol{\delta}|}$$
$$\times M(\boldsymbol{\alpha})M(\boldsymbol{\beta})M(\boldsymbol{\gamma})M(\boldsymbol{\delta}).$$

On the other hand, it is obvious that

$$I_4 = M(\boldsymbol{j})M(\boldsymbol{\ell})I_3.$$

Two different expressions for the number of multiple zeta values represented by I_4 lead to the following combinatorial identity.

Theorem 8.2.4. *For integers k, r, j, ℓ with $0 \leq j \leq k$ and $0 \leq \ell \leq r$, we have*

$$\sum_{\alpha+\beta=j} \sum_{\gamma+\delta=\ell} \binom{|\alpha|+k-j}{|\alpha|} \binom{|\beta|+|\gamma|}{|\beta|} \binom{|\delta|+r-\ell+1}{|\delta|}$$

$$\times M(\alpha)M(\beta)M(\gamma)M(\delta)$$

$$= M(j)M(\ell) \sum_{a+b=k} \binom{a}{k-j} \binom{b+r+3}{\ell+1}.$$

For nonnegative integers k, p, q, ℓ, the integral

$$(\star) \qquad I_5 = \frac{1}{k!p!q!\ell!} \int_{0<t_1<t_2<1} \left(\log \frac{1}{1-t_1}\right)^k \left(\log \frac{1}{1-t_2}\right)^p$$

$$\times \left(\log \frac{t_2}{t_1}\right)^q \left(\log \frac{1}{t_2}\right)^\ell \frac{dt_1}{1-t_1} \frac{dt_2}{t_2}$$

represents the sum of multiple zeta values

$$\sum_{a+b=p} \binom{k+a}{k} \sum_{|\alpha|=b+q+1} \zeta(\{1\}^{k+a}, \alpha_0, \alpha_1, \ldots, \alpha_b + \ell + 1)$$

if we replace the factor $\left(\log \frac{1}{1-t_2}\right)^p$ by

$$\sum_{a+b=p} \frac{p!}{a!b!} \left(\log \frac{1}{1-t_1}\right)^a \left(\log \frac{1-t_1}{1-t_2}\right)^b.$$

Therefore the number of multiple zeta values represented by the integral is

$$\sum_{a+b=p} \binom{k+a}{k} \binom{q+b}{q}$$

or

$$\binom{k+p+q+1}{p}.$$

Now we are going to investigate the combinatorial consequence from the shuffle product of such integral $(\star)$ and the multiple zeta value $\zeta(\{1\}^m, n+2)$ represented by the integral

$$\frac{1}{m!(n+1)!} \int_0^1 \left(\log \frac{1}{1-u}\right)^m \left(\log \frac{1}{u}\right)^{n+1} \frac{du}{1-u}.$$

It is clear that the number of multiple zeta values produced from the shuffle product is

$$\binom{k+p+q+1}{p}\binom{w_1+w_2}{w_1}$$

or

(8.2.4) $$\binom{k+p+q+1}{p}\sum_{\alpha=0}^{w_1}\binom{\alpha+m}{m}\binom{\alpha^*+n+1}{n+1}$$

with $w_1 = k+p+q+\ell+2$, $w_2 = m+n+2$ and $\alpha^* = w_1 - \alpha$.

In the following, we carry out the details of the shuffle process when $0 < t_1 < t_2 < u < 1$, the factors

$$\left(\log\frac{1}{1-t_2}\right)^p, \quad \left(\log\frac{1}{t_2}\right)^\ell \quad \text{and} \quad \left(\log\frac{1}{1-u}\right)^m$$

are replaced by

$$\left(\log\frac{1}{1-t_1}+\log\frac{1-t_1}{1-t_2}\right)^p, \quad \left(\log\frac{u}{t_2}+\log\frac{1}{u}\right)^\ell$$

and

$$\left(\log\frac{1}{1-t_1}+\log\frac{1-t_1}{1-t_2}+\log\frac{1-t_2}{1-u}\right)^m,$$

and then they are expanded into binomial expansion or trinomial expansions.

The number of multiple zeta values produced is given by

$$\sum_{|m|=m}\sum_{a+b=p}\sum_{c+d=\ell}\binom{m_1+k+a}{k+a}\binom{m_2+b+q}{b+q}\binom{m_3+c}{c}$$
$$\times\binom{k+a}{k}\binom{b+q}{q}\binom{d+n+1}{n+1}.$$

First summing over $m_1+m_2+m_3 = m$ and then over $a+b = p$, the number becomes

(8.2.5) $$\binom{k+p+q+1}{p}\sum_{c+d=\ell}\binom{c+k+p+q+m+2}{m}\binom{d+n+1}{n+1}.$$

This is a part of the total number of multiple zeta values with $0 \le \alpha^* \le \ell$.

In a similar manner, the numbers of multiple zeta values obtained from evaluations of integrals over the simplex $D_2 : 0 < t_1 < u < t_2 < 1$ and $D_3 : 0 < u < t_1 < t_2 < 1$ are

$$
(8.2.6) \qquad \sum_{a+b=p} \sum_{c+d=q} \binom{a+c+m+k+1}{m} \binom{a+c+k+1}{a}
$$
$$
\times \binom{b+d}{b} \binom{b+d+n+\ell+2}{n+1}
$$

and

$$
(8.2.7)
$$
$$
\sum_{a+b=p} \sum_{u+v=k} \binom{a+u+m}{m} \binom{a+u}{a} \binom{b+v+q+1}{b} \binom{b+v+n+q+\ell+3}{n+1}.
$$

As we have $(8.2.4) = (8.2.5) + (8.2.6) + (8.2.7)$, we conclude the following theorem.

Theorem 8.2.5. *For nonnegative integers k, p, q, ℓ, m, n, we have*

$$
(8.2.8) \qquad
\begin{aligned}
& \sum_{a+b=p} \sum_{c+d=q} \binom{a+c+m+k+1}{m} \binom{a+c+k+1}{a} \\
& \qquad \times \binom{b+d}{b} \binom{b+d+n+\ell+2}{n+1} \\
& + \sum_{a+b=p} \sum_{u+v=k} \binom{a+u+m}{m} \binom{a+u}{a} \\
& \qquad \times \binom{b+v+q+1}{b} \binom{b+v+n+q+\ell+3}{n+1} \\
& = \binom{k+p+q+1}{p} \sum_{\alpha=0}^{k+p+q+1} \binom{\alpha+m}{m} \binom{\alpha^*+n+\ell+2}{n+1}
\end{aligned}
$$
$$
(\alpha^* = k+p+q+1-\alpha).
$$

For the particular case $p = 0$, we conclude that the number of multiple zeta values in the integral

$$
I_6 = \frac{1}{k!q!\ell!m!(n+1)!} \int_{0<t_1<u<t_2<1} \left(\log \frac{1}{1-t_1} \right)^k \left(\log \frac{t_2}{t_1} \right)^q \left(\log \frac{1}{t_2} \right)^\ell
$$
$$
\times \left(\log \frac{1}{1-u} \right)^m \left(\log \frac{1}{u} \right)^{n+1} \frac{dt_1}{1-t_1} \frac{du}{1-u} \frac{dt_2}{t_2}
$$

is

$$\sum_{c+d=q} \binom{c+m+k+1}{m}\binom{d+n+\ell+2}{n+1}.$$

However if we replace $q!$ and $\left(\log\frac{t_2}{t_1}\right)^q$ in I_6 by $q_1!q_2!\cdots q_r!$ and $\left(\log\frac{t_2}{t_1}\right)^{q_1+q_2+\cdots+q_r}$, the number of multiple zeta values in the resulted integral is

$$\sum_{q=\alpha+\beta} \binom{|\alpha|+m+k+1}{m}M(\alpha)M(\beta)\binom{|\beta|+n+\ell+2}{n+1}.$$

But on the other hand, the resulted integral is $M(q)I_6$ and its number of multiple zeta values can be computed in other way. This leads to the following.

Theorem 8.2.6. *For nonnegative integers* $k, q_1, q_2, \ldots, q_r, \ell, m, n$ *with* $q = (q_1, q_2, \ldots, q_r)$, *we have*

(8.2.9)
$$\sum_{\alpha+\beta=q} \binom{|\alpha|+m+k+1}{m}M(\alpha)M(\beta)\binom{|\beta|+n+\ell+2}{n+1}$$
$$= M(q) \sum_{a+b=|q|} \binom{a+m+k+1}{m}\binom{b+n+\ell+2}{n+1}$$

with

$$M(g) = \binom{g_1+g_2+\cdots+g_r}{g_1, g_2, \ldots, g_r}$$

if $g = (g_1, g_2, \ldots, g_r)$.

When both p and q in I_5 are vectors, we obtain another vector version of Theorem 8.2.5.

Theorem 8.2.7. *For nonnegative integers* $k, p_1, p_2, \ldots, p_g, q_1, q_2, \ldots, q_r, \ell,$

m, n with $\boldsymbol{p} = (p_1, p_2, \ldots, p_g)$, $\boldsymbol{q} = (q_1, q_2, \ldots, q_r)$, we have

$$\sum_{\boldsymbol{\alpha}+\boldsymbol{\beta}=\boldsymbol{p}} \sum_{\boldsymbol{\gamma}+\boldsymbol{\delta}=\boldsymbol{q}} \binom{|\boldsymbol{\alpha}| + |\boldsymbol{\gamma}| + m + k + 1}{m} \binom{|\boldsymbol{\alpha}| + |\boldsymbol{\gamma}| + k + 1}{|\boldsymbol{\alpha}|}$$

$$\times \binom{|\boldsymbol{\beta}| + |\boldsymbol{\delta}|}{|\boldsymbol{\beta}|} \binom{|\boldsymbol{\beta}| + |\boldsymbol{\delta}| + n + \ell + 2}{n + 1} M(\boldsymbol{\alpha}) M(\boldsymbol{\beta}) M(\boldsymbol{\gamma}) M(\boldsymbol{\delta})$$

$$= M(\boldsymbol{p}) M(\boldsymbol{q}) \sum_{a+b=|\boldsymbol{p}|} \sum_{c+d=|\boldsymbol{q}|} \binom{a + c + m + k + 1}{m} \binom{a + c + k + 1}{a}$$

$$\times \binom{b + d}{b} \binom{b + d + n + \ell + 2}{n + 1}.$$

Here

$$M(\boldsymbol{b}) = \binom{b_1 + b_2 + \cdots + b_k}{b_1, b_2, \ldots, b_k}$$

if $\boldsymbol{b} = (b_1, b_2, \ldots, b_k)$.

Theorem 8.2.8. *Let* $\boldsymbol{p} = (p_1, p_2, \ldots, p_m)$ *and* $\boldsymbol{q} = (q_1, q_2, \ldots, q_n)$ *be a pair of vectors with components are nonnegative integers. Then for any positive integers* k*, we have*

$$\sum_{\boldsymbol{p}=\boldsymbol{\alpha}_1+\boldsymbol{\alpha}_2+\ldots+\boldsymbol{\alpha}_k} \sum_{\boldsymbol{q}=\boldsymbol{\beta}_1+\boldsymbol{\beta}_2+\ldots+\boldsymbol{\beta}_k} \prod_{j=1}^{k} \binom{|\boldsymbol{\alpha}_j| + |\boldsymbol{\beta}_j|}{|\boldsymbol{\alpha}_j|} M(\boldsymbol{\alpha}_j) M(\boldsymbol{\beta}_j)$$

$$= M(\boldsymbol{p}) M(\boldsymbol{q}) \binom{|\boldsymbol{p}| + |\boldsymbol{q}| + k - 1}{|\boldsymbol{p}|, |\boldsymbol{q}|, k - 1}.$$

8.3 Vector Versions of Pascal Identity

Through shuffle products of multiple zeta values, we had proved a generalization of Pascal identity read as

$$(8.3.1) \quad \binom{k + r + 4}{j + \ell + 2} = \sum_{a+b=k} \left\{ \binom{a}{j} \binom{b + r + 3}{\ell + 1} + \binom{a}{k - j} \binom{b + r + 3}{r - \ell + 1} \right\}$$

with integers k, r, j, ℓ satisfying $0 \le j \le k$ and $0 \le \ell \le r$. As

$$\binom{b + r + 3}{\ell + 1} = \binom{b + r + 3}{b + r - \ell + 2} = \sum_{m+p=r+2} \binom{b + m}{b} \binom{p}{r - \ell + 1}$$

and

$$\binom{b + r + 3}{r - \ell + 1} = \sum_{m+p=r+2} \binom{b + m}{b} \binom{p}{\ell + 1},$$

we are able to rewrite (8.3.1) as

$$
\binom{k+r+4}{j+\ell+2}
$$
(8.3.2)
$$
= \sum_{m+p=r+2} \left\{ \binom{m+k+1}{k-j}\binom{p}{r-\ell+1} + \binom{m+k+1}{j}\binom{p}{\ell+1} \right\}.
$$

In this section, we shall extend such a result to a more general form with k, j replaced by $k_1 + k_2 + \cdots + k_n$, $j_1 + j_2 + \cdots + j_n$ satisfying $0 \leq j_i \leq k_i$, $i = 1, 2, \ldots, n$. It is quite a surprise that the extension is just like (8.3.2), namely

$$
\binom{k_1 + k_2 + \cdots + k_n + r + 4}{j_1 + j_2 + \cdots + j_n + \ell + 2}
$$
$$
= \sum_{m+p=r+2} \left\{ \binom{m+|\boldsymbol{k}|+1}{|\boldsymbol{k}|-|\boldsymbol{j}|}\binom{p}{r-\ell+1} + \binom{m+|\boldsymbol{k}|+1}{|\boldsymbol{j}|}\binom{p}{\ell+1} \right\}.
$$

To accomplish such a mission, we need more basic identities among multinomial coefficients. Recall that the coefficient of $\mu^j (0 \leq j \leq k)$ of the polynomial

$$
(8.3.3) \qquad P(m, n, k; \mu) = \sum_{a+b=k} \binom{m+a}{m}\binom{n+b}{n}(\mu+1)^b
$$

is

$$
\binom{n+j}{j}\binom{m+n+k+1}{k-j}.
$$

Here is a multi-version of the above conclusion. However, we need some particular notations. As before, we let

$$
M(\boldsymbol{g}) = \binom{g_1 + g_2 + \cdots + g_n}{g_1, g_2, \ldots, g_n}
$$

if $\boldsymbol{g} = (g_1, g_2, \ldots, g_n)$. Also for vector variable $\boldsymbol{\mu} = (\mu_1, \mu_2, \ldots, \mu_n)$ and vector index $\boldsymbol{\alpha} = (\alpha_1, \alpha_2, \ldots, \alpha_n)$, we let $\boldsymbol{\mu}^{\boldsymbol{\alpha}}$ be the product of monomials $\mu_1^{\alpha_1} \mu_2^{\alpha_2} \cdots \mu_n^{\alpha_n}$.

Proposition 8.3.1. *If m is a nonnegative integer and vectors $\boldsymbol{k} = (k_1, k_2,$ $\ldots, k_n)$, $\boldsymbol{j} = (j_1, j_2, \ldots, j_n)$ with $0 \leq j_i \leq k_i$, $i = 1, 2, \ldots, n$, then the coefficient of $\boldsymbol{\mu^j} (= \mu_1^{j_1} \mu_2^{j_2} \cdots \mu_n^{j_n})$ of the polynomial*

$$(8.3.4) \qquad \sum_{\boldsymbol{\alpha}+\boldsymbol{\beta}=\boldsymbol{k}} M(\boldsymbol{\alpha}) M((m, \boldsymbol{\beta})) (\boldsymbol{\mu}+1)^{\boldsymbol{\alpha}}$$

is

$$\binom{m + |\boldsymbol{k}| + 1}{|\boldsymbol{k}| - |\boldsymbol{j}|} M(\boldsymbol{j}) M(\boldsymbol{k} - \boldsymbol{j}).$$

Here $(\boldsymbol{\mu} + 1)^{\boldsymbol{\alpha}} = (\mu_1 + 1)^{\alpha_1} (\mu_2 + 1)^{\alpha_2} \cdots (\mu_n + 1)^{\alpha_n}$.

Proof. Note that

$$M(\boldsymbol{\alpha}) = \binom{|\boldsymbol{\alpha}|}{\alpha_1} M((0, \alpha_2, \ldots, \alpha_n))$$

and

$$M((m, \boldsymbol{\beta})) = \binom{|\boldsymbol{\beta}| + m}{\beta_1} M((m, \beta_2, \ldots, \beta_n))$$

and the coefficient of $\mu_1^{j_1}$ of the polynomial

$$\sum_{\alpha_1+\beta_1=k_1} \binom{|\boldsymbol{\alpha}|}{\alpha_1} \binom{|\boldsymbol{\beta}| + m}{\beta_1} (\mu_1 + 1)^{\alpha_1}$$

is

$$\binom{m + |\boldsymbol{k}| + 1}{k_1 - j_1} \binom{j_1 + \alpha_2 + \cdots + \alpha_n}{j_1}.$$

Therefore the coefficient of $\mu_1^{j_1}$ in the polynomial (8.3.4) is

$$\binom{m + |\boldsymbol{k}| + 1}{k_1 - j_1} \sum (\mu_2 + 1)^{\alpha_2} \cdots (\mu_n + 1)^{\alpha_n}$$

$$\times M((j_1, \alpha_2, \ldots, \alpha_n)) M((m, \beta_2, \ldots, \beta_n)).$$

Repeat such a process $(n-1)$ times, the coefficient of $\boldsymbol{\mu^j}$ of the polynomial (8.3.4) is

$$\binom{m + |\boldsymbol{k}| + 1}{k_1 - j_1, k_2 - j_2, \ldots, k_n - j_n, m + |\boldsymbol{j}| + 1} M(\boldsymbol{j})$$

or

$$M(\boldsymbol{j}) M(\boldsymbol{k} - \boldsymbol{j}) \binom{m + |\boldsymbol{k}| + 1}{|\boldsymbol{k}| - |\boldsymbol{j}|}.$$

$\square$

Corollary 8.3.2. *If m is a nonnegative integer and $\boldsymbol{k} = (k_1, k_2, \ldots, k_n)$ with $k_1, k_2, \ldots, k_n \geq 0$, then*

$$\sum_{\boldsymbol{\alpha}+\boldsymbol{\beta}=\boldsymbol{k}} M(\boldsymbol{\alpha})M((m,\boldsymbol{\beta})) = M(\boldsymbol{k})\binom{m+|\boldsymbol{k}|+1}{|\boldsymbol{k}|}.$$

Proof. Such a sum of products of multinomial coefficients is just the constant of the polynomial (8.3.4). So we set $\mu_1 = \mu_2 = \cdots = \mu_n = 0$. This corresponds to the case when $j_1 = j_2 = \cdots = j_n = 0$. $\qquad\square$

Now we are ready to evaluate a generating function for

$$\binom{k_1 + k_2 + \cdots + k_n + r + 4}{j_1 + j_2 + \cdots + j_n + \ell + 2}$$

which can be used to determine the values of these multinomial coefficients.

Theorem 8.3.3. *For a nonnegative integer r and a vector $\boldsymbol{k} = (k_1, k_2, \ldots, k_n)$ with integers $k_1, k_2, \ldots, k_n \geq 0$, we have*

$$\sum_{\boldsymbol{\alpha}+\boldsymbol{\beta}=\boldsymbol{k}} \sum_{\ell=0}^{r} \boldsymbol{\mu}^{\boldsymbol{\alpha}} \lambda^{\ell+1} M(\boldsymbol{\alpha})M(\boldsymbol{\beta})\binom{|\boldsymbol{k}|+r+4}{|\boldsymbol{\alpha}|+\ell+2}$$

$$= \sum_{\boldsymbol{\alpha}+\boldsymbol{\beta}=\boldsymbol{k}} (\boldsymbol{\mu}+1)^{\boldsymbol{\alpha}} \sum_{m+p=r+2} \lambda^m (\lambda+1)^p M(\boldsymbol{\alpha})M((m,\boldsymbol{\beta}))$$

(8.3.5)
$$+ \sum_{\boldsymbol{\alpha}+\boldsymbol{\beta}=\boldsymbol{k}} (\boldsymbol{\mu}+1)^{\boldsymbol{\alpha}} \boldsymbol{\mu}^{\boldsymbol{\beta}} \sum_{m+p=r+2} (\lambda+1)^p M(\boldsymbol{\alpha})M((m,\boldsymbol{\beta}))$$

$$- \sum_{\boldsymbol{\alpha}+\boldsymbol{\beta}=\boldsymbol{k}} (\boldsymbol{\mu}+1)^{\boldsymbol{\alpha}} M(\boldsymbol{\alpha})M(\boldsymbol{\beta})\left\{\lambda^{r+2}\binom{|\boldsymbol{\beta}|+r+3}{r+2}+1\right\}$$

$$- \sum_{\boldsymbol{\alpha}+\boldsymbol{\beta}=\boldsymbol{k}} (\boldsymbol{\mu}+1)^{\boldsymbol{\alpha}} \boldsymbol{\mu}^{\boldsymbol{\beta}} M(\boldsymbol{\alpha})M(\boldsymbol{\beta})\left\{\binom{|\boldsymbol{\beta}|+r+3}{r+2}+\lambda^{r+2}\right\}.$$

Proof. Consider the integral

$$\frac{1}{k!r!} \iint_{R_1 \times R_2} \prod_{i=1}^{n} \left(\mu_i \log\frac{1}{1-t_1} + \log\frac{1}{1-u_1}\right)^{k_i} \left(\log\frac{1}{t_2} + \lambda\log\frac{1}{u_2}\right)^r$$

$$\times \frac{dt_1 dt_2}{(1-t_1)t_2} \frac{du_1 du_2}{(1-u_1)u_2}$$

with $R_1 : 0 < t_1 < t_2 < 1$ and $R_2 : 0 < u_1 < u_2 < 1$. After binomial expansions, the value of integral is simply given by

$$\sum_{\alpha+\beta=k} \sum_{\ell=0}^{r} \mu^{\alpha} \lambda^{\ell} M(\alpha) M(\beta) \zeta(\{1\}^{|\alpha|}, r - \ell + 2) \zeta(\{1\}^{|\beta|}, \ell + 2).$$

Therefore, after a shuffle product process, the generating function for number of multiple zeta values produced from the shuffle product is

$$(8.3.6) \qquad \sum_{\alpha+\beta=k} \sum_{\ell=0}^{r} \mu^{\alpha} \lambda^{\ell} M(\alpha) M(\beta) \binom{|\boldsymbol{k}| + r + 4}{|\boldsymbol{\alpha}| + r - \ell + 2}$$

which can be changed to left-hand side of (8.3.5) by changing λ into λ^{-1} and multiplying λ^{r+1}.

During the shuffle process, the generating function (8.3.6) becomes a sum of six partial generating functions obtained from integration over the following simplices:

$$D_1 : \; 0 < t_1 < t_2 < u_1 < u_2 < 1,$$
$$D_2 : \; 0 < u_1 < u_2 < t_1 < t_2 < 1,$$
$$D_3 : \; 0 < t_1 < u_1 < t_2 < u_2 < 1,$$
$$D_4 : \; 0 < t_1 < u_1 < u_2 < t_2 < 1,$$
$$D_5 : \; 0 < u_1 < t_1 < u_2 < t_2 < 1 \text{ and}$$
$$D_6 : \; 0 < u_1 < t_1 < t_2 < u_2 < 1.$$

On the first simplex $D_1 : \; 0 < t_1 < t_2 < u_1 < u_2 < 1$, the factors

$$\left(\mu_i \log \frac{1}{1 - t_1} + \log \frac{1}{1 - u_1} \right)^{k_i} \quad (i = 1, 2, \dots, n)$$

and

$$\left(\log \frac{1}{t_2} + \lambda \log \frac{1}{u_2} \right)^{r}$$

are replaced by

$$\left\{ (\mu_i + 1) \log \frac{1}{1 - t_1} + \log \frac{1 - t_1}{1 - t_2} + \log \frac{1 - t_2}{1 - u_1} \right\}^{k_i} \quad (i = 1, 2, \dots, n)$$

and

$$\left(\log \frac{u_1}{t_2} + \log \frac{u_2}{u_1} + (\lambda + 1) \log \frac{1}{u_2} \right)^{r},$$

and then expanded one by one into multinomial series, so that the generating function for number of multiple zeta values is

$$(8.3.7) \qquad \sum_{\alpha+\beta+\gamma=k} (\mu+1)^{\alpha} \sum_{m+p+g=r} (\lambda+1)^{g} M(\alpha) M(\beta) M((m,\gamma)).$$

Dummy variables β, γ are merged via Corollary 8.3.2 so that the above sum becomes

$$\lambda^{-1} \sum_{\alpha+\beta=k} (\mu+1)^{\alpha} \sum_{m+p=r} \left\{ (\lambda+1)^{p+1} - 1 \right\} M(\alpha) M((m+1,\beta))$$

or

$$(8.3.8)$$
$$\sum_{\alpha+\beta=k} (\mu+1)^{\alpha} \sum_{m+p=r+2} (\lambda+1)^{p} M(\alpha) M((m,\beta))$$
$$- \sum_{\alpha+\beta=k} (\mu+1)^{\alpha} M(\alpha) M(\beta) \left\{ \binom{|\beta|+r+3}{r+2} + (\lambda+1)^{r+2} - 1 \right\}.$$

Integrations over D_3 and D_4 only produce constant multiples of the multiple zeta value $\zeta(\{1\}^{|k|+1}, r+3)$ and the partial generating functions are

$$(8.3.9) \qquad \lambda^{-1} \sum_{\alpha+\beta=k} (\mu+1)^{\alpha} M(\alpha) M(\beta) \{(\lambda+1)^{r+1} - 1\}$$

and

$$(8.3.10) \qquad \sum_{\alpha+\beta=k} (\mu+1)^{\alpha} M(\alpha) M(\beta) \{(\lambda+1)^{r+1} - \lambda^{r+1}\}.$$

Consequently, the sum of (8.3.8), (8.3.9) and (8.3.10) is

$$(8.3.11)$$
$$\lambda^{-1} \sum_{\alpha+\beta=k} (\mu+1)^{\alpha} \sum_{m+p=r+2} (\lambda+1)^{m} M(\alpha) M((m,\beta))$$
$$- \lambda^{-1} \sum_{\alpha+\beta=k} (\mu+1)^{\alpha} M(\alpha) M(\beta) \left\{ \binom{|\beta|+r+3}{r+2} + \lambda^{r+2} \right\}.$$

In the same manner, we obtain the sum of generating functions from integrations over D_2, D_5 and D_6 as

(8.3.12)
$$\lambda^{-1} \sum_{\alpha+\beta=k} (\mu+1)^{\alpha} \sum_{m+p=r+2} \lambda^m (\lambda+1)^p M(\alpha) M((m, \beta))$$
$$- \lambda^{-1} \sum_{\alpha+\beta=k} (\mu+1)^{\alpha} \mu^{\beta} M(\alpha) M(\beta) \left\{ \lambda^{r+2} \binom{|\beta|+r+3}{r+2} + 1 \right\}.$$

Of course, we have
$$(8.3.6) = (8.3.11) + (8.3.12)$$

and hence our assertion follows.

$\square$

Separating the coefficient of $\mu^j \lambda^{\ell+1}$ on both sides of (8.3.5) with the help of Proposition 8.3.1, we obtain the value of the binomial coefficient
$$\binom{k_1 + k_2 + \cdots + k_n + r + 4}{j_1 + j_2 + \cdots + j_n + \ell + 2}.$$

Theorem 8.3.4. *For integers r, ℓ with $0 \le \ell \le r$ and vectors $\boldsymbol{k} = (k_1, k_2, \ldots, k_n)$, $\boldsymbol{j} = (j_1, j_2, \ldots, j_n)$ with $0 \le j_i \le k_i$, $i = 1, 2, \ldots, n$, we have*
$$\binom{k_1 + k_2 + \cdots + k_n + r + 4}{j_1 + j_2 + \cdots + j_n + \ell + 2}$$
$$= \sum_{m+p=r+2} \left\{ \binom{m + |\boldsymbol{k}| + 1}{|\boldsymbol{k}| - |\boldsymbol{j}|} \binom{p}{r - \ell + 1} + \binom{m + |\boldsymbol{k}| + 1}{|\boldsymbol{j}|} \binom{p}{\ell + 1} \right\}.$$

It is interesting to obtain the sum of an alternating series by setting $\mu_1 = \mu_2 = \cdots = \mu_n = -1$ and $\lambda = -1$.

Theorem 8.3.5. *For a nonnegative integer r and a vector $\boldsymbol{k} = (k_1, k_2, \ldots, k_n)$ with $k_1, k_2, \ldots, k_n \ge 0$, we have*
$$\frac{1}{2} \sum_{\alpha+\beta=k} \sum_{\ell=0}^{r} (-1)^{|\alpha|+\ell} M(\alpha) M(\beta) \binom{|\boldsymbol{k}| + r + 4}{|\alpha| + \ell + 2}$$
$$= (-1)^{r+1} \binom{k_1 + k_2 + \cdots + k_n + r + 2}{k_1, k_2, \ldots, k_n, r + 2}$$
$$+ \binom{k_1 + k_2 + \cdots + k_n}{k_1, k_2, \ldots, k_n} \left\{ (-1)^r \binom{|\boldsymbol{k}| + r + 3}{r + 2} + 1 \right\}$$

when $|\boldsymbol{k}| + r$ is even.

8.4 Problems on Combinatorial Identity

If a combinatorial identity can be interpreted as the number of multiple zeta values in a particular sum of multiple zeta values, then we can always produce a vector version of identity through shuffle products of multiple zeta values.

For example, for nonnegative integers p, q, r, we have

$$\sum_{a+b=p} \binom{q+a}{q}\binom{r+b}{r} = \binom{p+q+r+1}{p}$$

by Vandermonde's convolution. The vector version is

$$\sum_{\boldsymbol{\alpha}+\boldsymbol{\beta}=\boldsymbol{p}} \binom{q+\alpha_1+\alpha_2+\cdots+\alpha_n}{q,\alpha_1,\alpha_2,\ldots,\alpha_n}\binom{r+\beta_1+\beta_2+\cdots+\beta_n}{r,\beta_1,\beta_2,\ldots,\beta_n}$$

$$= \binom{p_1+p_2+\cdots+p_n}{p_1,p_2,\ldots,p_n}\binom{|\boldsymbol{p}|+q+r+1}{|\boldsymbol{p}|}$$

for $\boldsymbol{p} = (p_1, p_2, \ldots, p_n)$, an n-tuple of nonnegative integers.

In the following, we list some problems concerning applications of shuffle products of multiple zeta values in combinatorics. We employ notations from vectors such as

$$\boldsymbol{\mu^\alpha} = \mu_1^{\alpha_1}\mu_2^{\alpha_2}\cdots\mu_n^{\alpha_n}$$

if $\boldsymbol{\mu} = (\mu_1, \mu_2, \ldots, \mu_n)$ and $\boldsymbol{\alpha} = (\alpha_1, \alpha_2, \ldots, \alpha_n)$. Also

$$\boldsymbol{p}! = p_1!p_2!\cdots p_n!, \quad M(\boldsymbol{p}) = \binom{p_1+p_2+\cdots+p_n}{p_1,p_2,\ldots,p_n}$$

and

$$|\boldsymbol{p}| = p_1+p_2+\cdots+p_n$$

if $\boldsymbol{p} = (p_1, p_2, \ldots, p_n)$.

Problem 8.4.1. *For a pair of integers $m, n \geq 0$ and k-tuple of nonnegative integers $\boldsymbol{p} = (p_1, p_2, \ldots, p_k)$. Prove that*

$$\sum_{\boldsymbol{\alpha}+\boldsymbol{\beta}=\boldsymbol{p}} M\left((m, \boldsymbol{\alpha})\right) M\left((n, \boldsymbol{\beta})\right) = M(\boldsymbol{p})\binom{m+n+|\boldsymbol{p}|+1}{|\boldsymbol{p}|}.$$

Hint: Count the number of multiple zeta values represented by the integral

$$\frac{1}{m!p!n!} \int \int_{0<t_1<t_2<1} \left(\log \frac{1}{1-t_1}\right)^m \left(\log \frac{1}{1-t_2}\right)^{|p|} \left(\log \frac{t_2}{t_1}\right)^n \frac{dt_1 dt_2}{(1-t_1)t_2}$$

in two ways. If we substitute the factor

$$\left(\log \frac{1}{1-t_2}\right)^{|p|}$$

by

$$\sum_{\alpha+\beta=p} \frac{p!}{\alpha!\beta!} \left(\log \frac{1}{1-t_1}\right)^{|\alpha|} \left(\log \frac{1-t_1}{1-t_2}\right)^{|\beta|}$$

then the number of multiple zeta values is

$$\sum_{\alpha+\beta=p} M((m,\alpha)) M((n,\beta)).$$

Problem 8.4.2. *For a pair of k-tuples of integers $p = (p_1, p_2, \ldots, p_k)$ and $j = (j_1, j_2, \ldots, j_k)$ with $0 \leq j_i \leq p_i$, $i = 1, 2, \ldots, k$. Prove that the coefficient of μ^j of the multinomial*

$$\sum_{\alpha+\beta=p} (\mu+1)^\beta M((m,\alpha)) M((n,\beta))$$

is given by

$$M(j)M(p-j) \binom{n+|j|}{|j|} \binom{m+n+|p|+1}{|p|-|j|},$$

where $(\mu+1)^\beta = (\mu_1+1)^{\beta_1} (\mu_2+1)^{\beta_2} \cdots (\mu_k+1)^{\beta_k}$.

Problem 8.4.3. *For integers k, j, r, ℓ with $0 \leq j \leq k$ and $0 \leq \ell \leq r$, show that the coefficient of $\mu^j \lambda^\ell$ of the polynomial*

$$\sum_{a+b+c=k} \mu^{a+c}(\mu+1)^b \sum_{m+n+p=r} (\lambda+1)^n \binom{a+m}{a} \binom{b+n}{b} \binom{c+p}{c}$$

is given by

$$\binom{k+r+2}{j} \binom{k-j+r+2}{r-\ell} \binom{k-j+\ell}{\ell}.$$

Problem 8.4.4. *For integers* k, j, r, ℓ *with* $0 \le j \le k$ *and* $0 \le \ell \le r$, *show that the coefficient of* $\mu^j \lambda^\ell$ *of the polynomial*

$$\sum_{a+b+c=k} \mu^a (\mu+1)^b \sum_{m+n+p=r} (\lambda+1)^n \lambda^p \binom{a+m}{a} \binom{b+n}{b} \binom{c+p}{c}$$

is

$$\sum_{b+c=k} \sum_{m+p=r} \binom{b+m+1}{r-\ell} \binom{b+\ell-p+1}{j} \binom{b-j+\ell-p}{\ell-p} \binom{c+p}{c}.$$

Problem 8.4.5. *For* m-*tuple of nonnegative integers* $\boldsymbol{k} = (k_1, k_2, \ldots, k_m)$, n-*tuple of nonnegative integers* $\boldsymbol{r} = (r_1, r_2, \ldots, r_n)$ *and multi-variables* $\boldsymbol{\mu} = (\mu_1, \mu_2, \ldots, \mu_m), \boldsymbol{\lambda} = (\lambda_1, \lambda_2, \ldots, \lambda_n)$. *Find the coefficient of* $\boldsymbol{\mu}^j \boldsymbol{\lambda}^\ell$ *of the polynomial*

$$\sum_{\boldsymbol{\alpha}+\boldsymbol{\beta}+\boldsymbol{\gamma}=\boldsymbol{k}} \boldsymbol{\mu}^{\boldsymbol{\alpha}+\boldsymbol{\gamma}} (\boldsymbol{\mu}+1)^{\boldsymbol{\beta}} \sum_{\boldsymbol{u}+\boldsymbol{v}+\boldsymbol{w}=\boldsymbol{r}} (\boldsymbol{\lambda}+1)^{\boldsymbol{v}} M\left((\boldsymbol{\alpha}, \boldsymbol{u})\right) M\left((\boldsymbol{\beta}, \boldsymbol{v})\right) M\left((\boldsymbol{\gamma}, \boldsymbol{w})\right).$$

Problem 8.4.6. *For a pair of nonnegative integers* p, q, *prove that*

$$\sum_{a+b=p} \sum_{c+d=q} \binom{a+c}{a} \binom{b+d}{b} (-1)^{b+d} = \frac{1}{2} \left\{ 1 + (-1)^{p+q} \right\} \binom{p+q}{p}.$$

Hint: The inner sum is the coefficient of x^q *of the power series expansion at* $x = 0$ *of the rational function*

$$\frac{1}{(1-x)^{a+1}} \frac{(-1)^b}{(1+x)^{b+1}}.$$

Problem 8.4.7. *Let* $\boldsymbol{p} = (p_1, p_2, \ldots, p_m)$ *be an* m-*tuple of nonnegative integers and* $\boldsymbol{q} = (q_1, q_2, \ldots, q_n)$ *be an* n-*tuple of nonnegative integers. Prove that*

$$\sum_{\boldsymbol{\alpha}+\boldsymbol{\beta}=\boldsymbol{p}} \sum_{\boldsymbol{\gamma}+\boldsymbol{\delta}=\boldsymbol{q}} (-1)^{|\boldsymbol{\beta}|+|\boldsymbol{\delta}|} M\left((\boldsymbol{\alpha}, \boldsymbol{\gamma})\right) M\left((\boldsymbol{\beta}, \boldsymbol{\delta})\right)$$

$$= \frac{1}{2} M(\boldsymbol{p}) M(\boldsymbol{q}) \left\{ 1 + (-1)^{|\boldsymbol{p}|+|\boldsymbol{q}|} \right\} \binom{|\boldsymbol{p}| + |\boldsymbol{q}|}{|\boldsymbol{p}|}.$$

Hint: Count the number of multiple zeta values represented by the integral

$$\frac{1}{p!q!} \iint_{0<t_1<t_2<u_1<u_2<1} \left(\log \frac{1-t_1}{1-t_2} - \log \frac{1-u_1}{1-u_2} \right)^{|\boldsymbol{p}|} \left(\log \frac{t_2}{t_1} - \log \frac{u_2}{u_1} \right)^{|\boldsymbol{q}|}$$

$$\times \frac{dt_1 dt_2}{(1-t_1)t_2} \frac{du_1 du_2}{(1-u_1)u_2}.$$

Problem 8.4.8. *Suppose that k and r are nonnegative integers and $k+r$ is even. Prove that*

$$\frac{1}{2} \sum_{j=0}^{k} \sum_{\ell=0}^{r} (-1)^{j+\ell} \binom{j+r-\ell}{j} \binom{k-j+\ell}{\ell} \binom{k+r+4}{j+r-\ell+2}$$

$$= (k+r+3) \binom{k+r}{k}.$$

Problem 8.4.9. *For integers k, j with $0 \le j \le k$ and n-tuples of $\boldsymbol{r} = (r_1, r_2, \ldots, r_n)$, $\boldsymbol{\ell} = (\ell_1, \ell_2, \ldots, \ell_n)$ with $0 \le \ell_i \le r_i$, $i = 1, 2, \ldots, n$, prove that*

$$\binom{k+|\boldsymbol{r}|+4}{j+|\boldsymbol{\ell}|+2}$$

$$= \sum_{a+b=k+2} \left\{ \binom{a}{j+1} \binom{b+|\boldsymbol{r}|+1}{|\boldsymbol{\ell}|} + \binom{a}{k-j+1} \binom{b+|\boldsymbol{r}|+1}{|\boldsymbol{r}|-|\boldsymbol{\ell}|} \right\}$$

via the shuffle product with the integral

$$\frac{1}{k!r!} \iint_{R_1 \times R_2} \left(\mu \log \frac{1}{1-t_1} + \log \frac{1}{1-u_1} \right)^k \prod_{i=1}^{n} \left(\log \frac{1}{t_2} + \lambda_i \log \frac{1}{u_2} \right)^{r_i}$$

$$\times \frac{dt_1 dt_2}{(1-t_1)t_2} \frac{du_1 du_2}{(1-u_1)u_2}$$

where $R_1: 0 < t_1 < t_2 < 1$ and $R_2: 0 < u_1 < u_2 < 1$.

Problem 8.4.10. *For a nonnegative integer k and an n-tuple of nonnegative integers $\boldsymbol{r} = (r_1, r_2, \ldots, r_n)$, carry out the shuffle process with the* *integral*

$$\frac{1}{k!\boldsymbol{r}!} \iint_{R_1 \times R_2} \left(\mu \log \frac{1}{1-t_2} + \log \frac{1}{1-u_2}\right)^k \prod_{i=1}^{n} \left(\log \frac{1}{t_1} + \lambda_i \log \frac{1}{u_1}\right)^{r_i}$$

$$\times \frac{dt_1 dt_2}{(1-t_1)t_2} \frac{du_1 du_2}{(1-u_1)u_2}$$

and then evaluate the product of binomial coefficients

$$\left\{\binom{j+|\boldsymbol{r}|-|\boldsymbol{\ell}|+2}{j+1} - 1\right\}\left\{\binom{k-j+|\boldsymbol{\ell}|+2}{k-j+1} - 1\right\}\binom{k+|\boldsymbol{r}|+4}{j+|\boldsymbol{r}|-|\boldsymbol{\ell}|+2}$$

with $\boldsymbol{\ell} = (\ell_1, \ell_2, \ldots, \ell_n)$, $0 \le \ell_i \le r_i$, $i = 1, 2, \ldots, n$.

Problem 8.4.11. *Let q, m be integers such that $0 \le m \le q$ and $\boldsymbol{r} = (r_1, r_2, \ldots, r_n)$, $\boldsymbol{\ell} = (\ell_1, \ell_2, \ldots, \ell_n)$ be n-tuples of nonnegative integers with $0 \le \ell_j \le r_j$, $j = 1, 2, \ldots, n$. Prove the identity*

$$\binom{q+r_1+r_2+\cdots+r_n+4}{m+\ell_1+\ell_2+\cdots+\ell_n+2}$$
$$= \binom{m+|\boldsymbol{r}|+1}{|\boldsymbol{r}|-|\boldsymbol{\ell}|} + \binom{q-m+|\boldsymbol{\ell}|+1}{|\boldsymbol{\ell}|}$$
$$+ \sum_{a+b=q} \left\{\binom{a+1}{j} + \binom{a+1}{j+1}\right\}\left\{\binom{b+|\boldsymbol{r}|+1}{|\boldsymbol{r}|-|\boldsymbol{\ell}|} + \binom{b+|\boldsymbol{r}|+1}{|\boldsymbol{\ell}|}\right\}$$

through the shuffle products of multiple zeta values of the integral

$$\frac{1}{q!\boldsymbol{r}!} \iint_{R_1 \times R_2} \left(\mu \log \frac{t_2}{t_1} + \log \frac{u_2}{u_1}\right)^q \prod_{j=1}^{n} \left(\log \frac{1}{t_2} + \lambda_j \log \frac{1}{u_2}\right)^{r_j}$$

$$\times \frac{dt_1 dt_2}{(1-t_1)t_2} \frac{du_1 du_2}{(1-u_1)u_2}$$

with $R_1 : 0 < t_1 < t_2 < 1$ and $R_2 : 0 < u_1 < u_2 < 1$.

Problem 8.4.12. *Let $\boldsymbol{k} = (k_1, k_2, \ldots, k_m)$ be an m-tuple of nonnegative integers and $\boldsymbol{q} = (q_1, q_2, \ldots, q_n)$ be an n-tuple of nonnegative integers. Evaluate the generating function*

$$\sum_{\alpha+\beta=k} \sum_{\gamma+\delta=q} \mu^{\alpha} \lambda^{\delta} M(\alpha) M(\beta) M(\gamma) M(\delta) \binom{|\boldsymbol{k}|+|\boldsymbol{q}|+4}{|\alpha|+|\delta|+2}$$

via the shuffle product of multiple zeta values of the integral

$$\frac{1}{k!q!} \iint_{R_1 \times R_2} \prod_{i=1}^{m} \left(\mu_i \log \frac{1}{1-t_1} + \log \frac{1}{1-u_1} \right)^{k_i} \prod_{j=1}^{n} \left(\log \frac{1}{t_2} + \lambda_j \log \frac{1}{u_2} \right)^{q_j}$$

$$\times \frac{dt_1 dt_2}{(1-t_1)t_2} \frac{du_1 du_2}{(1-u_1)u_2}$$

with $\boldsymbol{\mu} = (\mu_1, \mu_2, \ldots, \mu_m)$ *and* $\boldsymbol{\lambda} = (\lambda_1, \lambda_2, \ldots, \lambda_n)$.

Problem 8.4.13. *Suppose that* k *and* r *are nonnegative integers with* k *even. Prove that*

$$\frac{1}{2} \sum_{j=0}^{k} \sum_{\ell=0}^{r} (-1)^j \binom{j+r-\ell}{j} \binom{k-j+\ell}{\ell} \binom{k+r+4}{j+r-\ell+2}$$

$$= \binom{k+r+1}{r} + 2 \sum_{\substack{a+b=k \\ b:even}} \sum_{m+n+p=r} 2^n \binom{a+m}{a} \binom{b+p}{b}.$$

Hint: Carry out of the shuffle product process for the integral

$$\frac{1}{k!r!} \iint_{R_1 \times R_2} \left(\log \frac{1-t_1}{1-t_2} - \log \frac{1-u_1}{1-u_2} \right)^k \left(\log \frac{t_2}{t_1} + \log \frac{u_2}{u_1} \right)^r$$

$$\times \frac{dt_1 dt_2}{(1-t_1)t_2} \frac{du_1 du_2}{(1-u_1)u_2}$$

and then count the number of multiple zeta values represented from the shuffle product.

Problem 8.4.14. *Suppose that* k *and* r *are nonnegative integers, prove that*

$$\frac{1}{2} \sum_{j=0}^{k} \sum_{\ell=0}^{r} \binom{j+\ell}{j} \binom{k-j+r-\ell}{k-j} \binom{k+r+4}{j+\ell+2}$$

$$= (k+r+1) \binom{k+r}{k} + \sum_{a+b+c=k} 2^b \sum_{m+n+p=r} 2^n \binom{a+m}{a} \binom{b+n}{b} \binom{c+p}{c}.$$

Problem 8.4.15. *Suppose that* k *and* r *are nonnegative integers. Show that*

$$
\sum_{a+b+c=k} 2^b \sum_{m+n+p=r} 2^n \binom{a+m}{a}\binom{b+n}{b}\binom{c+p}{c}
$$
$$
= \sum_{j=0}^{k}\sum_{\ell=0}^{r} \binom{k+r+2}{r-\ell}\binom{k+\ell+2}{k-j}\binom{j+\ell}{j}.
$$

Hint: Begin with the polynomial

$$
\sum_{a+b+c=k} (\mu+1)^b \sum_{m+n+p=r} \lambda^{m+p}(\lambda+1)^n \binom{a+m}{a}\binom{b+n}{b}\binom{c+p}{c}.
$$

Problem 8.4.16. *For a pair of nonnegative integers* q,r *and complex numbers* μ, λ*, prove that*

$$
\sum_{m+n=q} (\mu+1)^m \sum_{h+\ell=r} \binom{n+h}{n}(\lambda+1)^h
$$
$$
= \frac{(q+r+2)!}{q!r!} \int_0^1 \int_0^y [(\mu+1)(1-y)+x]^q [y+\lambda x]^r dx dy.
$$

Problem 8.4.17. *For nonnegative integers* k,p,q,ℓ*, prove that*

$$
\sum_{a+b=p} \binom{a+k}{k}\binom{b+q+\ell+1}{q} = \sum_{c+d=q} \binom{c+p+k+1}{p}\binom{d+\ell}{\ell}.
$$

Problem 8.4.18. *Suppose that* a,b *are nonnegative integers and* $\boldsymbol{k} = (k_1, k_2, \ldots, k_n)$ *and* $\boldsymbol{j} = (j_1, j_2, \ldots, j_n)$ *are* n*-tuples of nonnegative integers with* $0 \le j_\ell \le k_\ell$*,* $\ell = 1, 2, \ldots, n$*. Prove that the coefficient of* $\boldsymbol{\mu}^{\boldsymbol{j}} (= \mu_1^{j_1}\mu_2^{j_2}\cdots\mu_n^{j_n})$ *of the polynomial of* n *variables*

$$
\sum_{\boldsymbol{\alpha}+\boldsymbol{\beta}+\boldsymbol{\gamma}=\boldsymbol{k}} M(\boldsymbol{\alpha})M((a,\boldsymbol{\beta}))\, M((b,\boldsymbol{\gamma}))\, (\boldsymbol{\mu}+1)^{\boldsymbol{\beta}}
$$

is given by

$$
M(\boldsymbol{j})M(\boldsymbol{k}-\boldsymbol{j})\binom{a+|\boldsymbol{j}|}{|\boldsymbol{j}|}\binom{a+b+|\boldsymbol{k}|+2}{|\boldsymbol{k}-\boldsymbol{j}|}.
$$

Also the coefficient of $\boldsymbol{\mu}^{\boldsymbol{j}}$ of the polynomial

$$\sum_{\boldsymbol{\alpha}+\boldsymbol{\beta}+\boldsymbol{\gamma}=\boldsymbol{k}} M(\boldsymbol{\alpha})M\left((a,\boldsymbol{\beta})\right) M\left((b,\boldsymbol{\gamma})\right) \boldsymbol{\mu}^{\boldsymbol{\alpha}+\boldsymbol{\gamma}}(\boldsymbol{\mu}+1)^{\boldsymbol{\beta}}$$

is given by

$$M(\boldsymbol{j})M(\boldsymbol{k}-\boldsymbol{j})\binom{a+|\boldsymbol{k}-\boldsymbol{j}|}{|\boldsymbol{k}-\boldsymbol{j}|}\binom{a+b+|\boldsymbol{k}|+2}{|\boldsymbol{j}|}.$$

Problem 8.4.19. *Let a, b be nonnegative integers and $\boldsymbol{k} = (k_1, k_2, \ldots, k_n)$, $\boldsymbol{j} = (j_1, j_2, \ldots, j_n)$ be n-tuples of nonnegative integers with $0 \le j_\ell \le k_\ell$, $\ell = 1, 2, \ldots, n$. Show that the coefficient of $\boldsymbol{\mu}^{\boldsymbol{j}}$ of the polynomial of n variables*

$$\sum_{\boldsymbol{\alpha}+\boldsymbol{\beta}+\boldsymbol{\gamma}=\boldsymbol{k}} M(\boldsymbol{\alpha})M\left((a,\boldsymbol{\beta})\right) M\left((b,\boldsymbol{\gamma})\right) (\boldsymbol{\mu}+1)^{\boldsymbol{\beta}}\boldsymbol{\mu}^{\boldsymbol{\gamma}}$$

is given by

$$M(\boldsymbol{j})M(\boldsymbol{k}-\boldsymbol{j}) \sum_{c+d=|\boldsymbol{j}|} \binom{c+b}{b}\binom{d+a}{a}\binom{d+a+|\boldsymbol{k}-\boldsymbol{j}|+1}{|\boldsymbol{k}-\boldsymbol{j}|}.$$

Hint: Let $\boldsymbol{\gamma} = (\gamma_1, \gamma_2, \ldots, \gamma_n)$. The coefficient of $\boldsymbol{\mu}^{\boldsymbol{j}}$ of the polynomial $(\boldsymbol{\mu}+1)^{\boldsymbol{\beta}}\boldsymbol{\mu}^{\boldsymbol{\gamma}}$ is zero unless $0 \le \gamma_1 \le j_1$, $0 \le \gamma_2 \le j_2$, $\ldots$, $0 \le \gamma_n \le j_n$. Restrict those $\boldsymbol{\gamma}$ with the above property and then find that the coefficient of $\boldsymbol{\mu}^{\boldsymbol{j}-\boldsymbol{\gamma}}$ of the polynomial

$$\sum_{\boldsymbol{\alpha}+\boldsymbol{\beta}=\boldsymbol{k}-\boldsymbol{\gamma}} M(\boldsymbol{\alpha})M\left((a,\boldsymbol{\beta})\right) (\boldsymbol{\mu}+1)^{\boldsymbol{\beta}}$$

is given by

$$M(\boldsymbol{k}-\boldsymbol{j})M(\boldsymbol{j}-\boldsymbol{\gamma})\binom{a+|\boldsymbol{j}-\boldsymbol{\gamma}|}{|\boldsymbol{j}-\boldsymbol{\gamma}|}\binom{a+|\boldsymbol{k}-\boldsymbol{\gamma}|+1}{|\boldsymbol{k}-\boldsymbol{j}|}.$$

For n-tuples of nonnegative integers $\boldsymbol{k} = (k_1, k_2, \ldots, k_n)$, $\boldsymbol{j} = (j_1, j_2, \ldots, j_n)$ with $0 \le j_i \le k_i$, $i = 1, 2, \ldots, n$, and integers ℓ, r with $0 \le \ell \le r$, we are going to evaluate

$$\binom{|\boldsymbol{j}|+\ell}{\ell}\binom{|\boldsymbol{k}|+r+4}{|\boldsymbol{j}|+\ell+2}$$

through the shuffle product process of multiple zeta values appearing in the integral

$$(*) \quad \frac{1}{k!r!} \iint_{R_1 \times R_2} \prod_{i=1}^{n} \left(\mu_i \log \frac{1-t_1}{1-t_2} + \log \frac{1}{1-u_1} \right)^{k_i} \left(\lambda \log \frac{t_2}{t_1} + \log \frac{1}{u_2} \right)^r$$

$$\times \frac{dt_1 dt_2}{(1-t_1)t_2} \frac{du_1 du_2}{(1-u_1)u_2}$$

with $R_1 : \; 0 < t_1 < t_2 < 1$ and $R_2 : \; 0 < u_1 < u_2 < 1$. The value of the integral is

$$\sum_{\alpha+\beta=k} \sum_{\ell=0}^{r} \mu^\alpha \lambda^\ell M(\alpha) M(\beta) \zeta \left(|\alpha| + \ell + 2 \right) \zeta(\{1\}^{|\beta|}, r - \ell + 2)$$

so that the generating function for numbers of multiple zeta values produced from the shuffle product process is

$$\sum_{\alpha+\beta=k} \sum_{\ell=0}^{r} \mu^\alpha \lambda^\ell M(\alpha) M(\beta) \binom{|\alpha| + \ell}{\ell} \binom{|k| + r + 4}{|\alpha| + \ell + 2}.$$

The following problems are devoted to separate the coefficients of $\mu^j \lambda^\ell$ in 6 partial generating functions.

Problem 8.4.20. *Show that the partial generating function from integration* $(*)$ *over* $D_1 : \; 0 < t_1 < t_2 < u_1 < u_2 < 1$ *is*

$$\sum_{\alpha+\beta+\gamma=k} (\mu+1)^\beta \sum_{m+n=r} \lambda^m M(\alpha) M((m,\beta)) M(\gamma),$$

and its coefficient of $\mu^j \lambda^\ell$ *is given by*

$$M(j) M(k-j) \binom{|j| + \ell}{\ell} \binom{|k| + \ell + 2}{|k-j|}.$$

Problem 8.4.21. *Show that the partial generating function from integration* $(*)$ *over* $D_2 : \; 0 < u_1 < u_2 < t_1 < t_2 < 1$ *is*

$$\sum_{\alpha+\beta=k} \mu^\beta \sum_{m+n+p=r} (\lambda+1)^n M(\alpha) M((n,\beta)),$$

and its coefficient of $\mu^j \lambda^\ell$ is given by

$$M(j)M(k-j)\binom{|j|+\ell}{\ell}\binom{|j|+r+2}{r-\ell}.$$

Problem 8.4.22. *Show that the partial generating function from integration* (*) *over* $D_3 :$ $0 < t_1 < u_1 < t_2 < u_2 < 1$ *is*

$$\sum_{\alpha+\beta+\gamma=k} (\mu+1)^\beta \mu^\gamma \sum_{m+n+p=r} \lambda^{m+n} M(\alpha) M\left((m,\beta)\right) M\left((n,\gamma)\right),$$

and its coefficient of $\mu^j \lambda^\ell$ is

$$M(j)M(k-j)\sum_{g=0}^{|k-j|}\binom{g+\ell+1}{g}\binom{g+\ell+|j|+1}{|j|}.$$

Problem 8.4.23. *The partial generating function from integration* (*) *over* $D_4 :$ $0 < t_1 < u_1 < u_2 < t_2 < 1$ *is*

$$\sum_{\alpha+\beta+\gamma+\delta=k} (\mu+1)^\beta \mu^{\gamma+\delta} \sum_{m+n+p+q=r} \lambda^{m+n}(\lambda+1)^p$$

$$\times M(\alpha) M\left((m,\beta)\right) M\left((n,\gamma)\right) M(\delta),$$

or after careful manipulation

$$\sum_{\alpha+\beta+\gamma=k} (\mu+1)^\beta \mu^\gamma \sum_{m+n+q=r} \lambda^m \left[(\lambda+1)^{n+1} - \lambda^{n+1}\right]$$

$$\times M(\alpha) M\left((m,\beta)\right) M(n+1,\gamma),$$

and its coefficient of $\mu^j \lambda^\ell$ is

$$\sum_{m=0}^{r}\sum_{g=0}^{|k-j|}\binom{m+g+2}{\ell}\binom{m+g+|j|+2}{|j|}.$$

Problem 8.4.24. *The partial generating function from integration* (*) *over* $D_5 : 0 < u_1 < t_1 < u_2 < t_2 < 1$ *is*

$$\sum_{\boldsymbol{\alpha}+\boldsymbol{\beta}+\boldsymbol{\gamma}=\boldsymbol{k}} \boldsymbol{\mu}^{\boldsymbol{\beta}+\boldsymbol{\gamma}} \sum_{m+n+p=r} \lambda^m (\lambda+1)^n M(\boldsymbol{\alpha}) M(m,\boldsymbol{\beta}) M(n,\boldsymbol{\gamma}),$$

and its coefficient of $\boldsymbol{\mu}^{\boldsymbol{j}} \lambda^\ell$ *is given by*

$$M(\boldsymbol{j}) M(\boldsymbol{k}-\boldsymbol{j}) \binom{|\boldsymbol{j}|+\ell}{\ell} \left\{ \binom{|\boldsymbol{j}|+r+2}{r-\ell+1} - 1 \right\}.$$

Problem 8.4.25. *The partial generating function from integration* (*) *over* $D_6 : 0 < u_1 < t_1 < t_2 < u_2 < 1$ *is*

$$\sum_{\boldsymbol{\alpha}+\boldsymbol{\beta}=\boldsymbol{k}} \boldsymbol{\mu}^{\boldsymbol{\beta}} \sum_{m+n=r} \lambda^m M(\boldsymbol{\alpha}) M\left((m,\boldsymbol{\beta})\right),$$

and its coefficient of $\boldsymbol{\mu}^{\boldsymbol{j}} \lambda^\ell$ *is given by*

$$M(\boldsymbol{j}) M(\boldsymbol{k}-\boldsymbol{j}) \binom{|\boldsymbol{j}|+\ell}{\ell}.$$

Problem 8.4.26. *Let* $\boldsymbol{k} = (k_1, k_2, \ldots, k_n)$ *be an* n-*tuple of nonnegative integers and* r *be a nonnegative integer. Carry out the shuffle product process with the integral*

$$\frac{1}{k!r!} \iint_{R_1 \times R_2} \prod_{i=1}^{n} \left(\mu_i \log \frac{1-t_1}{1-t_2} + \log \frac{1}{1-u_1} \right)^{k_i} \left(\lambda \log \frac{t_2}{t_1} + \log \frac{u_2}{u_1} \right)^r$$

$$\times \frac{dt_1 dt_2}{(1-t_1)t_2} \frac{du_1 du_2}{(1-u_1)u_2}$$

and then the evaluation of

$$\binom{j_1 + j_2 + \cdots + j_n + \ell}{j_1, j_2, \ldots, j_n, \ell} \binom{|\boldsymbol{k}|+r+4}{|\boldsymbol{j}|+\ell+2}$$

with $0 \leq j_i \leq k_i$, $i = 1, 2, \ldots, n$, *and* $0 \leq \ell \leq r$.

Suppose that k is a nonnegative integer and $\boldsymbol{r} = (r_1, r_2, \ldots, r_n)$ is an n-tuple of nonnegative integers. The next three problems are related to the shuffle product process of the integral

$$(\star) \quad \frac{1}{k! \boldsymbol{r}!} \iint_{R_1 \times R_2} \left(\mu \log \frac{1}{1-t_2} + \log \frac{1}{1-u_2} \right)^k \prod_{i=1}^n \left(\lambda_i \log \frac{1}{t_1} + \log \frac{1}{u_1} \right)^{r_i}$$

$$\times \frac{dt_1 dt_2}{(1-t_1)t_2} \frac{du_1 du_2}{(1-u_1)u_2},$$

which has the value

$$\sum_{j=0}^k \sum_{\alpha+\beta=r} \mu^j \lambda^\alpha M(\alpha) M(\beta) \left\{ \binom{j+|\alpha|+2}{j+1} - 1 \right\} \left\{ \binom{k-j+|\beta|+2}{k-j+1} - 1 \right\}$$

$$\times \zeta(\{1\}^j, |\alpha|+2) \zeta(\{1\}^{k-j}, |\beta|+2).$$

Consequently, the generating function for number of multiple zeta values produced from the shuffle product process is

$$\sum_{j=0}^k \sum_{\alpha+\beta=r} \mu^j \lambda^\alpha M(\alpha) M(\beta) \left\{ \binom{j+|\alpha|+2}{j+1} - 1 \right\}$$

$$\times \left\{ \binom{k-j+|\beta|+2}{k-j+1} - 1 \right\} \binom{k+|\boldsymbol{r}|+4}{j+|\alpha|+2}.$$

Problem 8.4.27. *Prove that the partial generating function from integration* $(\star)$ *over* $D_1 : 0 < t_1 < t_2 < u_1 < u_2 < 1$ *is*

$$\sum_{a+b+c+d=k} (\mu+1)^{a+b} \sum_{\alpha+\beta+\gamma+\delta=r} \lambda^{\alpha+\beta} (\lambda+1)^{\gamma+\delta}$$

$$\times M((b, \alpha)) M((c, \beta)) M((d, \gamma)) M(\delta)$$

or after manipulation

$$\sum_{a+b+c+d=k} (\mu+1)^{a+b} \sum_{\alpha+\beta=r} \lambda^\alpha (\lambda+1)^\beta M((b+c+1, \alpha)) M((d+1, \beta))$$

and its coefficient of $\mu^j \boldsymbol{\lambda}^{\boldsymbol{\ell}}$ $(= \mu^j \lambda_1^{\ell_1} \lambda_2^{\ell_2} \cdots \lambda_n^{\ell_n})$ is given by

$$M(\boldsymbol{\ell})M(\boldsymbol{r} - \boldsymbol{\ell}) \sum_{a+b=k} \binom{a}{j} \left\{ \binom{b + |\boldsymbol{r} - \boldsymbol{\ell}| + 2}{|\boldsymbol{r} - \boldsymbol{\ell}| + 1} - 1 \right\}$$
$$\times \left\{ \binom{k + |\boldsymbol{r}| + 4}{|\boldsymbol{\ell}| + 1} - \binom{b + |\boldsymbol{r}| + 3}{|\boldsymbol{\ell}| + 1} \right\}.$$

Problem 8.4.28. *Prove that the partial generating function from integration $(\star)$ over D_3 : $0 < t_1 < u_1 < t_2 < u_2 < 1$ is*

$$\sum_{a+b+c+d=k} (\mu + 1)^{a+b+c} \sum_{\alpha+\beta+\gamma+\delta=r} \boldsymbol{\lambda}^{\boldsymbol{\alpha}}(\boldsymbol{\lambda} + 1)^{\beta+\gamma+\delta}$$
$$\times M\left((b, \boldsymbol{\alpha})\right) M\left((c, \boldsymbol{\beta})\right) M\left((d, \boldsymbol{\gamma})\right) M(\boldsymbol{\delta})$$

or

$$\sum_{a+b+c+d=k} (\mu + 1)^{a+b+c} \sum_{\alpha+\beta=r} \boldsymbol{\lambda}^{\boldsymbol{\alpha}}(\boldsymbol{\lambda} + 1)^{\beta} M\left((b, \boldsymbol{\alpha})\right) M\left((c + d + 2, \boldsymbol{\beta})\right)$$

and its coefficient of $\mu^j \boldsymbol{\lambda}^{\boldsymbol{\ell}}$ is given by

$$M(\boldsymbol{\ell})M(\boldsymbol{r} - \boldsymbol{\ell}) \sum_{a+b=k} \left\{ \binom{k+1}{j+1} - \binom{a}{j+1} \right\} \binom{b + |\boldsymbol{r} - \boldsymbol{\ell}| + 2}{|\boldsymbol{r} - \boldsymbol{\ell}|}$$
$$\times \left\{ \binom{k + |\boldsymbol{r}| + 4}{|\boldsymbol{\ell}| + 1} - \binom{b + |\boldsymbol{r}| + 3}{|\boldsymbol{\ell}| + 1} \right\}.$$

Problem 8.4.29. *Let*

$$H_{a,b}(j, \boldsymbol{\ell}) = \binom{a}{j} \left\{ \binom{b + |\boldsymbol{r} - \boldsymbol{\ell}| + 2}{|\boldsymbol{r} - \boldsymbol{\ell}| + 1} - 1 \right\}$$
$$\times \left\{ \binom{k + |\boldsymbol{r}| + 4}{|\boldsymbol{\ell}| + 1} - \binom{b + |\boldsymbol{r}| + 3}{|\boldsymbol{\ell}| + 1} \right\}$$

and

$$K_{a,b}(j, \boldsymbol{\ell}) = \left\{ \binom{k+1}{j+1} - \binom{a}{j+1} \right\} \binom{b + |\boldsymbol{r} - \boldsymbol{\ell}| + 2}{|\boldsymbol{r} - \boldsymbol{\ell}|}$$
$$\times \left\{ \binom{k + |\boldsymbol{r}| + 4}{|\boldsymbol{\ell}| + 1} - \binom{b + |\boldsymbol{r}| + 3}{|\boldsymbol{\ell}| + 1} \right\},$$

prove that

$$\left\{\binom{j+|\boldsymbol{\ell}|+2}{j+1}-1\right\}\left\{\binom{k-j+|\boldsymbol{r}-\boldsymbol{\ell}|+2}{k-j+1}-1\right\}\binom{k+|\boldsymbol{r}|+4}{j+|\boldsymbol{\ell}|+2}$$

$$=\sum_{a+b=k}\left\{H_{a,b}(j,\boldsymbol{\ell})+H_{a,b}(k-j,\boldsymbol{r}-\boldsymbol{\ell})\right.$$

$$\left.+K_{a,b}(j,\boldsymbol{\ell})+K_{a,b}(j,\boldsymbol{r}-\boldsymbol{\ell})+K_{a,b}(k-j,\boldsymbol{\ell})+K_{a,b}(k-j,\boldsymbol{r}-\boldsymbol{\ell})\right\}.$$

Problem 8.4.30. *Let $\boldsymbol{k}=(k_1,k_2,\ldots,k_n)$ be an n-tuple of nonnegative integers and r be a nonnegative integer. Carry out the shuffle product process with the integral*

$$\frac{1}{\boldsymbol{k}!r!}\iint_{R_1\times R_2}\prod_{i=1}^{n}\left(\mu_i\log\frac{1}{1-t_2}+\log\frac{1}{1-u_2}\right)^{k_i}\left(\lambda\log\frac{1}{t_1}+\log\frac{1}{u_1}\right)^{r}$$

$$\times\frac{dt_1dt_2}{(1-t_1)t_2}\frac{du_1du_2}{(1-u_1)u_2},$$

which has the value

$$\sum_{\boldsymbol{\alpha}+\boldsymbol{\beta}=\boldsymbol{k}}\sum_{\ell=0}^{r}\boldsymbol{\mu}^{\boldsymbol{\alpha}}\lambda^{\ell}M(\boldsymbol{\alpha})M(\boldsymbol{\beta})\left\{\binom{|\boldsymbol{\alpha}|+\ell+2}{\ell+1}-1\right\}$$

$$\times\left\{\binom{|\boldsymbol{\beta}|+r-\ell+2}{r-\ell+1}-1\right\}\zeta\left(\{1\}^{|\boldsymbol{\alpha}|},\ell+2\right)\zeta\left(\{1\}^{|\boldsymbol{\beta}|},r-\ell+2\right).$$

Assume that $\boldsymbol{k}=(k_1,k_2,\ldots,k_n)$ is an n-tuple of nonnegative integers and r is a nonnegative integer. In the following considerations, we carry out a shuffle product process with the integral

$$(\dagger)\quad\frac{1}{\boldsymbol{k}!r!}\iint_{R_1\times R_2}\prod_{i=1}^{n}\left(\mu_i\log\frac{1-t_1}{1-t_2}+\log\frac{1-u_1}{1-u_2}\right)^{k_i}\left(\log\frac{t_2}{t_1}+\lambda\log\frac{u_2}{u_1}\right)^{r}$$

$$\times\frac{dt_1dt_2}{(1-t_1)t_2}\frac{du_1du_2}{(1-u_1)u_2}.$$

Note that the value of the above integral is

$$\sum_{\boldsymbol{\alpha}+\boldsymbol{\beta}=\boldsymbol{k}}\sum_{\ell=0}^{r}\boldsymbol{\mu}^{\boldsymbol{\alpha}}\lambda^{\ell}M(\boldsymbol{\alpha})M(\boldsymbol{\beta})\zeta\left(|\boldsymbol{\alpha}|+r-\ell+2\right)\zeta\left(|\boldsymbol{\beta}|+\ell+2\right),$$

so that the generating function for the number of multiple zeta values produced from the shuffle product process is

$$\sum_{\alpha+\beta=k}\sum_{\ell=0}^{r}\mu^{\alpha}\lambda^{\ell}M(\alpha)M(\beta)\binom{|\alpha|+r-\ell}{r-\ell}\binom{|\beta|+\ell}{\ell}\binom{|k|+r+4}{|\alpha|+r-\ell+2}.$$

Problem 8.4.31. *Prove that the partial generating function from integration* (†) *over* $D_1: \ 0 < t_1 < t_2 < u_1 < u_2 < 1$ *is*

$$\sum_{\alpha+\beta=k}\mu^{\alpha}\sum_{m+n=r}\lambda^{n}M\left((m,\alpha)\right)M\left((n,\beta)\right)$$

and its coefficient of $\mu^{j}\lambda^{\ell}$ *is*

$$M\left((r-\ell,j)\right)M\left((\ell,k-j)\right)$$

or

$$M(j)M(k-j)\binom{|j|+r-\ell}{r-\ell}\binom{|k-j|+\ell}{\ell}.$$

Problem 8.4.32. *Prove that the partial generating function from integration* (†) *over* $D_2: \ 0 < u_1 < u_2 < t_1 < t_2 < 1$ *is*

$$\sum_{\alpha+\beta=k}\mu^{\beta}\sum_{m+n=r}\lambda^{m}M\left((m,\alpha)\right)M\left((n,\beta)\right)$$

and its coefficient of $\mu^{j}\lambda^{\ell}$ *is*

$$M\left((r-\ell,j)\right)M\left((\ell,k-j)\right),$$

or

$$M(j)M(k-j)\binom{|j|+r-\ell}{r-\ell}\binom{|k-j|+\ell}{\ell}.$$

Problem 8.4.33. *Prove that the partial generating function from integration* (†) *over* $D_3 :$ $0 < t_1 < u_1 < t_2 < u_2 < 1$ *is*

$$\sum_{\alpha+\beta+\gamma=k} \mu^{\alpha}(\mu+1)^{\beta} \sum_{m+n+p=r} (\lambda+1)^n \lambda^p M\left((m,\alpha)\right) M\left((n,\beta)\right) M\left((p,\gamma)\right),$$

and its coefficient of $\mu^j \lambda^\ell$ *is given by*

$$M(j)M(k-j) \sum_{c+d=|j|} \sum_{m+n=r} \binom{c+m}{m} \binom{d+n+1}{\ell}$$
$$\times \binom{d+n-\ell}{d} \binom{d+n+|k-j|+1}{|k-j|}.$$

Problem 8.4.34. *Prove that the partial generating function from integration* (†) *over* $D_5 :$ $0 < u_1 < t_1 < u_2 < t_2 < 1$ *is*

$$\sum_{\alpha+\beta+\gamma=k} (\mu+1)^{\beta} \mu^{\gamma} \sum_{m+n+p=r} \lambda^m (\lambda+1)^n M\left((m,\alpha)\right) M\left((n,\beta)\right) M\left((p,\gamma)\right),$$

which is the same generating function as the previous problem if we exchange α, m *with* γ, p, *so that they have the same coefficient of* $\mu^j \lambda^\ell$.

Problem 8.4.35. *Prove that the partial generating function from integration* (†) *over* $D_4 :$ $0 < t_1 < u_1 < u_2 < t_2 < 1$ *is*

$$\sum_{\alpha+\beta+\gamma=k} \mu^{\alpha+\gamma}(\mu+1)^{\beta} \sum_{m+n+p=r} (\lambda+1)^n M\left((m,\alpha)\right) M\left((n,\beta)\right) M\left((p,\gamma)\right),$$

and its coefficient of $\mu^j \lambda^\ell$ *is given by*

$$M(j)M(k-j)\binom{|k|+r+2}{|j|} \binom{|k-j|+\ell}{\ell} \binom{|k-j|+r+2}{r-\ell}.$$

Problem 8.4.36. *Prove that the partial generating function from integration* (†) *over* $D_6 :$ $0 < u_1 < t_1 < t_2 < u_2 < 1$ *is*

$$\sum_{\alpha+\beta+\gamma=k} (\mu+1)^{\beta} \sum_{m+n+p=r} \lambda^{m+p}(\lambda+1)^n M\left((m,\alpha)\right) M\left((n,\beta)\right) M\left((p,\gamma)\right),$$

and its coefficient of $\mu^j \lambda^\ell$ is given by

$$M(\boldsymbol{j})M(\boldsymbol{k}-\boldsymbol{j})\binom{|\boldsymbol{k}|+r+2}{|\boldsymbol{k}-\boldsymbol{j}|}\binom{|\boldsymbol{j}|+r-\ell}{r-\ell}\binom{|\boldsymbol{j}|+r+2}{\ell}.$$

Problem 8.4.37. *Carry out the shuffle product process with the integral*

$$\frac{1}{k!r!}\iint_{R_1\times R_2}\left(\mu\log\frac{1-t_1}{1-t_2}+\log\frac{1-u_1}{1-u_2}\right)^k\prod_{i=1}^n\left(\log\frac{t_2}{t_1}+\lambda_i\log\frac{u_2}{u_1}\right)^{r_i}$$

$$\times\frac{dt_1dt_2}{(1-t_1)t_2}\frac{du_1du_2}{(1-u_1)u_2}$$

with $\boldsymbol{r}=(r_1,r_2,\ldots,r_n)$, $R_1:\ 0<t_1<t_2<1$ and $R_2:\ 0<u_1<u_2<1$. In particular, we obtain an evaluation of the product of binomial coefficient

$$\binom{j+|\boldsymbol{\ell}|}{j}\binom{k-j+|\boldsymbol{r}-\boldsymbol{\ell}|}{k-j}\binom{k+|\boldsymbol{r}|+4}{j+|\boldsymbol{\ell}|+2}$$

when $0\le j\le k$ and $0\le \ell_i\le r_i$, $i=1,2,\ldots,n$.

Problem 8.4.38. *Let $\boldsymbol{p}=(p_1,p_2,\ldots,p_m)$ be m-tuples of nonnegative integers and $\boldsymbol{q}=(q_1,q_2,\ldots,q_n)$ be n-tuples of nonnegative integers. Show that*

$$\sum_{\boldsymbol{\alpha}+\boldsymbol{\beta}=\boldsymbol{p}}\sum_{\boldsymbol{u}+\boldsymbol{v}=\boldsymbol{q}}M\left((\boldsymbol{\alpha},\boldsymbol{u})\right)M\left((\boldsymbol{\beta},\boldsymbol{v})\right)=M(\boldsymbol{p})M(\boldsymbol{q})\binom{|\boldsymbol{p}|+|\boldsymbol{q}|+1}{|\boldsymbol{p}|,|\boldsymbol{q}|,1}.$$

Problem 8.4.39. *Let $\boldsymbol{\alpha},\boldsymbol{\beta}$ be an m-tuple of nonnegative integers and $\boldsymbol{q}$ be an n-tuple of nonnegative integers. Show that*

$$\sum_{\boldsymbol{u}+\boldsymbol{v}=\boldsymbol{q}}M\left((\boldsymbol{\alpha},\boldsymbol{u})\right)M\left((\boldsymbol{\beta},\boldsymbol{v})\right)=M(\boldsymbol{\alpha})M(\boldsymbol{\beta})M\left((|\boldsymbol{\alpha}|+|\boldsymbol{\beta}|+1,\boldsymbol{q})\right).$$

Problem 8.4.40. *Suppose that $\boldsymbol{k} = (k_1, k_2, \ldots, k_m)$ and $\boldsymbol{j} = (j_1, j_2, \ldots, j_m)$, are m-tuples of nonnegative integers with $0 \le j_i \le k_i$, $i = 1, 2, \ldots, m$ and $\boldsymbol{r} = (r_1, r_2, \ldots, r_n)$, $\boldsymbol{\ell} = (\ell_1, \ell_2, \ldots, \ell_n)$ are n-tuples of nonnegative integers with $0 \le \ell_p \le r_p$, $p = 1, 2, \ldots, n$. Prove that the coefficient of $\boldsymbol{\mu^j \lambda^\ell}$ of the polynomial of $m + n$ variables*

$$\sum_{\boldsymbol{\alpha}+\boldsymbol{\beta}+\boldsymbol{\gamma}=\boldsymbol{k}} (\boldsymbol{\mu}+1)^{\boldsymbol{\beta}} \sum_{\boldsymbol{u}+\boldsymbol{v}+\boldsymbol{w}=\boldsymbol{r}} (\boldsymbol{\lambda}+1)^{\boldsymbol{v}} M\left((\boldsymbol{\alpha}, \boldsymbol{u})\right) M\left((\boldsymbol{\beta}, \boldsymbol{v})\right) M\left((\boldsymbol{\gamma}, \boldsymbol{w})\right)$$

is given by

$$M(\boldsymbol{j}) M(\boldsymbol{k}-\boldsymbol{j}) M(\boldsymbol{\ell}) M(\boldsymbol{r}-\boldsymbol{\ell}) \binom{|\boldsymbol{j}|+|\boldsymbol{\ell}|}{|\boldsymbol{j}|} \binom{|\boldsymbol{k}|+|\boldsymbol{r}|+2}{|\boldsymbol{r}-\boldsymbol{\ell}|} \binom{|\boldsymbol{k}|+|\boldsymbol{\ell}|+2}{|\boldsymbol{k}-\boldsymbol{j}|}.$$

Problem 8.4.41. *Notations as introduced in the previous problem. Prove that*

$$\sum_{\boldsymbol{\alpha}+\boldsymbol{\beta}+\boldsymbol{\gamma}=\boldsymbol{k}} \sum_{\boldsymbol{u}+\boldsymbol{v}+\boldsymbol{w}=\boldsymbol{r}} M\left((\boldsymbol{\alpha}, \boldsymbol{u})\right) M\left((\boldsymbol{\beta}, \boldsymbol{v})\right) M\left((\boldsymbol{\gamma}, \boldsymbol{w})\right)$$

$$= M(\boldsymbol{k}) M(\boldsymbol{r}) \binom{|\boldsymbol{k}|+|\boldsymbol{r}|+2}{|\boldsymbol{k}|, |\boldsymbol{r}|, 2}.$$

Problem 8.4.42. *Suppose that a, b and r are nonnegative integers, prove that the coefficient of λ^ℓ, $0 \le \ell \le r$, of the polynomial in λ*

$$\sum_{m+n+p=r} \lambda^m \left\{ (\lambda+1)^{n+1} - \lambda^{n+1} \right\} \binom{m+a}{a} \binom{n+1+b}{b}$$

is given by

$$\sum_{m=0}^{\ell} \binom{m+a}{a} \binom{b+\ell-m}{b} \left\{ \binom{b+r-m+2}{r-\ell+1} - 1 \right\}.$$

Hint: The coefficient of λ^ℓ is zero unless $0 \le m \le \ell$. Fix m, then the coefficient of $\lambda^{\ell-m}$ of

$$\sum_{n'=0}^{r-m+1} \binom{b+n'}{b} \left\{ (\lambda+1)^{n'} - \lambda^{n'} \right\}$$

is

$$\binom{b+\ell-m}{b}\left\{\binom{b+r-m+2}{r-\ell+1}-1\right\}.$$

Problem 8.4.43. *Let k, j be m-tuples of nonnegative integers with $0 \leq j_i \leq k_i$, $i = 1, 2, \ldots, m$ and r, ℓ be n-tuples of nonnegative integers with $0 \leq \ell_p \leq r_p$, $p = 1, 2, \ldots, n$. Also $\boldsymbol{\mu} = (\mu_1, \mu_2, \ldots, \mu_m)$ and $\boldsymbol{\lambda} = (\lambda_1, \lambda_2, \ldots, \lambda_n)$. Prove that the coefficient of $\boldsymbol{\mu}^j \boldsymbol{\lambda}^\ell$ of the polynomial of $m + n$ variables*

$$\sum_{\boldsymbol{\alpha}+\boldsymbol{\beta}+\boldsymbol{\gamma}=\boldsymbol{k}} (\boldsymbol{\mu}+1)^{\boldsymbol{\alpha}} M(\boldsymbol{\alpha}) M(\boldsymbol{\beta}) \sum_{\boldsymbol{u}+\boldsymbol{v}+\boldsymbol{w}=\boldsymbol{r}} \boldsymbol{\lambda}^{\boldsymbol{u}+\boldsymbol{v}} (\boldsymbol{\lambda}+1)^{\boldsymbol{w}} M\left((\boldsymbol{\gamma}, \boldsymbol{u})\right) M(\boldsymbol{v}) M(\boldsymbol{w})$$

is given by

$$\sum_{a+b=|\boldsymbol{k}|} \binom{a+1}{|\boldsymbol{j}|+1} \binom{b+|\boldsymbol{r}|+2}{|\boldsymbol{\ell}|} M(\boldsymbol{j}) M(\boldsymbol{k}-\boldsymbol{j}) M(\boldsymbol{\ell}) M(\boldsymbol{r}-\boldsymbol{\ell}).$$

Here

$$M(\boldsymbol{g}) = \binom{g_1 + g_2 + \cdots + g_n}{g_1, g_2, \ldots, g_n}$$

if $\boldsymbol{g} = (g_1, g_2, \ldots, g_n)$.

Problem 8.4.44. *Let k, j be integers with $0 \leq j \leq k$ and $\boldsymbol{r} = (r_1, r_2, \ldots, r_n)$, $\boldsymbol{\ell} = (\ell_1, \ell_2, \ldots, \ell_n)$ be n-tuples of nonnegative integers with $0 \leq \ell_p \leq r_p$, $p = 1, 2, \ldots, n$. Show that the coefficient of $\mu^j \boldsymbol{\lambda}^\ell$ of the polynomial*

$$\sum_{a+b+c=k} (\mu+1)^b \sum_{\boldsymbol{\alpha}+\boldsymbol{\beta}=\boldsymbol{r}} \boldsymbol{\lambda}^{\boldsymbol{\alpha}} M\left((b, \boldsymbol{\alpha})\right) M(\boldsymbol{\beta})$$

is given by

$$\binom{j+|\boldsymbol{\ell}|}{j} \binom{k+|\boldsymbol{\ell}|+2}{k-j} M(\boldsymbol{\ell}) M(\boldsymbol{r}-\boldsymbol{\ell}).$$

Problem 8.4.45. *Let k, j, r, ℓ be given as in the previous problem. Show that the coefficient of $\mu^j \boldsymbol{\lambda}^\ell$ of the polynomial*

$$\sum_{a+b=k} \mu^b \sum_{\boldsymbol{\alpha}+\boldsymbol{\beta}=\boldsymbol{r}} \boldsymbol{\lambda}^{\boldsymbol{\beta}} M(\boldsymbol{\alpha}) M\left((b, \boldsymbol{\beta})\right)$$

is given by

$$\binom{j + |\boldsymbol{\ell}|}{j} M(\boldsymbol{\ell}) M(\boldsymbol{r} - \boldsymbol{\ell}).$$

Problem 8.4.46. *Let* $k, j, \boldsymbol{r}, \boldsymbol{\ell}$ *be given as in* Problem 8.4.44. *Show that the coefficient of* $\mu^j \boldsymbol{\lambda}^{\boldsymbol{\ell}}$ *of the polynomial*

$$\sum_{a+b+c=k} (\mu+1)^b \mu^c \sum_{\boldsymbol{\alpha}+\boldsymbol{\beta}+\boldsymbol{\gamma}=\boldsymbol{r}} \boldsymbol{\lambda}^{\boldsymbol{\alpha}} (\boldsymbol{\lambda}+1)^{\boldsymbol{\beta}} M\left((b, \boldsymbol{\alpha})\right) M\left((c, \boldsymbol{\beta})\right) M(\boldsymbol{\gamma})$$

is given by

$$\sum_{a=0}^{k} \sum_{g=0}^{|\boldsymbol{r}-\boldsymbol{\ell}|} \binom{a+g+1}{j} \binom{a+g+|\boldsymbol{\ell}|+1}{|\boldsymbol{\ell}|} M(\boldsymbol{\ell}) M(\boldsymbol{r} - \boldsymbol{\ell}).$$

Problem 8.4.47. *Let* $k, j, \boldsymbol{r}, \boldsymbol{\ell}$ *be given as in* Problem 8.4.44. *Show that the coefficient of* $\mu^j \boldsymbol{\lambda}^{\boldsymbol{\ell}}$ *of the polynomial*

$$\sum_{a+b+c+d=k} (\mu+1)^b \mu^{c+d} \sum_{\boldsymbol{\alpha}+\boldsymbol{\beta}+\boldsymbol{\gamma}=\boldsymbol{r}} \boldsymbol{\lambda}^{\boldsymbol{\alpha}+\boldsymbol{\gamma}} (\boldsymbol{\lambda}+1)^{\boldsymbol{\beta}} M\left((b, \boldsymbol{\alpha})\right) M\left((c, \boldsymbol{\beta})\right) M\left((d, \boldsymbol{\gamma})\right)$$

is given by

$$\sum_{b=0}^{k} \binom{b+|\boldsymbol{r}-\boldsymbol{\ell}|+2}{j} \binom{b+|\boldsymbol{r}|+2}{|\boldsymbol{\ell}|} M(\boldsymbol{\ell}) M(\boldsymbol{r} - \boldsymbol{\ell}).$$

Problem 8.4.48. *Let* $k, j, \boldsymbol{r}, \boldsymbol{\ell}$ *be given as in* Problem 8.4.44. *Show that the coefficient of* $\mu^j \boldsymbol{\lambda}^{\boldsymbol{\ell}}$ *of the polynomial*

$$\sum_{a+b+c=k} \mu^{b+c} \sum_{\boldsymbol{\alpha}+\boldsymbol{\beta}+\boldsymbol{\gamma}=\boldsymbol{r}} (\boldsymbol{\lambda}+1)^{\boldsymbol{\beta}} \boldsymbol{\lambda}^{\boldsymbol{\gamma}} M(\boldsymbol{\alpha}) M\left((b, \boldsymbol{\beta})\right) M\left((c, \boldsymbol{\gamma})\right)$$

is given by

$$\sum_{b+c=j} \sum_{m+n=|\boldsymbol{\ell}|} \binom{b+m}{b} \binom{b+|\boldsymbol{r}-\boldsymbol{\ell}|+m+1}{|\boldsymbol{r}-\boldsymbol{\ell}|} \binom{c+n}{c} M(\boldsymbol{\ell}) M(\boldsymbol{r} - \boldsymbol{\ell}).$$

Problem 8.4.49. *Let* k, j, r, ℓ *be given as in* Problem 8.4.44. *Show that the coefficient of* $\mu^j \lambda^\ell$ *of the polynomial*

$$\sum_{a+b=k} \mu^b \sum_{\alpha+\beta+\gamma=r} (\lambda+1)^\beta M(\alpha) M((b,\beta)) M(\gamma)$$

is given by

$$\binom{j+|\ell|}{j} \binom{j+|r|+2}{|r-\ell|} M(\ell) M(r-\ell).$$

Problem 8.4.50. *Let* k *be a nonnegative integer and* $r = (r_1, r_2, \ldots, r_n)$ *be an* n-*tuple of nonnegative integers. Carry out the shuffle product process with the integral*

$$\frac{1}{k! \, r!} \iint_{R_1 \times R_2} \left(\mu \log \frac{1-t_1}{1-t_2} + \log \frac{1}{1-u_1} \right)^k \prod_{j=1}^n \left(\lambda_j \log \frac{t_2}{t_1} + \log \frac{u_2}{u_1} \right)^{r_j}$$

$$\times \frac{dt_1 dt_2}{(1-t_1)t_2} \frac{du_1 du_2}{(1-u_1)u_2}.$$

Problem 8.4.51. *Let* a, b, c *be nonnegative integers and* $k = (k_1, k_2, \ldots, k_n)$, $j = (j_1, j_2, \ldots, j_n)$ *be* n-*tuples of nonnegative integers with* $0 \le j_i \le k_i$, $i = 1, 2, \ldots, n$. *Prove that*

$$\sum_{m+p=|j|} \binom{b+m}{b} \binom{c+p}{c} \binom{a+b+|k-j|+m+1}{|k-j|}$$

$$= \sum_{m+p=|k-j|} \binom{b+m}{b} \binom{a+p}{a} \binom{b+c+|j|+m+1}{|j|}.$$

Hint: Up to a constant, both sides are coefficients of μ^j *of the polynomial*

$$\sum_{\alpha+\beta+\gamma=k} (\mu+1)^\beta \mu^\gamma M((a,\alpha)) M((b,\beta)) M((c,\gamma)).$$

Problem 8.4.52. *Let k, j be integers with $0 \le j \le k$ and $\boldsymbol{r} = (r_1, r_2, \ldots, r_n)$, $\boldsymbol{\ell} = (\ell_1, \ell_2, \ldots, \ell_n)$ be n-tuples of nonnegative integers with $0 \le \ell_i \le r_i$, $i = 1, 2, \ldots, n$. Show that the coefficient of $\mu^j \boldsymbol{\lambda}^{\boldsymbol{\ell}}$ of the polynomial*

$$\sum_{a+b+c+d=k} (\mu + 1)^b \mu^{c+d} \sum_{\alpha+\beta+\gamma+\delta=r} \lambda^{\alpha+\beta}(\lambda + 1)^\gamma$$

is given by

$$\sum_{b=0}^{k} \sum_{m=0}^{|r-\ell|} \binom{b + m + 2}{j} \binom{b + m + |\boldsymbol{\ell}| + 2}{|\boldsymbol{\ell}|}.$$

Problem 8.4.53. *Let $\boldsymbol{p} = (p_1, p_2, \ldots, p_m)$, $\boldsymbol{q} = (q_1, q_2, \ldots, q_n)$ be m-tuple and n-tuple of nonnegative integers, respectively. Show that*

$$\sum_{\alpha+\beta=p} \sum_{u+v=q} \binom{m + |\alpha| + |u| + 1}{m} \binom{|\alpha| + |u| + 1}{|\alpha|} \binom{|\beta| + |v| + 1}{|v|}$$

$$\times \binom{n + |\beta| + |v| + 2}{n + 1} M(\alpha)M(\beta)M(u)M(v)$$

$$= M(\boldsymbol{p})M(\boldsymbol{q}) \sum_{a+b=|p|} \sum_{c+d=|q|} \binom{m + a + c + 1}{m} \binom{a + c + 1}{a}$$

$$\times \binom{b + d + 1}{d} \binom{n + b + d + 2}{n + 1}$$

for all integers $m, n \ge 0$.

Problem 8.4.54. *Let $a_1, a_2, \ldots, a_n, k, q$ be nonnegative integers with $a_1 + a_2 + \cdots + a_n = p$. Show that for $0 \le j \le k$,*

$$\sum_{|b|=q} \binom{a_1 + b_1}{a_1} \binom{a_2 + b_2}{a_2} \cdots \binom{a_n + b_n}{a_n} \binom{b_n}{j}$$

$$= \binom{a_n + j}{j} \binom{p + q + n - 1}{k - j}.$$

Problem 8.4.55. *Let $\boldsymbol{k}, \boldsymbol{j}$ be n-tuples of nonnegative integers with $0 \leq j_i \leq k_i$, $i = 1, 2, \ldots, n$, and r, ℓ be integers with $0 \leq \ell \leq r$. Prove that*

$$
\binom{k_1 + k_2 + \cdots + k_n + r + 4}{j_1 + j_2 + \cdots + j_n + \ell + 2}
$$
$$
= \binom{|\boldsymbol{k} - \boldsymbol{j}| + r + 2}{\ell} + \binom{|\boldsymbol{j}| + r + 2}{r - \ell}
$$
$$
+ \sum_{m+p=r} \left\{ \binom{|\boldsymbol{k}| + m + 2}{|\boldsymbol{j}|}\binom{p}{\ell} + \binom{|\boldsymbol{k}| + m + 2}{|\boldsymbol{k} - \boldsymbol{j}|}\binom{p}{r - \ell} \right\}
$$
$$
+ \sum_{a+b=|\boldsymbol{j}|} \binom{a + |\boldsymbol{k} - \boldsymbol{j}| + 1}{|\boldsymbol{k} - \boldsymbol{j}|}\binom{b + r + 1}{r - \ell}
$$
$$
+ \sum_{a+b=|\boldsymbol{k}-\boldsymbol{j}|} \binom{a + |\boldsymbol{j}| + 1}{|\boldsymbol{j}|}\binom{b + r + 1}{\ell}
$$

via the consideration of the integral

$$
\frac{1}{k!r!} \iint_{R_1 \times R_2} \prod_{i=1}^{n} \left(\mu_i \log \frac{t_2}{t_1} + \log \frac{u_2}{u_1} \right)^{k_i} \left(\lambda \log \frac{1}{t_2} + \log \frac{1}{u_2} \right)^{r}
$$
$$
\times \frac{dt_1 dt_2}{(1 - t_1)t_2} \frac{du_1 du_2}{(1 - u_1)u_2}
$$

with $R_1 : 0 < t_1 < t_2 < 1$ and $R_2 : 0 < u_1 < u_2 < 1$. Note that the value of integral in terms of multiple zeta values is

$$
\sum_{\boldsymbol{\alpha}+\boldsymbol{\beta}=\boldsymbol{k}} \sum_{\ell=0}^{r} \boldsymbol{\mu}^{\boldsymbol{\alpha}} \lambda^{\ell} M(\boldsymbol{\alpha}) M(\boldsymbol{\beta}) \zeta(|\boldsymbol{\alpha}| + \ell + 2) \zeta(|\boldsymbol{\beta}| + r - \ell + 2)
$$

so that the generating function for number of multiple zeta values produced from the shuffle product process is

$$
\sum_{\boldsymbol{\alpha}+\boldsymbol{\beta}=\boldsymbol{k}} \sum_{\ell=0}^{r} \boldsymbol{\mu}^{\boldsymbol{\alpha}} \lambda^{\ell} M(\boldsymbol{\alpha}) M(\boldsymbol{\beta}) \binom{|\boldsymbol{k}| + r + 4}{|\boldsymbol{\alpha}| + \ell + 2}.
$$

Problem 8.4.56. *Let $\boldsymbol{k}$ be an n-tuple of nonnegative integers and $b_1, b_2, \ldots, b_p$ be nonnegative integers with $b_1 + b_2 + \cdots + b_p = q$. Show that*

$$
\sum_{\boldsymbol{\alpha}_1+\boldsymbol{\alpha}_2+\cdots+\boldsymbol{\alpha}_p=\boldsymbol{k}} M((b_1, \boldsymbol{\alpha}_1)) M((b_2, \boldsymbol{\alpha}_2)) \cdots M((b_p, \boldsymbol{\alpha}_p)) = M((q+p-1, \boldsymbol{k})).
$$

Problem 8.4.57. *Let $\boldsymbol{k}$ be an n-tuple of nonnegative integers and $\boldsymbol{u}_1$, $\boldsymbol{u}_2, \ldots, \boldsymbol{u}_p$ be m-tuples of nonnegative integers with $\boldsymbol{u}_1 + \boldsymbol{u}_2 + \cdots + \boldsymbol{u}_p = \boldsymbol{q}$. Prove that*

$$\sum_{\boldsymbol{\alpha}_1+\boldsymbol{\alpha}_2+\cdots+\boldsymbol{\alpha}_p=\boldsymbol{k}} M((\boldsymbol{u}_1,\boldsymbol{\alpha}_1))M((\boldsymbol{u}_2,\boldsymbol{\alpha}_2))\cdots M((\boldsymbol{u}_p,\boldsymbol{\alpha}_p))$$

$$= M(\boldsymbol{u}_1)M(\boldsymbol{u}_2)\cdots M(\boldsymbol{u}_p)M(\boldsymbol{k})\binom{|\boldsymbol{q}|+|\boldsymbol{k}|+p-1}{|\boldsymbol{k}|}$$

and

$$\sum_{\boldsymbol{u}_1+\boldsymbol{u}_2+\cdots+\boldsymbol{u}_p=\boldsymbol{q}}\sum_{\boldsymbol{\alpha}_1+\boldsymbol{\alpha}_2+\cdots+\boldsymbol{\alpha}_p=\boldsymbol{k}} M((\boldsymbol{u}_1,\boldsymbol{\alpha}_1))M((\boldsymbol{u}_2,\boldsymbol{\alpha}_2))\cdots M((\boldsymbol{u}_p,\boldsymbol{\alpha}_p))$$

$$= M(\boldsymbol{q})M(\boldsymbol{k})\binom{|\boldsymbol{q}|+|\boldsymbol{k}|+p-1}{|\boldsymbol{q}|,|\boldsymbol{k}|,p-1}.$$

Problem 8.4.58. *Let $\boldsymbol{p}, \boldsymbol{q}$ be m-tuple and n-tuple of nonnegative integers, respectively. Show that for nonnegative integers k, m, n, ℓ,*

$$\sum_{\boldsymbol{\alpha}+\boldsymbol{\beta}=\boldsymbol{p}}\sum_{\boldsymbol{\gamma}+\boldsymbol{\delta}=\boldsymbol{q}}\binom{m+|\boldsymbol{\alpha}|+|\boldsymbol{\gamma}|+k+1}{m}M((\boldsymbol{\alpha},\boldsymbol{\gamma}))M((\boldsymbol{\beta},\boldsymbol{\delta}))$$

$$\times \binom{n+|\boldsymbol{\beta}|+|\boldsymbol{\delta}|+\ell+2}{n+1}$$

$$= M((\boldsymbol{p},\boldsymbol{q}))\sum_{a+b=|\boldsymbol{p}|+|\boldsymbol{q}|}\binom{a+m+k+1}{m}\binom{b+n+\ell+2}{n}.$$

Here

$$M((\boldsymbol{p},\boldsymbol{q})) = \binom{p_1+p_2+\cdots+p_m+q_1+q_2+\cdots+q_n}{p_1,p_2,\ldots,p_m,q_1,q_2,\ldots,q_n}$$

$$= \frac{(|\boldsymbol{p}|+|\boldsymbol{q}|)!}{|\boldsymbol{p}|!|\boldsymbol{q}|!}M(\boldsymbol{p})M(\boldsymbol{q}).$$

Problem 8.4.59. *Notations as given in the previous problem. Show that*

$$\sum_{\boldsymbol{\alpha}+\boldsymbol{\beta}+\boldsymbol{\gamma}=\boldsymbol{p}} 2^{|\boldsymbol{\beta}|} \sum_{\boldsymbol{u}+\boldsymbol{v}+\boldsymbol{w}=\boldsymbol{q}} 2^{|\boldsymbol{v}|} M((\boldsymbol{\alpha},\boldsymbol{u}))M((\boldsymbol{\beta},\boldsymbol{v}))M((\boldsymbol{\gamma},\boldsymbol{w}))$$

$$= M(\boldsymbol{p})M(\boldsymbol{q})\sum_{j=0}^{|\boldsymbol{p}|}\sum_{\ell=0}^{|\boldsymbol{q}|}\binom{|\boldsymbol{p}|+|\boldsymbol{q}|+2}{|\boldsymbol{p}|-j}\binom{j+\ell}{j}\binom{|\boldsymbol{q}|+j+2}{|\boldsymbol{q}|-\ell}.$$

Problem 8.4.60. *Let a, b, c be nonnegative integers and $\boldsymbol{k} = (k_1, k_2, \ldots, k_n)$ be an n-tuple of nonnegative integers. Show that*

$$\sum_{\boldsymbol{\alpha}+\boldsymbol{\beta}=\boldsymbol{k}} (-1)^{|\boldsymbol{\alpha}|} \sum_{m+n=|\boldsymbol{\beta}|} \binom{b+m}{b} \binom{c+n}{c} \binom{a+b+m+|\boldsymbol{\alpha}|+1}{|\boldsymbol{\alpha}|}$$

$$\times M(\boldsymbol{\alpha})M(\boldsymbol{\beta})$$

$$= M((a+c+1, \boldsymbol{k})) = \frac{(|\boldsymbol{k}|+a+c+1)!}{k_1! k_2! \cdots k_n!(a+c+1)!}.$$

Problem 8.4.61. *Notations as given in the previous problem. Show that*

$$\sum_{\boldsymbol{\alpha}+\boldsymbol{\beta}=\boldsymbol{k}} (-1)^{|\boldsymbol{\alpha}|} \sum_{m+n=|\boldsymbol{\alpha}|} \binom{b+m}{b} \binom{a+n}{a} \binom{b+c+m+|\boldsymbol{\beta}|+1}{|\boldsymbol{\beta}|}$$

$$= M((a+c+1, \boldsymbol{k})) = \frac{(|\boldsymbol{k}|+a+c+1)!}{k_1! k_2! \cdots k_n!(a+c+1)!}.$$

Problem 8.4.62. *Notations as given in* Problem 8.4.60. *Show that*

$$\sum_{\boldsymbol{\alpha}+\boldsymbol{\beta}=\boldsymbol{k}} (-1)^{|\boldsymbol{\alpha}|} \binom{b+|\boldsymbol{\beta}|}{b} \binom{a+b+c+|\boldsymbol{k}|+2}{|\boldsymbol{\alpha}|} M(\boldsymbol{\alpha})M(\boldsymbol{\beta})$$

$$= M((a+c+1, \boldsymbol{k})) = \frac{(|\boldsymbol{k}|+a+c+1)!}{k_1! k_2! \cdots k_n!(a+c+1)!}.$$

Problem 8.4.63. *Let $\boldsymbol{q}$ be an n-tuple of nonnegative integers. Prove that*

$$\sum_{\boldsymbol{\alpha}_1+\boldsymbol{\alpha}_2+\cdots+\boldsymbol{\alpha}_m=\boldsymbol{q}} M(\boldsymbol{\alpha}_1)M(\boldsymbol{\alpha}_2)\cdots M(\boldsymbol{\alpha}_m) = M(\boldsymbol{q})\binom{|\boldsymbol{q}|+m-1}{|\boldsymbol{q}|}.$$

Problem 8.4.64. *Let $\boldsymbol{u}_1, \boldsymbol{u}_2, \ldots, \boldsymbol{u}_m$ be n-tuples of nonnegative integers with $\boldsymbol{u}_1 + \boldsymbol{u}_2 + \cdots + \boldsymbol{u}_m = \boldsymbol{q}$ and $\boldsymbol{p}$ be an another r-tuple of nonnegative integers. Prove that*

$$\sum_{\boldsymbol{\alpha}_1+\boldsymbol{\alpha}_2+\cdots+\boldsymbol{\alpha}_m=\boldsymbol{p}} M((\boldsymbol{u}_1, \boldsymbol{\alpha}_1))M((\boldsymbol{u}_2, \boldsymbol{\alpha}_2))\cdots M((\boldsymbol{u}_m, \boldsymbol{\alpha}_m))$$

$$= M(\boldsymbol{u}_1)M(\boldsymbol{u}_2)\cdots M(\boldsymbol{u}_m)M(\boldsymbol{p})\binom{|\boldsymbol{p}|+|\boldsymbol{q}|+m-1}{|\boldsymbol{p}|}$$

and

$$\sum_{\boldsymbol{u}_1+\boldsymbol{u}_2+\cdots+\boldsymbol{u}_m=\boldsymbol{q}} \sum_{\boldsymbol{\alpha}_1+\boldsymbol{\alpha}_2+\cdots+\boldsymbol{\alpha}_m=\boldsymbol{p}} M((\boldsymbol{u}_1,\boldsymbol{\alpha}_1))M((\boldsymbol{u}_2,\boldsymbol{\alpha}_2))\cdots M((\boldsymbol{u}_m,\boldsymbol{\alpha}_m))$$
$$= M(\boldsymbol{p})M(\boldsymbol{q})\binom{|\boldsymbol{p}|+|\boldsymbol{q}|+m-1}{|\boldsymbol{p}|,|\boldsymbol{q}|,m-1}.$$

Here

$$M((\boldsymbol{\alpha},\boldsymbol{\beta})) = \binom{\alpha_1+\cdots+\alpha_m+\beta_1+\cdots+\beta_n}{\alpha_1,\ldots,\alpha_m,\beta_1,\ldots,\beta_n}$$

if

$$\boldsymbol{\alpha} = (\alpha_1,\alpha_2,\ldots,\alpha_m) \quad and \quad \boldsymbol{\beta} = (\beta_1,\beta_2,\ldots,\beta_n).$$

Let $\boldsymbol{k} = (k_1,k_2,\ldots,k_m)$ and $\boldsymbol{r} = (r_1,r_2,\ldots,r_n)$ be m-tuple and n-tuple of nonnegative integers, respectively. In the following considerations, we are going to evaluate the triple product

$$M((\boldsymbol{j},\boldsymbol{r}-\boldsymbol{\ell})) M((\boldsymbol{k}-\boldsymbol{j},\boldsymbol{\ell}))\binom{|\boldsymbol{k}|+|\boldsymbol{r}|+4}{|\boldsymbol{j}|+|\boldsymbol{r}-\boldsymbol{\ell}|+2}$$

with

$$\boldsymbol{j} = (j_1,j_2,\ldots,j_m), \quad 0 \le j_i \le k_i, \quad i = 1,2,\ldots,m$$

and

$$\boldsymbol{\ell} = (\ell_1,\ell_2,\ldots,\ell_n), \quad 0 \le \ell_g \le r_g, \quad g = 1,2,\ldots,n.$$

We begin with the integral

(‡)

$$\frac{1}{\boldsymbol{k}!\boldsymbol{r}!} \iint_{R_1\times R_2} \prod_{i=1}^{m} \left(\mu_i \log\frac{1-t_1}{1-t_2} + \log\frac{1-u_1}{1-u_2}\right)^{k_i} \prod_{g=1}^{n} \left(\log\frac{t_2}{t_1} + \lambda_g \log\frac{u_2}{u_1}\right)^{r_g}$$
$$\times \frac{dt_1 dt_2}{(1-t_1)t_2}\frac{du_1 du_2}{(1-u_1)u_2}$$

which has the value

$$\sum_{\boldsymbol{\alpha}+\boldsymbol{\beta}=\boldsymbol{k}} \sum_{\boldsymbol{\gamma}+\boldsymbol{\delta}=\boldsymbol{r}} \boldsymbol{\mu}^{\boldsymbol{\alpha}}\boldsymbol{\lambda}^{\boldsymbol{\delta}} M(\boldsymbol{\alpha})M(\boldsymbol{\beta})M(\boldsymbol{\gamma})M(\boldsymbol{\delta})\zeta(|\boldsymbol{\alpha}|+|\boldsymbol{\gamma}|+2)\zeta(|\boldsymbol{\beta}|+|\boldsymbol{\delta}|+2)$$

so that the generating function for the number of multiple zeta values produced from shuffle product process is

$$\sum_{\alpha+\beta=k} \sum_{\gamma+\delta=r} \mu^\alpha \lambda^\delta M((\alpha,\gamma)) M((\beta,\delta)) \binom{|k|+|r|+4}{|\alpha|+|\gamma|+2}.$$

The following problems are concern with separating the coefficient of $\mu^j \lambda^\ell$ from 6 partial generating functions whose sum is equal to the above generating function.

Problem 8.4.65. *Prove that the partial generation function from the integration ($\ddagger$) over $D_1: 0 < t_1 < t_2 < u_1 < u_2 < 1$ is*

$$\sum_{\alpha+\beta=k} \sum_{u+v=r} \mu^\alpha \lambda^v M((\alpha,u)) M((\beta,v))$$

and the coefficient of $\mu^j \lambda^\ell$ is

$$M((j, r-\ell)) M((k-j, \ell)).$$

Also, prove that the integration ($\ddagger$) over $D_2: 0 < t_1 < u_1 < t_2 < u_2 < 1$ produced the same generating function with the same coefficient of $\mu^j \lambda^\ell$.

Problem 8.4.66. *Prove that the partial generating function from the integration ($\ddagger$) over $D_3: 0 < t_1 < u_1 < t_2 < u_2 < 1$ is*

$$\sum_{\alpha+\beta+\gamma=k} \mu^\alpha (\mu+1)^\beta \sum_{u+v+w=r} (\lambda+1)^v \lambda^w M((\alpha,u)) M((\beta,v)) M((\gamma,w))$$

and the coefficient of $\mu^j \lambda^\ell$ is given by

$$M(j) M(k-j) M(\ell) M(r-\ell)$$
$$\times \sum_{a+b=|k-j|} \sum_{m+n=|\ell|} \binom{a+m}{a} \binom{b+n}{b} \binom{a+m+|j|+|r-\ell|+1}{a+m+1, |j|, |r-\ell|}.$$

Also, prove that the integration ($\ddagger$) over $D_5: 0 < u_1 < t_1 < u_2 < t_2 < 1$ produced the same generating function.

Problem 8.4.67. *Prove that the partial generating function from the integration* (‡) *over D_4 : $0 < t_1 < u_1 < u_2 < t_2 < 1$ is*

$$\sum_{\alpha+\beta+\gamma=k} \mu^{\alpha+\delta}(\mu+1)^\beta \sum_{u+v+w=r} (\lambda+1)^v M((\alpha,u))M((\beta,v))M((\gamma,w))$$

and its coefficient of $\mu^j \lambda^\ell$ is given by

$$M(j)M(k-j)M(\ell)M(r-\ell)$$

$$\times \binom{|k|+|r|+2}{|j|}\binom{|k-j|+|\ell|}{|\ell|}\binom{|r|+|k-j|+2}{|r-\ell|}.$$

Also, prove that the partial generating function produced from the integration (‡) *over D_6 : $0 < u_1 < t_1 < t_2 < u_2 < 1$ is*

$$\sum_{\alpha+\beta+\gamma=k} (\mu+1)^\beta \sum_{u+v+w=r} \lambda^{u+w}(\lambda+1)^v M((\alpha,u))M((\beta,v))M((\gamma,w))$$

and its coefficient of $\mu^j \lambda^\ell$ is given by

$$M(j)M(k-j)M(\ell)M(r-\ell)$$

$$\times \binom{|k|+|r|+2}{|k-j|}\binom{|j|+|r-\ell|}{|r-\ell|}\binom{|r|+|j|+2}{|\ell|}.$$

Problem 8.4.68. *Notations as shown above, prove that*

$$\binom{|j|+|r-\ell|}{|j|}\binom{|k-j|+|\ell|}{|\ell|}\left\{\binom{|k|+|r|+4}{|j|+|r-\ell|+2}-2\right\}$$

$$= \binom{|k|+|r|+2}{|j|}\binom{|k-j|+|\ell|}{|\ell|}\binom{|r|+|k-j|+2}{|r-\ell|}$$

$$+ \binom{|k|+|r|+2}{|k-j|}\binom{|j|+|r-\ell|}{|r-\ell|}\binom{|r|+|j|+2}{|\ell|}$$

$$+ 2 \sum_{a+b=|k-j|}\sum_{m+n=|\ell|} \binom{a+m}{a}\binom{b+n}{b}\binom{a+m+|j|+|r-\ell|+1}{a+m+1,|j|,|r-\ell|}.$$

Problem 8.4.69. *Let* k, j *be* m-*tuples of nonnegative integers with* $0 \leq j_i \leq k_i$, $i = 1, 2, \ldots, m$, *and* r, ℓ *be* n-*tuples of nonnegative integers with* $0 \leq \ell_g \leq r_g$, $g = 1, 2, \ldots, n$. *Prove that*

$$\sum_{a+b=|k-j|} \sum_{m+n=|\ell|} \binom{a+m}{a}\binom{b+n}{b}\binom{a+m+|j|+|r-\ell|+1}{a+m+1, |j|, |r-\ell|}$$

$$= \sum_{a+b=|j|} \sum_{m+n=|r-\ell|} \binom{a+m}{a}\binom{b+n}{b}\binom{a+m+|k-j|+|\ell|+1}{a+m+1, |k-j|, |\ell|}.$$

Problem 8.4.70. *Notations as in the previous problem. Show that the coefficient of* $\boldsymbol{\mu}^j \boldsymbol{\lambda}^\ell$ $(= \mu_1^{j_1}\mu_2^{j_2}\cdots\mu_m^{j_m}\lambda_1^{\ell_1}\lambda_2^{\ell_2}\cdots\lambda_n^{\ell_n})$ *of the polynomial function*

$$\sum_{\alpha+\beta+\gamma=k} (\boldsymbol{\mu}+1)^\alpha \sum_{u+v+w=r} (\boldsymbol{\lambda}+1)^w M(\alpha)M(\beta)M((\gamma, u))M(v)M(w)$$

is given by

$$M(j)M(k-j)M(\ell)M(r-\ell) \sum_{m+n=|r-\ell|} \binom{m+|k|+2}{|k-j|}\binom{n+|\ell|+1}{|\ell|+1}$$

or

$$M(j)M(k-j)M(\ell)M(r-\ell) \sum_{a+b=|k-j|} \binom{a+|j|+1}{|j|+1}\binom{b+|r|+2}{|r-\ell|}.$$

Appendices

APPENDIX A

Singular Modular Forms on the Exceptional Domain

*A modular form f with the Fourier expansion

$$f(Z) = \sum_{T \geq 0} a(T) e^{2\pi i(T, Z)}$$

is called a singular modular form if $a(T) = 0$ when T is of full rank.

According to a general theorem by E. Freitag, all singular modular forms on the Siegel upper half plane

$$\mathscr{H}_n = \left\{ Z = X + iY \mid {}^t X = X \in M_n(\mathbb{R}), {}^t Y = Y \in M_n(\mathbb{R}), Y > 0 \right\},$$

are nothing but finite linear combinations of theta series of the form

$$\vartheta(S, Z) = \sum_{G \in \mathbb{Z}^{m \times n}} e^{\pi i(S[G], Z)} \quad (m < n)$$

where S is a $m \times m$ positive definite integral symmetric matrix and $S[G] = {}^t G S G$.

---

*This is a lecture note of the talk with the same title, given at KIAS of Korea in August 24, 2011 on modular forms and their geometric applications.

Such an assertion is not quite true on the exceptional domain of 27 dimensions. Indeed, it is a long-standing open problem to construct possible theta series on the exceptional domain since Baily initiated the theory of modular forms on such domain in 1970.

A.1   Cayley Numbers and Integral Cayley Numbers

Let $\mathfrak{f}$ be a field. The Cayley numbers $\mathscr{C}_\mathfrak{f}$ over $\mathfrak{f}$ is an eight-dimensional vector space over $\mathfrak{f}$ with a standard basis $e_0, e_1, e_2, e_3, e_4, e_5, e_6, e_7$:

$$(A.1.1) \qquad \mathscr{C}_\mathfrak{f} = \left\{ \sum_{j=0}^{n} x_j e_j \ \middle|\ x_j \in \mathfrak{f}, \ j = 0, 1, \ldots, 7 \right\}.$$

In $\mathscr{C}_\mathfrak{f}$, there is a multiplication satisfying the following rules:

(i) $x e_0 = e_0 x = x$ for all $x \in \mathscr{C}_\mathfrak{f}$,

(ii) $e_j^2 = -e_0$, $j = 1, 2, \ldots, 7$ and

(iii) $e_1 e_2 e_4 = e_2 e_3 e_5 = e_3 e_4 e_6 = e_4 e_5 e_7 = e_5 e_6 e_1 = e_6 e_7 e_2 = e_7 e_1 e_3 = -e_0$.

Remark A.1.1. In general, the multiplication of $\mathscr{C}_\mathfrak{f}$ is not associative. However, for some triads such as e_1, e_2, e_4 have the same property as i, j, k in quaternions i.e. $e_1 e_2 = -e_2 e_1 = e_4$, $e_2 e_4 = -e_4 e_2 = e_1$, $e_4 e_1 = -e_1 e_4 = e_2$.

For $x = \sum_{j=0}^{7} x_j e_j$ and $y = \sum_{j=0}^{7} y_j e_j$ in $\mathscr{C}_\mathfrak{f}$, we define the following operations on $\mathscr{C}_\mathfrak{f}$.

(1) Involution: $x \mapsto \overline{x} = x_0 e_0 - \sum_{j=1}^{7} x_j e_j$;

(2) Trace operator: $T(x) = x + \overline{x} = 2x_0$;

(3) Norm operator: $N(x) = x\overline{x} = \overline{x}x = \sum_{j=0}^{7} x_j^2$;

(4) Inner product: $\sigma : \mathscr{C}_\mathfrak{f} \times \mathscr{C}_\mathfrak{f} \to \mathfrak{f}$

$$\sigma(x, y) = T(x\overline{y}) = T(y\overline{x}) = 2 \sum_{j=0}^{7} x_j y_j.$$

Note that we have the following further property.

$$N(x + y) = N(x) + N(y) + \sigma(x, y).$$

The integral Cayley numbers $\mathscr{O}$ is the $\mathbb{Z}$-module in $\mathscr{C}_{\mathbb{Q}}$ generated by α_j $(j = 0, 1, 2, \ldots, 7)$, as follows:

$$\alpha_0 = e_0, \quad \alpha_1 = e_1, \quad \alpha_2 = e_2, \quad \alpha_3 = -e_4,$$

$$\alpha_4 = \frac{1}{2}(e_1 + e_2 + e_3 - e_4), \quad \alpha_5 = \frac{1}{2}(-e_0 - e_1 - e_4 + e_5),$$

$$\alpha_6 = \frac{1}{2}(-e_0 + e_1 - e_2 + e_6), \quad \alpha_7 = \frac{1}{2}(-e_0 + e_2 + e_4 + e_7).$$

Remark A.1.2. $\mathscr{O}$ is the maximal $\mathbb{Z}$-module of $\mathscr{C}_{\mathbb{Q}}$ satisfying the following properties:

(1) $N(x) \in \mathbb{Z}$ and $T(x) \in \mathbb{Z}$ for each x in the set;

(2) The set is closed under substraction and multiplication;

(3) The set contains 1.

For z in the upper half plane, the theta series

$$\vartheta(z) = \sum_{t \in \mathscr{O}} e^{2\pi i N(t) z}$$

is a modular form of weight 4. Therefore $\vartheta(z)$ is just the normalized Eisenstein series

$$E_4(z) = 1 + 240 \sum_{n=1}^{\infty} \left(\sum_{d|n} d^3 \right) e^{2\pi i n z}$$

and it follows that the number of solutions to the equation

$$N(t) = n, \quad t \in \mathscr{O}$$

is given by

$$240 \sum_{d|n} d^3.$$

A.2   The Exceptional Domain

The exceptional domain of dimension 27 is a tube domain in 3×3 Hermitian matrices over Cayley numbers. Let $\mathscr{T}_{\mathbb{R}}$ be the set of 3×3 Hermitian matrices over real Cayley numbers:

$$\mathscr{T}_{\mathbb{R}} = \left\{ \begin{bmatrix} \xi_1 & x_{12} & x_{13} \\ \overline{x}_{12} & \xi_2 & x_{23} \\ \overline{x}_{13} & \overline{x}_{23} & \xi_3 \end{bmatrix} \middle| \ \xi_1, \xi_2, \xi_3 \in \mathbb{R}, \ x_{12}, x_{13}, x_{23} \in \mathscr{C}_{\mathbb{R}} \right\}.$$

We supply $\mathscr{T}_{\mathbb{R}}$ with a product $X \circ Y$ defined by

$$X \circ Y = \frac{1}{2} \left(XY + YX \right),$$

where XY is the ordinary matrix product. Then $\mathscr{T}_{\mathbb{R}}$ becomes a real Jordan algebra with this product. Define an inner product on $\mathscr{T}_{\mathbb{R}}$ by

$$(X, Y) = \operatorname{trace} (X \circ Y).$$

Finally, we let $\mathscr{K}$ be the set of squares $X \circ X$ of $\mathscr{T}_{\mathbb{R}}$ and $\mathscr{K}^+$ be the interior of $\mathscr{K}$. The exceptional domain $\mathscr{H}$ in $\mathbb{C}^{27}$ is then defined by

$$\mathscr{H} = \left\{ Z = X + iY \mid X, Y \in \mathscr{T}_{\mathbb{R}}, \ Y \in \mathscr{K}^+ \right\}.$$

Set $\mathscr{T}_{\mathscr{O}} = \mathscr{T}_{\mathbb{R}} \cap M_3(\mathscr{O})$. Here $M_3(\mathscr{O})$ is the set of 3×3 matrices over integral Cayley numbers. For $1 \leq i, j \leq 3$, let e_{ij} be the 3×3 matrix with 1 at the (i, j)-entry and 0 elsewhere. When $i \neq j$ and $t \in \mathscr{C}_{\mathbb{R}}$, we let $U_{ij}(t) = E + te_{ij}$, E being the 3×3 identity matrix.

The set of one-to-one and onto bi-holomorphic functions of $\mathscr{H}$ form a group $\mathscr{G}$ which is a Lie group of type E_7. Let Γ be the discrete subgroup of $\mathscr{G}$ generated by the following automorphisms of $\mathscr{H}$.

(1) $\iota : Z \mapsto -Z^{-1}$;

(2) $p_B : Z \mapsto Z + B, \ B \in \mathscr{T}_{\mathscr{O}}$;

(3) $t_U : Z \mapsto {}^t\overline{U} Z U, \ U = U_{ij}(t), \ t \in \mathscr{O}$.

Also let Γ_0 be the subgroup of Γ generated by (2) and (3). Suppose that $J(\gamma, Z)$ is the determinant of the Jacobian matrix of γ at Z. Then it has the following properties:

(a) $J(p_B, Z) = 1$ for all $B \in \mathscr{T}_{\mathbb{Q}}$;

(b) $J(t_U, Z) = 1$ for all $U = U_{ij}(t)$, $t \in \mathscr{O}$;

(c) $J(\iota, Z) = \det(-Z)^{-18}$.

Baily was the first to consider the Eisenstein series

$$E_\ell(Z) = \sum_{\gamma \in \Gamma/\Gamma_0} J(\gamma, Z)^\ell, \qquad Z \in \mathscr{H}.$$

He proved that $E_\ell(Z)$ is a modular form of weight 18ℓ and it has rational Fourier coefficients.

In 1993 Kim (a student of Baily) constructed a singular modular form of weight 4 through the analytic continuation of a non-holomorphic Eisenstein series

$$E_{k,s}(Z) = \sum_{\gamma \in \Gamma/\Gamma_0} j(\gamma, Z)^{-k} |j(\gamma, Z)|^{-s},$$

where $j(\gamma, Z)$ is a factor of $J(\gamma, Z)$ with the following properties:

(a) $j(p_B, Z) = 1$ for all $B \in \mathscr{T}_{\mathbb{Q}}$;

(b) $j(t_U, Z) = 1$ for all $U = U_{ij}(t)$, $t \in \mathscr{O}$;

(c) $j(\iota, Z) = \det(-Z)$ and

(d) $j(g_1 g_2, Z) = j(g_1, g_2(Z)) j(g_2, Z))$ for all $g_1, g_2 \in \Gamma$.

He conclude that $E_{4,0}(Z)$ is a singular modular form of weight 4 while $E_{8,0}(Z) = \{E_{4,0}(Z)\}^2$ is a singular modular form of weight 8. Furthermore, the Fourier coefficient $a(T)$ of $E_{4,0}(Z)$ is given by

$$a\left(\begin{bmatrix} m & 0 & 0 \\ 0 & 0 & 0 \\ 0 & 0 & 0 \end{bmatrix}\right) = 240 \sum_{d|m} d^3$$

if m is a positive integer. Also $a(T) = 0$ unless rank $T = 0$ or 1.

A.3   The Theory of Jacobi Forms

The study of the theory of Jacobi forms began in 1985 with a textbook written by M. Eichler and D. Zagier in an attempt to extend Maaß's work on the Saito-Kurokawa conjecture which asserted the existence of a lifting from modular forms of one variable of weight $2k-2$ to Siegel modular forms of weight k and degree two.

Let k and m be a pair of nonnegative integers. A holomorphic function $\varphi : \mathscr{H} \times \mathbb{C} \to \mathbb{C}$ is called a Jacobi form of weight k and index m with respect to Jacobi group $\mathrm{SL}_2(\mathbb{Z}) \ltimes \mathbb{Z}^2$ if it satisfies the following conditions:

(J-1) For all $\begin{bmatrix} a & b \\ c & d \end{bmatrix}$ in $\mathrm{SL}_2(\mathbb{Z})$

$$\varphi\left(\frac{az+b}{cz+d}, \frac{w}{cz+d}\right) = (cz+d)^k \exp\left\{2\pi i m c w^2/(cz+d)\right\} \varphi(z,w).$$

(J-2) For all integers λ and μ

$$\varphi\left(z, w + \lambda z + \mu\right) = \exp\left\{-2\pi i m(\lambda^2 z + 2\lambda w)\right\} \varphi(z,w).$$

(J-3) $\varphi(z,w)$ has a Fourier expansion of the form

$$\varphi(z,w) = \sum_{n=0}^{\infty} \sum_{r^2 \leq 4mn} \alpha(n,r) e^{2\pi i(nz+rw)}.$$

A main source of Jacobi forms come from the coefficient of Fourier-Jacobi expansions of Siegel modular forms of degree two. Let

$$f(Z) = \sum_{T \geq 0} a(T) e^{2\pi i(T,Z)}$$

be a modular form of weight k with respect to $\mathrm{Sp}(2,\mathbb{Z})$. Set

$$T = \begin{bmatrix} n & t/2 \\ t/2 & m \end{bmatrix} \quad \text{and} \quad Z = \begin{bmatrix} z & w \\ w & z^* \end{bmatrix}.$$

Then

$$(T,Z) = nz + tw + mz^*$$

so that

$$f(Z) = \sum_{m=0}^{\infty} \varphi_m(z, w) e^{2\pi i m z^*},$$

with

$$\varphi(z, w) = \sum_{n=0}^{\infty} \sum_{t^2 \leq 4mn} a \left(\begin{bmatrix} n & t/2 \\ t/2 & m \end{bmatrix} \right) e^{2\pi i (nz + tw)}.$$

Theorem A.3.1. $\varphi_m(z, w)$ *is a Jacobi form of weight k and index m.*

We develop the theory of Jacobi forms over Cayley numbers in 1992 and then the theory of Jacobi forms of degree two over Cayley numbers in 2000.

Remark A.3.2. The theta series

$$\sum_{h \in \mathcal{O}^2} e^{2\pi i (h^t \overline{h}, Z)}$$

is a singular modular form of weight 4 on the 10 dimension domain

$$\mathscr{H}_2 = \left\{ Z = \begin{bmatrix} x_1 & x_{12} \\ \overline{x}_{12} & x_2 \end{bmatrix} + i \begin{bmatrix} y_1 & y_{12} \\ \overline{y}_{12} & y_2 \end{bmatrix} \;\middle|\; x_1, x_2, y_1, y_2 \in \mathbb{R}, \right.$$
$$\left. x_{12}, y_{12} \in \mathscr{C}_{\mathbb{R}}, \; y_1 > 0, \; y_1 y_2 > N(y_{12}) \right\}.$$

However, it was proved by Krieg that the theta series

$$\sum_{h \in \mathcal{O}^3} e^{2\pi i (h^t \overline{h}, Z)}, \quad Z \in \mathscr{H}$$

is not a singular modular on the exceptional domain for several reasons. Indeed, the rank of $h^t \overline{h}$ maybe be 3 and we are unable to prove the theta series is a modular form!

If

$$f(Z) = \sum_{T \geq 0} a(T) e^{2\pi i (T, Z)}$$

is a modular form of weight k on the exceptional domain. Let

$$T = \begin{bmatrix} T_1 & q \\ {}^t q & m \end{bmatrix} \quad \text{and} \quad Z = \begin{bmatrix} Z_1 & W \\ {}^t W & z^* \end{bmatrix}.$$

Then

$$(T, Z) = (T_1, Z_1) + \sigma(q_1, w_1) + \sigma(q_2, w_2) + mz^*$$

so that

$$f(Z) = \sum_{m=0}^{\infty} \varphi_m(Z_1, W) e^{2\pi i m z^*},$$

with

$$\varphi_m(Z_1, W) = \sum_{T_1 \geq 0} \sum_{T_1 \geq q^t q/m} a(T) e^{2\pi i [(T_1, Z_1) + \sigma(q_1, w_1) + \sigma(q_2, w_2)]}.$$

Theorem A.3.3. $\varphi_m(Z_1, W)$ *is a Jacobi form of weight k and index m on* $\mathscr{H}_2 \times (\mathscr{C}_{\mathbb{C}})^2$.

On the other hand, a Jacobi form is just a vector-valued modular form:

$$f(Z, W) = \sum_{q:(\mathscr{O}/m\mathscr{O})^2} F_q(Z) \vartheta_{m,q}(Z, W)$$

with

$$F_q(Z) = \sum_{T \geq q^t \overline{q}/m} a(T, q) e^{2\pi i (T - q^t \overline{q}/m, Z)}$$

and

$$\vartheta_{m,q}(Z, W) = \sum_{\substack{h = \lambda + q/m \\ \lambda \in \mathscr{O}^2}} \exp \left\{ 2\pi i m \left[(h^t \overline{h}, Z) + \sigma(h_1, w_1) + \sigma(h_2, w_2) \right] \right\}.$$

A.4   A Final Application

Now suppose that $f(Z)$ is a singular modular form of weight 4 with

$$f(Z) = \sum_{m=0}^{\infty} \varphi_m(Z_1, W) e^{2\pi i m z^*}.$$

Then $\varphi_m(Z, W)$ is a Jacobi form of weight 4 and index m and hence

$$\varphi_m(Z, W) = \sum_{q:(\mathscr{O}/m\mathscr{O})^2} C_m(q) \vartheta_{m,q}(Z, W)$$

with $C_m(q)$ constant. Of course the vector $(C_m(q))_{q:(\mathscr{O}/m\mathscr{O})^2}$ must satisfy certain condition so that $\varphi_m(Z, W)$ is a Jacobi form. Here is the definition of $C_m(q)$.

For a positive integer m and $q = {}^t(q_1, q_2) \in \mathscr{O}^2$, let

$$T = \begin{bmatrix} q^t\bar{q}/m & q \\ {}^t\bar{q} & m \end{bmatrix}.$$

Define

$$C_m(q) = \begin{cases} 240 \sum_{d|\epsilon(T)} d^3 & \text{if } T \in \mathscr{T}_{\mathscr{O}}; \\ 0 & \text{otherwise;} \end{cases}$$

where $\epsilon(T)$ is the greatest positive integer such that $[\epsilon(T)]^{-1}T \in \mathscr{T}_{\mathscr{O}}$.

Theorem A.4.1. [8] *Notations as above, then*

$$C_m(q) = \frac{1}{m^8} \sum_{p:(\mathscr{O}/m\mathscr{O})^2} e^{-2\pi i[\sigma(q_1,p_1)+\sigma(q_2,p_2)]} C_m(p).$$

Such a condition can be proved directly from the definition and then it implies that $\varphi_m(Z, W)$ is a Jacobi form of weight $k - 4$ and index m. Therefore the singular modular forms of weight 4 is given by

$$E(Z) = f(Z_1) + \sum_{m=1}^{\infty} \varphi_m(Z_1, W)e^{2\pi imz^*}$$

with

$$f(Z_1) = \sum_{h \in \mathscr{O}} e^{2\pi i(h^t\bar{h}, Z_1)}$$

a modular form of weight 4 on the exceptional domain of 27 dimensions.

Appendix (i): Jacobi Forms over Cayley Numbers

Let k and m be a pair of nonnegative integers. A holomorphic function $\varphi : \mathscr{H} \times \mathscr{C}_{\mathbb{C}} \to \mathbb{C}$ is called a Jacobi form of weight k and index m with respect to $\mathrm{SL}_2(\mathbb{Z}) \ltimes \mathscr{O}^2$ if it satisfies the following conditions:

(J-1) For all $\begin{bmatrix} a & b \\ c & d \end{bmatrix}$ in $\mathrm{SL}_2(\mathbb{Z})$

$$\varphi\left(\frac{az+b}{cz+d}, \frac{w}{cz+d}\right) = (cz+d)^k \exp\left\{2\pi imcN(w)/(cz+d)\right\} \varphi(z,w);$$

(J-2) For all λ and μ in $\mathcal{O}$

$$\varphi\left(z, w + \lambda z + \mu\right) = \exp\left\{-2\pi im[N(\lambda)z + \sigma(\lambda w)]\right\} \varphi(z, w)$$

and

(J-3) $\varphi(z, w)$ has a Fourier expansion of the form

$$\varphi(z, w) = \sum_{n=0}^{\infty} \sum_{t \in \mathcal{O}, N(t) \leq mn} \alpha(n, t) e^{2\pi i[nz + \sigma(t, w)]}.$$

Appendix (ii): Basic Properties of a Set of Theta Series

Given a positive integer m and an integral Cayley number q, we define the theta series $\vartheta_{m,q}(z, w)$ as

$$\vartheta_{m,q}(z, w) = \sum_{\lambda \in \mathcal{O}} e^{2\pi im[N(\lambda + q/m)z + \sigma(\lambda + q/m, w)]}.$$

Directly from the above definition, it is easy to see that

(1) $\vartheta_{m,q}(z + 1, w) = e^{2\pi iN(q)/m}\vartheta_{m,q}(z, w)$,

(2) $\vartheta_{m,q}(z, w + \lambda z + \mu) = e^{-2\pi im[N(\lambda)z + \sigma(\lambda, w)]}\vartheta_{m,q}(z, w)$,

(3) $\vartheta_{m,q_1}(z, w) = \vartheta_{m,q_2}(z, w)$ if $q_1 \equiv q_2 \pmod{m}$.

Furthermore, we have

(4) $\vartheta_{m,q}(-\frac{1}{z}, \frac{w}{z}) = \left(\frac{z}{m}\right)^4 e^{2\pi imN(w)/z} \sum_{p:\mathcal{O}/m\mathcal{O}} e^{-2\pi i\sigma(p,q)/m}\vartheta_{m,p}(z, w)$.

Fix a set of representatives $q_1, q_2, \ldots, q_{m^8}$ of $\mathcal{O}/m\mathcal{O}$, we let

$$\Theta = {}^t(\vartheta_{m,q_1}, \vartheta_{m,q_2}, \ldots, \vartheta_{m,q_{m^8}}).$$

There is unique group homomorphism $\psi : \mathrm{SL}_2(\mathbb{Z}) \to U(m^8)$, unitary group of size m^8, such that

$$\Theta\left(\frac{az + b}{cz + d}\right) = (cz + d)^k e^{2\pi imcN(w)/(cz+d)}\psi\left(\begin{bmatrix} a & b \\ c & d \end{bmatrix}\right)\Theta(z, w)$$

for all $\begin{bmatrix} a & b \\ c & d \end{bmatrix}$ in $\mathrm{SL}_2(\mathbb{Z})$. In particular, for $T = \begin{bmatrix} 1 & 1 \\ 0 & 1 \end{bmatrix}$ and $J = \begin{bmatrix} 0 & -1 \\ 1 & 0 \end{bmatrix}$

we have

(a) $\psi(T) = \text{diag}\left[e^{2\pi i N(q_1)/m}, e^{2\pi i N(q_2)/m}, \ldots, e^{2\pi i N(q_{m^8})/m}\right]$;

(b) $\psi(J) = m^{-4}\left[e^{-2\pi i \sigma(q_\mu, q_\nu)/m}\right]_{\mu,\nu=1,2,\ldots,m^8}$.

APPENDIX B

Shuffle Product Formulas of Multiple Zeta Values

*A multiple zeta value or r-fold Euler sum defined by

$$\zeta(\alpha_1, \alpha_2, \ldots, \alpha_r) = \sum_{1 \le n_1 < n_2 < \cdots < n_r} n_1^{-\alpha_1} n_2^{-\alpha_2} \cdots n_r^{-\alpha_r}$$

with $\alpha = (\alpha_1, \alpha_2, \ldots, \alpha_r)$ an r-tuple of positive integers and $\alpha_r \ge 2$, is a natural generalization of the classical Euler sum

$$S_{p,q} = \sum_{k=1}^{\infty} \frac{1}{k^q} \sum_{j=1}^{k} \frac{1}{j^p}.$$

It is a problem proposed by Goldbach to Euler in an attempt to evaluate $S_{p,q}$ in terms of the special values at positive integers of the Riemann zeta function defined by

$$\zeta(s) = \sum_{n=1}^{\infty} \frac{1}{n^s}, \quad \text{Re } s > 1.$$

---

*This is a lecture note of the talk with the same title, given at KIAS of Korea in August 25, 2011 on modular forms and their geometric applications.

The shuffle product formula of two multiple zeta values then express the product of two multiple zeta values as a linear combination of multiple zeta values with integral coefficients.

B.1   Introduction

For an r-tuple of positive integers $\alpha = (\alpha_1, \alpha_2, \ldots, \alpha_r)$ with $\alpha_r \geq 2$, the multiple zeta values or r-fold Euler sum is defined as

$$\zeta(\alpha) = \sum_{1 \leq n_1 < n_2 < \cdots < n_r} n_1^{-\alpha_1} n_2^{-\alpha_2} \cdots n_r^{-\alpha_r},$$

or in free dummy variables as

$$\zeta(\alpha) = \sum_{\mathbf{k} \in \mathbb{N}^r} k_1^{-\alpha_1} (k_1 + k_2)^{-\alpha_2} \cdots (k_1 + k_2 + \cdots + k_r)^{-\alpha_r},$$

or in traditional nested form as

$$\sum_{n_r=1}^{\infty} \frac{1}{n_r^{\alpha_r}} \cdots \sum_{n_2=1}^{n_3-1} \frac{1}{n_2^{\alpha_2}} \sum_{n_1=1}^{n_2-1} \frac{1}{n_1^{\alpha_1}}.$$

The numbers r and $|\alpha| = \alpha_1 + \alpha_2 + \cdots + \alpha_r$ are the depth and the weight of $\zeta(\alpha)$, respectively.

     The case $r = 2$ went back to 1742 as a problem proposed by Goldbach to Euler. For a pair of positive integers p, q with $q \geq 2$, the classical Euler double sum [4, 15, 20, 28] is defined as

$$S_{p,q} = \sum_{k=1}^{\infty} \frac{1}{k^q} \sum_{j=1}^{k} \frac{1}{j^p}.$$

The purpose of the problem is to evaluate $S_{p,q}$ in terms of the special values at positive integers of the Riemann zeta function defined by

$$\zeta(s) = \sum_{n=1}^{\infty} \frac{1}{n^s}, \quad \mathrm{Re}\, s > 1.$$

     In 1775, Euler proved the case $p = 1$ and gave a general formula for $S_{p,q}$ when the weight is odd without any proof. N. Nielsen was the first to fill the gap by giving the correct version and a proof by solving a system of linear equations.

By the definition of Euler double sums, we see immediately that

$$S_{p,q} + S_{q,p} = \zeta(p)\zeta(q) + \zeta(p+q)$$

for all $p, q \geq 2$. This is called the reflection formula of Euler sums. Here we mention two well-known results concerning the evaluation of $S_{p,q}$.

Theorem B.1.1. *For each positive integer $n \geq 2$, we have*

$$S_{1,n} = \frac{n+2}{2}\zeta(n+1) - \frac{1}{2}\sum_{r=2}^{n-1}\zeta(r)\zeta(n+1-r).$$

Theorem B.1.2. *For an odd weight $w = p + q$ with $p, q \geq 2$, we have*

$$\begin{aligned}
S_{p,q} = &\frac{1}{2}\zeta(w) + \frac{1 - (-1)^p}{2}\zeta(p)\zeta(q) \\
&+ (-1)^p \sum_{k=0}^{[p/2]} \binom{w - 2k - 1}{q - 1}\zeta(2k)\zeta(w - 2k) \\
&+ (-1)^p \sum_{k=0}^{[q/2]} \binom{w - 2k - 1}{p - 1}\zeta(2k)\zeta(w - 2k).
\end{aligned}$$

B.2   The Shuffle Product Formula of Two Multiple Zeta Values

For our convenience, we let $\{1\}^k$ be the k repetitions of 1, or more general, let $\{a\}^k$ be the k repetitions of a. For example, we have

$$\zeta(\{1\}^5, 6) = \zeta(1, 1, 1, 1, 1, 6) \quad \text{and} \quad \zeta(\{2\}^4) = \zeta(2, 2, 2, 2).$$

Multiple zeta value $\zeta(\alpha_1, \alpha_2, \ldots, \alpha_r)$ can be represented by Drinfeld iterated integrals over the simplex

$$E_{|\boldsymbol{\alpha}|} : 0 < t_1 < t_2 < \cdots < t_{|\boldsymbol{\alpha}|} < 1$$

as

$$\int_{E_{|\boldsymbol{\alpha}|}} \Omega_1 \Omega_2 \cdots \Omega_{|\boldsymbol{\alpha}|}$$

with

$$\Omega_j = \begin{cases} \frac{dt_j}{1-t_j}, & \text{if } j = 1, \alpha_1 + 1, \alpha_1 + \alpha_2 + 1, \ldots, \alpha_1 + \alpha_2 + \cdots + \alpha_{r-1} + 1; \\ \frac{dt_j}{t_j}, & \text{otherwise.} \end{cases}$$

For examples:

$$\zeta(3) = \int_{E_3} \frac{dt_1}{1-t_1} \frac{dt_2}{t_2} \frac{dt_3}{t_3}$$

and

$$\zeta(\{1\}^2, 3) = \int_{E_5} \frac{dt_1}{1-t_1} \frac{dt_2}{1-t_2} \frac{dt_3}{1-t_3} \frac{dt_4}{t_4} \frac{dt_5}{t_5}.$$

Usually, we write the integral representation symbolically as

$$\int_0^1 \Omega_1 \Omega_2 \cdots \Omega_{|\alpha|}.$$

It always begins with $\Omega_1 = \frac{dt_1}{1-t_1}$ and end up with $\Omega_{|\alpha|} = \frac{dt_{|\alpha|}}{t_{|\alpha|}}$. In particular, for all integers $m, n \geq 0$, we have

$$\zeta(\{1\}^m, n+2) = \int_{E_{m+n+2}} \prod_{j=1}^{m+1} \frac{dt_j}{1-t_j} \prod_{k=m+2}^{m+n+2} \frac{dt_k}{t_k}.$$

Note that the change of variables:

$$u_1 = 1 - t_{m+n+2}, u_2 = 1 - t_{m+n+1}, \ldots, u_{m+n+2} = 1 - t_1$$

yields the duality theorem:

$$\zeta(\{1\}^m, n+2) = \zeta(\{1\}^n, m+2).$$

This is just a special case of Drinfeld duality theorem.

Theorem B.2.1 (Drinfeld duality theorem). *Suppose that $a_1, b_1; a_2, b_2; \ldots; a_n, b_n$ are n pairs of nonnegative integers,*

$$\mathbf{k} = (\{1\}^{a_1}, b_1 + 2, \{1\}^{a_2}, b_2 + 2, \ldots, \{1\}^{a_n}, b_n + 2)$$

and

$$\mathbf{k}' = \left(\{1\}^{b_n}, a_n + 2, \{1\}^{b_{n-1}}, a_{n-1} + 2, \ldots, \{1\}^{b_1}, a_1 + 2 \right).$$

Then $\zeta(\mathbf{k}) = \zeta(\mathbf{k}')$.

Some multiple zeta values can be evaluated through the duality theorem. For example,

$$\zeta(\{1\}^m, 2) = \zeta(m + 2).$$

Once multiple zeta values are expressed in Drinfeld iterated integrals, the shuffle product formula of two multiple zeta values take the form

$$\int_0^1 \Omega_1\Omega_2\cdots\Omega_m \int_0^1 \Omega_{m+1}\Omega_{m+2}\cdots\Omega_{m+n} = \sum_\sigma \int_0^1 \Omega_{\sigma(1)}\Omega_{\sigma(2)}\cdots\Omega_{\sigma(m+n)},$$

where σ ranges over all $(m + n)!/(m!n!)$ permutations of the set $\{1, 2, \ldots, m + n\}$ which preserves of orderings of $\Omega_1\Omega_2\cdots\Omega_m$ and $\Omega_{m+1}\Omega_{m+2}\cdots$ Ω_{m+n}. More precisely, for $1 \le i < j \le m$ or $m + 1 \le i < j \le m + n$, we have

$$\sigma^{-1}(i) < \sigma^{-1}(j).$$

In general, it is not easy to perform the shuffle precess for two arbitrary multiple zeta values. Instead we choose special multiple zeta values of the form $\zeta(\{1\}^m, n + 2))$ or a sum of multiple zeta values as our candidates. However, we first have to reduce the number of variables of our candidates.

Proposition B.2.2. *For nonnegative integers m, n, we have*

$$\zeta(\{1\}^m, n + 2) = \frac{1}{m!n!} \int_{0<t_1<t_2<1} \left(\log\frac{1}{1-t_1}\right)^m \left(\log\frac{1}{t_2}\right)^n \frac{dt_1}{1-t_1}\frac{dt_2}{t_2}$$

$$= \frac{1}{m!n!} \int_{0<t_1<t_2<1} \left(\log\frac{1}{1-t_1}\right)^m \left(\log\frac{t_2}{t_1}\right)^n \frac{dt_1}{1-t_1}\frac{dt_2}{t_2}$$

$$= \frac{1}{m!(n+1)!} \int_0^1 \left(\log\frac{1}{1-t}\right)^m \left(\log\frac{1}{t}\right)^{n+1} \frac{dt}{1-t}.$$

Proposition B.2.3. *For nonnegative integers p, m, n, r, we have*

$$\sum_{|\alpha|=m+r+1} \zeta(\{1\}^p, \alpha_0, \alpha_1, \ldots, \alpha_r + n + 1)$$

$$= \frac{1}{p!m!r!n!} \int_{0<t_1<t_2<1} \left(\log\frac{1}{1-t_1}\right)^p \left(\log\frac{1-t_1}{1-t_2}\right)^r \left(\log\frac{t_2}{t_1}\right)^m$$

$$\times \left(\log\frac{1}{t_2}\right)^n \frac{dt_1}{1-t_1}\frac{dt_2}{t_2}.$$

B.3 Some Basic Shuffle Relations

In this section, we perform the shuffle process of a product of two single zeta values to reproof the classical Euler's decomposition theorem. Then we perform the shuffle process of a product of two multiple zeta values of the form $\zeta(\{1\}^m, n+2)$ to give a more general decomposition theorem.

Proposition B.3.1 (Euler's Decomposition Theorem). *For a pair of positive integers p, q,*

$$\zeta(p+1)\zeta(q+1) = \sum_{|\boldsymbol{\alpha}|=p+q+1} \zeta(\alpha_1, \alpha_2+1)\left\{\binom{\alpha_2}{p} + \binom{\alpha_2}{q}\right\}.$$

Proof. By Proposition B.2.2, we have

$$\zeta(p+1)\zeta(q+1) = \frac{1}{p!q!}\int_0^1\int_0^1 \left(\log\frac{1}{t}\right)^p \left(\log\frac{1}{u}\right)^q \frac{dt}{1-t}\frac{du}{1-u}.$$

Subdivide the region of integration into simplices as

$$D_1: \ 0 < t < u < 1 \quad \text{and} \quad D_2: \ 0 < u < t < 1.$$

On the simplex D_1, we replace the factor $\left(\log\frac{1}{t}\right)^p$ by its binomial expression

$$\sum_{a=0}^p \frac{p!}{a!(p-a)!}\left(\log\frac{u}{t}\right)^a \left(\log\frac{1}{u}\right)^{p-a},$$

so that the value of integration over D_1 is

$$\sum_{a=0}^p \binom{p+q-a}{q}\zeta(a+1, p+q-a+1)$$

or

$$\sum_{|\boldsymbol{\alpha}|=p+q+1} \zeta(\alpha_1, \alpha_2+1)\binom{\alpha_2}{q}.$$

On the other hand, the value of the integration over D_2 is

$$\sum_{b=0}^q \binom{p+q-b}{p}\zeta(b+1, p+q-b+1)$$

or

$$\sum_{|\boldsymbol{\alpha}|=p+q+1} \zeta(\alpha_1, \alpha_2+1)\binom{\alpha_2}{p}.$$

Of course, the value of the integral is equal to the sums of the values of integration over D_1 and D_2. Consequently, our assertion follows. $\qquad\square$

Here is a immediate application. We let $p = \ell + 1$ and $q = (r - \ell) + 1$ with $0 \leq \ell \leq r$. Then Euler's decomposition theorem read as

$$\zeta(\ell+2)\zeta(r-\ell+2) = \sum_{|\alpha|=r+3} \zeta(\alpha_1, \alpha_2 + 1) \left\{ \binom{\alpha_2}{\ell+1} + \binom{\alpha_2}{r-\ell+1} \right\}.$$

Sum over all ℓ with $0 \leq \ell \leq r$ then yields

$$\sum_{\ell=0}^{r} \zeta(\ell+2)\zeta(r-\ell+2) = \sum_{|\alpha|=r+3} (2^{\alpha_2+1} - 2)\zeta(\alpha_1, \alpha_2 + 1) - 2\zeta(1, r+3).$$

In light of the evaluation of $\zeta(1, r+3)$ given by

$$\zeta(1, r+3) = \frac{r+3}{2}\zeta(r+4) - \frac{1}{2}\sum_{\ell=0}^{r} \zeta(\ell+2)\zeta(r-\ell+2),$$

and the sum formula, we get that

$$\sum_{|\alpha|=r+3} 2^{\alpha_2}\zeta(\alpha_1, \alpha_2 + 1) = \frac{r+5}{2}\zeta(r+4).$$

This is just the weighted Euler sum formula proved by Y. Ohno and W. Zudilin in 2008.

Here we produce a generalization of Euler's decomposition theorem and a few applications on double weighted sum formulas.

Theorem B.3.2 (A new theorem by the author). *For a pair of nonnegative integers k and r, we have*

$$\sum_{p+q=k} \binom{p}{j} \sum_{|\alpha|=q+r+3} \zeta(\{1\}^p, \alpha_0, \ldots, \alpha_q, \alpha_{q+1} + 1)\binom{\alpha_{q+1}}{\ell+1}$$

$$+ \sum_{p+q=k} \binom{p}{k-j} \sum_{|\alpha|=q+r+3} \zeta(\{1\}^p, \alpha_0, \ldots, \alpha_q, \alpha_{q+1} + 1)\binom{\alpha_{q+1}}{r-\ell+1}$$

$$= \zeta(\{1\}^j, r-\ell+2)\zeta(\{1\}^{k-j}, \ell+2)$$

for all $0 \leq j \leq k$ and $0 \leq \ell \leq r$.

Proof. By Proposition B.2.2, we have

$$\zeta(\{1\}^j, r-\ell+2)\zeta(\{1\}^{k-j}, \ell+2)$$

$$= \frac{1}{j!(k-j)!(r-\ell+1)!(\ell+1)!} \int_0^1 \int_0^1 \left(\log\frac{1}{1-t}\right)^j \left(\log\frac{1}{t}\right)^{r-\ell+1}$$

$$\times \left(\log\frac{1}{1-u}\right)^{k-j} \left(\log\frac{1}{u}\right)^{\ell+1} \frac{dt\,du}{(1-t)(1-u)}.$$

Again, we decompose the region of integration into two simplices as

$$D_1 : 0 < t < u < 1 \quad \text{and} \quad D_2 : 0 < u < t < 1.$$

On the simplex D_1, we replace the factors

$$\left(\log \frac{1}{1-u} \right)^{k-j} \quad \text{and} \quad \left(\log \frac{1}{t} \right)^{r-\ell+1},$$

by their binomial expansions

$$\sum_{a+b=k-j} \frac{(k-j)!}{a!b!} \left(\log \frac{1}{1-t} \right)^a \left(\log \frac{1-t}{1-u} \right)^b$$

and

$$\sum_{m+n=r-\ell+1} \frac{(r-\ell+1)!}{m!n!} \left(\log \frac{u}{t} \right)^m \left(\log \frac{1}{u} \right)^n.$$

So that the value of integration over D_1 is

$$\sum_{a+b=k-j} \sum_{m+n=r-\ell+1} \binom{j+a}{j} \binom{n+\ell+1}{\ell+1}$$
$$\times \sum_{|\alpha|=b+m+1} \zeta(\{1\}^{j+a}, \alpha_0, \alpha_1, \ldots, \alpha_b, n+\ell+2),$$

or

$$\sum_{p+q=k} \binom{p}{j} \sum_{|\alpha|=q+r+3} \zeta(\{1\}^p, \alpha_0, \alpha_1, \ldots, \alpha_q, \alpha_{q+1}+1) \binom{\alpha_{q+1}}{\ell+1}.$$

In a similar way, we evaluate the value of the integration over D_2 and our assertion follows. $\qquad\square$

Here we mention a few applications of our previous theorem.

(1) Just sum over all $0 \le j \le k$ and $0 \le \ell \le r$, we get

$$\sum_{p+q=k} 2^p \sum_{|\alpha|=q+r+3} (2^{\alpha_{q+1}} - 1)\zeta(\{1\}^p, \alpha_0, \ldots, \alpha_q, \alpha_{q+1}+1)$$
$$= (2^{k+1} - 1)\zeta(\{1\}^{k+1}, r+3)$$
$$+ \frac{1}{2} \sum_{j=0}^{k} \sum_{\ell=0}^{r} \zeta(\{1\}^j, r-\ell+2)\zeta(\{1\}^{k-j}, \ell+2).$$

(2) When r is even, sum over all $0 \le j \le k$, and $0 \le \ell \le r$ with ℓ even yields

$$\sum_{p+q=k} 2^p \sum_{|\alpha|=q+r+3} 2^{\alpha_{q+1}+1} \zeta(\{1\}^p, \alpha_0, \ldots, \alpha_q, \alpha_{q+1}+1)$$

$$= \sum_{j=0}^{k} \sum_{\substack{\ell=0 \\ \ell:\text{even}}}^{r} \zeta(\{1\}^j, r-\ell+2)\zeta(\{1\}^{k-j}, \ell+2).$$

(3) When k is even, multiply both sides by $(-1)^j$ and then sum over all $0 \le j \le k$ and $0 \le \ell \le r$, we get

$$\sum_{|\alpha|=k+r+3} \zeta(\alpha_0, \ldots, \alpha_k, \alpha_{k+1}+1)2^{\alpha_{k+1}+1}$$

$$= \zeta(k+r+4) + \zeta(\{1\}^{k+1}, r+3)$$

$$+ \frac{1}{2}\sum_{j=0}^{k}\sum_{\ell=0}^{r}(-1)^j\zeta(\{1\}^j, r-\ell+2)\zeta(\{1\}^{k-j}, \ell+2).$$

(4) Multiply both sides of the identity by $(-1)^{j+\ell}$ and then sum over all $0 \le j \le k$ and $0 \le \ell \le r$, we get

$$\{1+(-1)^{k+r}\} \sum_{|\alpha|=k+r+3} \zeta(\alpha_0, \alpha_1, \ldots, \alpha_k, \alpha_{k+1}+1)$$

$$+ \{(-1)^k + (-1)^r\} \zeta(\{1\}^{k+1}, r+3)$$

$$= \sum_{j=0}^{k}\sum_{\ell=0}^{r}(-1)^{j+\ell}\zeta(\{1\}^j, r-\ell+2)\zeta(\{1\}^{k-j}, \ell+2).$$

(5) When k is even, we have

$$\sum_{|\alpha|=k+r+3} \zeta(\alpha_0, \ldots, \alpha_k, \alpha_{k+1}+1)\binom{\alpha_{k+1}+r}{r+1}$$

$$= \frac{1}{2}\sum_{j=0}^{k}\sum_{\ell=0}^{r}(-1)^j\binom{r}{\ell}\zeta(\{1\}^j, r-\ell+2)\zeta(\{1\}^{k-j}, \ell+2).$$

Here we need the following identity:

$$\sum_{\ell=0}^{r}\binom{\alpha_{k+1}}{\ell+1}\binom{r}{r-\ell} = \binom{\alpha_{k+1}+r}{r+1}.$$

B.4   Shuffle Relations of Two Sums of Multiple Zeta Values

To obtain the shuffle relations of two sums of multiple zeta values we need to express a sum

$$\sum_{|\alpha|=p+q+1} \zeta(\alpha_0, \alpha_1, \ldots, \alpha_p + 1)$$

as a double integral

$$\frac{1}{p!q!} \int_{0<t_1<t_2<1} \left(\log \frac{1-t_1}{1-t_2}\right)^p \left(\log \frac{t_2}{t_1}\right)^q \frac{dt_1 dt_2}{(1-t_1)t_2}.$$

Besides, we need the following special case of Ohno's generalization of the duality and sum formula.

Proposition B.4.1. *For a pair of integers $k, r \geq 0$, we have*

$$\sum_{j=0}^{k}\sum_{\ell=0}^{r}(-1)^j \sum_{\substack{|\alpha|=j+r-\ell+1 \\ |\beta|=k-j+\ell+1}} \zeta(\alpha_0, \alpha_1, \ldots, \alpha_j + 1, \beta_0, \beta_1, \ldots, \beta_{k-j} + 1)$$

$$=\sum_{j=0}^{k}\sum_{\ell=0}^{r}(-1)^j\zeta(k-j+\ell+2, j+(r-\ell)+2).$$

For a pair of nonnegative integers k, r with k even, we consider the integral

$$\frac{1}{k!r!} \iint_{R_1 \times R_2} \left(\log \frac{1-u_1}{1-u_2} - \log \frac{1-t_1}{1-t_2}\right)^k \left(\log \frac{t_2}{t_1} + \log \frac{u_2}{u_1}\right)^r$$

$$\times \frac{dt_1 dt_2}{(1-t_1)t_2} \frac{du_1 du_2}{(1-u_1)u_2},$$

where $R_1 : 0 < t_1 < t_2 < 1$ and $R_2 : 0 < u_1 < u_2 < 1$. Such an integral is separable and its value is equal to

$$\sum_{j=0}^{k}\sum_{\ell=0}^{r}(-1)^j\zeta(j+r-\ell+2)\zeta(k-j+\ell+2).$$

As a replacement of the standard shuffle process, we decompose $R_1 \times R_2$

into 6 simplices obtained from all possible interlacings of t_1, t_2 and u_1, u_2:.

$$D_1 : \ 0 < t_1 < t_2 < u_1 < u_2 < 1,$$
$$D_2 : \ 0 < u_1 < u_2 < t_1 < t_2 < 1,$$
$$D_3 : \ 0 < t_1 < u_1 < t_2 < u_2 < 1,$$
$$D_4 : \ 0 < t_1 < u_1 < u_2 < t_2 < 1,$$
$$D_5 : \ 0 < u_1 < t_1 < u_2 < t_2 < 1 \text{ and}$$
$$D_6 : \ 0 < u_1 < t_1 < t_2 < u_2 < 1.$$

As the integrand is invariant if exchange t_1, t_2 and u_1, u_2. We only need to evaluate the integrations over D_1, D_3 and D_4. The value of the integration over D_1 is given by Proposition B.4.1. On the simplex D_3, we substitute the factors

$$\left(\log \frac{1-u_1}{1-u_2} - \log \frac{1-t_1}{1-t_2} \right)^k \quad \text{and} \quad \left(\log \frac{t_2}{t_1} + \log \frac{u_2}{u_1} \right)^r$$

by

$$\sum_{j=0}^{k} \binom{k}{j} (-1)^j \left(\log \frac{1-t_1}{1-u_1} \right)^j \left(\log \frac{1-t_2}{1-u_2} \right)^{k-j}$$

and

$$\sum_{m+n+\ell=r} \frac{r!}{m!n!\ell!} \left(\log \frac{u_1}{t_1} \right)^m \left(2 \log \frac{t_2}{u_1} \right)^n \left(\log \frac{u_2}{t_2} \right)^{\ell}.$$

So that the value of integration over D_3 is

$$\sum_{j=0}^{k} (-1)^j \sum_{m+n+\ell=r} 2^n$$
$$\times \sum_{\substack{|\alpha|=j+m+1 \\ |\beta|=k-j+\ell+1}} \zeta(\alpha_0, \alpha_1, \ldots, \alpha_j, \beta_0 + n + 1, \beta_1, \ldots, \beta_{k-j} + 1),$$

which is rewritten as

$$\sum_{j=0}^{k} (-1)^j \sum_{|\alpha|=k+r+3} \left(2^{\alpha_{j+1}-1} - 1 \right) \zeta(\alpha_0, \ldots, \alpha_k, \alpha_{k+1} + 1).$$

In a similar way, we get the value of integration over D_4 as

$$\sum_{j=0}^{k} \sum_{|\alpha|=k+r+3} \left(2^{\alpha_{j+1}-1} - 1\right) \zeta(\alpha_0, \ldots, \alpha_k, \alpha_{k+1} + 1).$$

This leads to the following theorem.

Theorem B.4.2. *For a pair of nonnegative integers k, r with k even, we have*

$$\sum_{\substack{1 \le j \le k+1 \\ j:odd}} \sum_{|\alpha|=k+r+3} \left(2^{\alpha_j-1} - 1\right) \zeta(\alpha_0, \ldots, \alpha_k, \alpha_{k+1}+1) = \frac{1}{4}(r+1)\zeta(k+r+4).$$

In light of the sum formula, we rewrite the above theorem as the following.

Theorem B.4.3. *For a pair of positive integers n, k with $n \ge k$ and k is even, we have*

$$\sum_{|\alpha|=n} \left(\sum_{\substack{1 \le j \le k \\ j:even}} 2^{\alpha_j} \right) \zeta(\alpha_1, \alpha_2, \ldots, \alpha_k + 1) = \frac{n+k}{2}\zeta(n+1).$$

When k is odd, we have the following theorem.

Theorem B.4.4. *For a pair of positive integers n, k with k odd and $n > k \ge 3$, we have*

$$(k-2) \sum_{\substack{j=2 \\ j:even}}^{k} \sum_{|\alpha|=n} 2^{\alpha_j} \zeta(\alpha_1, \ldots, \alpha_{k-1}, \alpha_k + 1)$$

$$+ \sum_{j=3}^{k} (j-2)(-1)^{j+1} \sum_{|\alpha|=n} 2^{\alpha_j} \zeta(\alpha_1, \ldots, \alpha_{k-1}, \alpha_k + 1)$$

$$+ \sum_{\substack{j=2 \\ j:even}}^{k-1} \sum_{|\alpha|=n} 2^{\alpha_j} \left(2^{\alpha_{j+1}} - 2\right) \zeta(\alpha_1, \ldots, \alpha_{k-1}, \alpha_k + 1)$$

$$= \frac{(k-1)(n+k-2)}{2}\zeta(n+1).$$

B.5 The Generating Function of Height One

The height of a multiple zeta value $\zeta(\alpha_1, \alpha_2, \ldots, \alpha_r)$ is defined as the number of element of the set

$$\{j \mid 1 \leq j \leq r, \alpha_j \geq 2\}.$$

In particular, multiple zeta values of height one has the form $\zeta(\{1\}^m, n+2)$. Here we give a new proof for the generating function of multiple zeta values of height one.

Theorem B.5.1. *The generating function of* $\zeta(\{1\}^m, n+2)$ *is given by*

$$\sum_{m=0}^{\infty} \sum_{n=0}^{\infty} \zeta(\{1\}^m, n+2) x^{m+1} y^{n+1}$$

$$= 1 - \frac{\Gamma(1-x)\Gamma(1-y)}{\Gamma(1-x-y)}$$

$$= 1 - \exp\left\{\sum_{k=2}^{\infty} (x^k + y^k - (x+y)^k) \frac{\zeta(k)}{k}\right\}.$$

Proof. For the time being, we assume that $x < 0$. By Proposition B.2.2, we have

$$\sum_{m=0}^{\infty} \sum_{n=0}^{\infty} \zeta(\{1\}^m, n+2) x^m y^{n+1}$$

$$= \sum_{m=0}^{\infty} \sum_{n=0}^{\infty} \frac{x^m y^{n+1}}{m!(n+1)!} \int_0^1 \left(\log \frac{1}{1-t}\right)^m \left(\log \frac{1}{t}\right)^{n+1} \frac{dt}{1-t}$$

$$= \int_0^1 (1-t)^{-x-1}(t^{-y} - 1) dt.$$

Under the condition $x < 0$, the above integral can be separated into two convergent improper integrals and the value is

$$\frac{1}{x} + \frac{\Gamma(-x)\Gamma(1-y)}{\Gamma(1-x-y)}$$

provides that $0 \leq y < 1$ and $0 \leq x + y < 1$. Multiply both sides by x and employ the functional equation

$$(-x)\Gamma(-x) = \Gamma(1-x),$$

we concluded that the generating function is

$$1 - \frac{\Gamma(1-x)\Gamma(1-y)}{\Gamma(1-x-y)}.$$

The condition $x < 0$ disappears automatically. The second part of the identity follows from the infinite product of gamma function

$$\frac{1}{\Gamma(1+s)} = e^{\gamma s} \prod_{n=1}^{\infty} (1 + \frac{s}{n}) e^{-s/n}.$$

$\square$

Remark B.5.2. As the generating function is invariant under the interexchange of x and y, this implies the Drinfeld dual theorem:

$$\zeta(\{1\}^m, n+2) = \zeta(\{1\}^n, m+2).$$

Besides, multiple zeta values of the form $\zeta(\{1\}^m, n+2)$ can be evaluated in terms of single zeta values.

Appendix (i): Double Weighted Sum Formulas

For a pair of nonnegative integers k, r and complex numbers μ, λ, we let

$$E(\mu, \lambda) = \sum_{p+q=k} \mu^p \sum_{|\alpha|=q+r+3} \zeta(\{1\}^p, \alpha_0, \ldots, \alpha_q, \alpha_{q+1} + 1)\lambda^{\alpha_{q+1}}.$$

The following theorems come from shuffle relations of some particular integrals such as

$$\frac{1}{k!r!} \iint_{R_1 \times R_2} \left(\mu \log \frac{1}{1-t_1} + \log \frac{1}{1-u_1} \right)^k \left(\log \frac{1}{t_2} + \lambda \log \frac{1}{u_2} \right)^r$$
$$\times \frac{dt_1 dt_2}{(1-t_1)t_2} \frac{du_1 du_2}{(1-u_1)u_2}$$

or producing from the basic shuffle relations

$$\sum_{p+q=k} \binom{p}{j} \sum_{|\alpha|=q+r+3} \zeta(\{1\}^p, \alpha_0, \ldots, \alpha_q, \alpha_{q+1} + 1) \binom{\alpha_{q+1}}{\ell+1}$$
$$+ \sum_{p+q=k} \binom{p}{k-j} \sum_{|\alpha|=q+r+3} \zeta(\{1\}^p, \alpha_0, \ldots, \alpha_q, \alpha_{q+1} + 1) \binom{\alpha_{q+1}}{r-\ell+1}.$$

Theorem A. *Notation as above, then we have*

$$E(\mu + 1, \lambda + 1) - E(\mu + 1, 1)$$

$$+ \mu^k \lambda^{r+2} \left\{ E\left(\frac{\mu+1}{\mu}, \frac{\lambda+1}{\lambda}\right) - E\left(\frac{\mu+1}{\mu}, 1\right) \right\}$$

$$= \left\{ \frac{1}{\mu}[(\mu+1)^{k+1} - 1]\lambda^{r+2} + [(\mu+1)^{k+1} - \mu^{k+1}] \right\} \zeta(\{1\}^{k+1}, r+3)$$

$$+ \sum_{j=0}^{k} \sum_{\ell=0}^{r} \mu^j \lambda^{\ell+1} \zeta(\{1\}^j, r - \ell + 2) \zeta(\{1\}^{k-j}, \ell + 2).$$

Let

$$E_1(\mu, \lambda) = \frac{\partial}{\partial \mu} E(\mu, \lambda)$$

$$= \sum_{p+q=k} p\mu^{p-1} \sum_{|\alpha|=q+r+3} \zeta(\{1\}^p, \alpha_0, \ldots, \alpha_q, \alpha_{q+1} + 1)\lambda^{\alpha_{q+1}}.$$

Theorem B. *Notation as above, then we have*

$$E(\mu + 1, \lambda + 1) - E(\mu + 1, 1)$$

$$+ (k+1)\mu^k \lambda^{r+2} \left\{ E\left(\frac{\mu+1}{\mu}, \frac{\lambda+1}{\lambda}\right) - E\left(\frac{\mu+1}{\mu}, 1\right) \right\}$$

$$+ \mu\left[E_1(\mu+1, \lambda+1) - E_1(\mu+1, 1)\right]$$

$$- \mu^{k-1}\lambda^{r+2} \left\{ E_1\left(\frac{\mu+1}{\mu}, \frac{\lambda+1}{\lambda}\right) - E_1\left(\frac{\mu+1}{\mu}, 1\right) \right\}$$

$$= \left\{ (k+1)(\mu+1)^k \lambda^{r+2} + (\mu+1)^k[(k+2)\mu + 1] - (k+2)\mu^{k+1} \right\}$$

$$\times \zeta(\{1\}^{k+1}, r+3)$$

$$+ \sum_{j=0}^{k} \sum_{\ell=0}^{r} (j+1)\mu^j \lambda^{\ell+1} \zeta(\{1\}^j, r - \ell + 2) \zeta(\{1\}^{k-j}, \ell + 2).$$

In Theorem A, if we let $\mu = -1$ and $\lambda = 1$, we get

$$\sum_{|\alpha|=k+r+3} (2^{\alpha_{k+1}} - 1)\zeta(\alpha_0, \ldots, \alpha_k, \alpha_{k+1} + 1)$$

$$= \zeta(\{1\}^{k+1}, r+3) + \frac{1}{2} \sum_{j=0}^{k} \sum_{\ell=0}^{r} (-1)^j \zeta(\{1\}^j, r - \ell + 2)\zeta(\{1\}^{k-j}, \ell + 2)$$

when k is even.

On the other hand, when k is odd, we let $\mu = -1$ and $\lambda = 1$ in Theorem B to obtain

$$k \sum_{|\alpha|=k+r+3} (2^{\alpha_{k+1}} - 1)\zeta(\alpha_0, \ldots, \alpha_k, \alpha_{k+1} + 1)$$

$$+ 2 \sum_{|\alpha|=k+r+2} (2^{\alpha_k} - 1)\zeta(1, \alpha_0, \ldots, \alpha_{k-1}, \alpha_k + 1)$$

$$= (k+2)\zeta(\{1\}^{k+1}, r+3)$$

$$+ \sum_{j=0}^{k} \sum_{\ell=0}^{r} (-1)^{j+1}(j+1)\zeta(\{1\}^j, r-\ell+2)\zeta(\{1\}^{k-j}, \ell+2).$$

Appendix (ii): Evaluations of Some Particular Integrals

Let

$$E(\lambda) = E(2, \lambda) = \sum_{p+q=k} 2^p \sum_{|\alpha|=q+r+3} \zeta(\{1\}^p, \alpha_0, \ldots, \alpha_q, \alpha_{q+1} + 1)\lambda^{\alpha_{q+1}}$$

and

$$P(\mu, \lambda, \nu) = \frac{1}{k!r!} \iint_{R_1 \times R_2} \left(\log \frac{1}{1-t_1} + \log \frac{1}{1-u_1} \right)^k$$
$$\times \left(\mu \log \frac{t_2}{t_1} + \lambda \log \frac{u_1}{t_2} + \nu \log \frac{u_2}{u_1} \right)^r \frac{dt_1 dt_2}{(1-t_1)t_2} \frac{du_1 du_2}{(1-u_1)u_2}.$$

In terms of multiple zeta values, $P(\mu, \lambda, \nu)$ can be expressed as

$$\sum_{a+b+c=k} 2^a \sum_{m+n+p=r} \mu^m \lambda^n \nu^p$$

$$\times \sum_{\substack{|\alpha|=b+m+1 \\ |\beta|=c+n+1}} \zeta(\{1\}^a, \alpha_0, \ldots, \alpha_{b-1}, \alpha_b + \beta_0, \beta_1, \ldots, \beta_c, p+2).$$

Theorem C. *For complex numbers* μ, λ, ν *with* $\mu \neq \lambda$ *and* $\mu\lambda\nu \neq 0$, *we have*

$$P(\mu, \lambda, \nu) = \frac{1}{\nu(\lambda-\mu)} \left\{ \lambda^{r+2} E\left(\frac{\nu}{\lambda}\right) - \mu^{r+2} E\left(\frac{\nu}{\mu}\right) \right\}.$$

APPENDIX C

The Sum Formula and Their Generalizations

*The sum formula

$$\sum_{\substack{|\boldsymbol{\alpha}|=m+r+1 \\ \alpha_j \geq 1}} \zeta(\alpha_0, \alpha_1, \ldots, \alpha_r + 1) = \zeta(m+r+2)$$

asserted that the sum of all multiple zeta values of the same depth $r+1$ and weight $m+r+2$ is equal to a single zeta values $\zeta(m+r+2)$. In particular, when $m = 0$, it implies that

$$\zeta(\{1\}^r, 2) = \zeta(r+2)$$

which is a special case of duality. It was originally conjectured by C. Moen and M. Schmidt independently around 1994. It was proved for the case of depth 2 ($r = 1$) by Euler a long time ago and for the case of depth 3 ($r = 2$) by Hoffman and Moen in 1996. A. Granville proved the general cases in 1996 and he mentioned that it was also proved independently by

---

*This is a lecture note of the talk with the same title, given at KIAS of Korea in August 26, 2011 on modular forms and their geometric applications.

D. Zagier in one of his unpublished papers. Zagier had also made a remark: Although this proof is not very long, it seems too complicated compared with the elegance of the statement.

C.1   The Double Integral Representation of a Sum

A multiple zeta value of depth $r+1$ and weight $m+r+2$ can be expressed in iterated integral as

$$(\text{C.1.1}) \qquad \int_{0<t_1<t_2<\cdots<t_{m+r+2}<1} \Omega_1\Omega_2\cdots\Omega_{m+r+2}$$

with

$$\Omega_j = \begin{cases} \frac{dt_j}{1-t_j}, & \text{if } j = 1,\ \alpha_0+1,\ \alpha_0+\alpha_1+1,\ \ldots,\ \alpha_0+\alpha_1+\cdots+\alpha_{r-1}+1; \\ \frac{dt_j}{t_j}, & \text{otherwise.} \end{cases}$$

For example, we have

$$\zeta(1,1,2) = \int_{0<t_1<t_2<t_3<t_4<1} \frac{dt_1}{1-t_1}\frac{dt_2}{1-t_2}\frac{dt_3}{1-t_3}\frac{dt_4}{t_4}$$

and

$$\zeta(2,1,3) = \int_{0<t_1<\cdots<t_6<1} \frac{dt_1}{1-t_1}\frac{dt_2}{t_2}\frac{dt_3}{1-t_3}\frac{dt_4}{1-t_4}\frac{dt_5}{t_5}\frac{dt_6}{t_6}.$$

Fix $t_1, t_{\alpha_0+1}, \ldots, t_{\alpha_0+\cdots+\alpha_{r-1}+1}$ and t_{m+r+2} and integrate with respect to the remaining variables, we obtain another integral representation as

$$(\text{C.1.2})$$

$$\int_{0<t_1<t_2<\cdots<t_{r+2}<1} \left\{ \prod_{j=1}^{r+1} \frac{dt_j}{1-t_j}\frac{1}{(\alpha_{j-1}-1)!} \left(\log\frac{t_{j+1}}{t_j}\right)^{\alpha_{j-1}-1} \right\} \frac{dt_{r+2}}{t_{r+2}}.$$

Let $\beta_0 = \alpha_0 - 1, \beta_1 = \alpha_1 - 1, \ldots, \beta_r = \alpha_r - 1$. Sum over all nonnegative integers $\beta_0, \beta_1, \ldots, \beta_r$ with $|\boldsymbol{\beta}| = m$ and in light of the multinomial theorem:

$$\sum_{|\boldsymbol{\beta}|=m} \frac{1}{\beta_0!\beta_1!\cdots\beta_r!} \left(\log\frac{t_2}{t_1}\right)^{\beta_0} \left(\log\frac{t_3}{t_2}\right)^{\beta_1} \cdots \left(\log\frac{t_{r+2}}{t_{r+1}}\right)^{\beta_r}$$

$$= \frac{1}{m!} \left(\log\frac{t_2}{t_1} + \log\frac{t_3}{t_2} + \cdots + \log\frac{t_{r+2}}{t_{r+1}}\right)^m$$

$$= \frac{1}{m!} \left(\log\frac{t_{r+2}}{t_1}\right)^m,$$

we conclude that

$$\sum_{|\alpha|=m+r+1} \zeta(\alpha_0, \alpha_1, \ldots, \alpha_r + 1)$$

$$= \frac{1}{m!} \int_{0<t_1<\cdots<t_{r+2}<1} \left(\log \frac{t_{r+2}}{t_1}\right)^m \left(\prod_{j=1}^{r+1} \frac{dt_j}{1-t_j}\right) \frac{dt_{r+2}}{t_{r+2}}.$$

Fix t_1 and t_{r+2} and integrate with respect to $t_2, \ldots, t_{r+2}$, we obtain the following theorem.

Theorem C.1.1. *For a pair of integers $m, r \geq 0$, we have*

(C.1.3)
$$\sum_{|\alpha|=m+r+1} \zeta(\alpha_0, \alpha_1, \ldots, \alpha_r + 1)$$

$$= \frac{1}{m!r!} \int_{0<t_1<t_2<1} \left(\log \frac{1-t_1}{1-t_2}\right)^r \left(\log \frac{t_2}{t_1}\right)^m \frac{dt_1}{1-t_1} \frac{dt_2}{t_2}.$$

Once a sum of multiple zeta values is expressed as a double integral over a simplex, the sum formula follows with a change of variables. Proof of the sum formula: Let

$$x_1 = \log \frac{1}{1-t_1} \quad \text{and} \quad x_2 = \log \frac{t_2}{t_1}$$

be new variables in place of t_1 and t_2. A direct calculation shows

$$dx_1 dx_2 = \frac{dt_1}{1-t_1} \frac{dt_2}{t_2} \quad \text{and} \quad \log \frac{1-t_1}{1-t_2} = \log \frac{1}{e^{x_1} + e^{x_2} - e^{x_1+x_2}},$$

and the double integral is transformed into the integral

$$\frac{1}{m!r!} \int_{D_2} x_2^m \left(\log \frac{1}{e^{x_1} + e^{x_2} - e^{x_1+x_2}}\right)^r dx_1 dx_2$$

where

$$D_2 = \{(x_1, x_2) | x_1 > 0, x_2 > 0, e^{x_1} + e^{x_2} > e^{x_1+x_2}\}.$$

Note that D_2 is invariant under the exchange of x_1 and x_2, so that the integral is also equal to

$$\frac{1}{m!r!} \int_{D_2} x_1^m \left(\log \frac{1}{e^{x_1} + e^{x_2} - e^{x_1+x_2}}\right)^r dx_1 dx_2$$

or

$$\frac{1}{m!r!} \int_{0<t_1<t_2<1} \left(\log \frac{1}{1-t_1}\right)^m \left(\log \frac{1-t_1}{1-t_2}\right)^r \frac{dt_1}{1-t_1} \frac{dt_2}{t_2}$$

which is $\zeta(\{1\}^{m+r}, 2)$. By Drinfeld duality theorem, it is $\zeta(m+r+2)$.

C.2   Euler Sums with Two Branches

Euler sums with two branches first appeared in the author's proof of the restricted sum formula. For a pair of integers $p, q \geq 0$ and a positive integer $n \geq 2$, an Euler sum with two branches is defined as

$$(C.2.1) \qquad G_n(p,q) = \sum_{k_0=1}^{\infty} \frac{1}{k_0^n} \sum_{\mathbf{k}} \frac{1}{k_1 k_2 \cdots k_q} \sum_{\boldsymbol{\ell}} \frac{1}{\ell_1 \ell_2 \cdots \ell_p},$$

where $\mathbf{k}$ ranges over all positive integers $k_1, k_2, \ldots, k_q$ with

$$k_0 \geq k_1 \geq k_2 \geq \cdots \geq k_q$$

and $\boldsymbol{\ell}$ ranges over all positive integers $\ell_1, \ell_2, \ldots, \ell_p$ with

$$k_0 > \ell_1 > \ell_2 > \cdots > \ell_p \geq 1.$$

$G_n(p,q)$ has another expression in free variables as

$$(C.2.2) \qquad \sum_{k \in \mathbb{N}^{p+1}} \sum_{j \in \mathbb{N}^q} \frac{1}{\sigma_1 \sigma_2 \cdots \sigma_p \sigma_{p+1}^{n-1} \tau_1 \cdots \tau_q (\sigma_{p+1} + \tau_q)}$$

with $\sigma_j = k_1 + k_2 + \cdots + k_j$ and $\tau_\ell = j_1 + j_2 + \cdots + j_\ell$. Therefore it has iterated integral representation as

$$(C.2.3) \qquad \int_D \prod_{j=1}^{p+1} \frac{dt_j}{1 - t_j} \prod_{k=p+2}^{p+n-1} \frac{dt_k}{t_k} \left(\prod_{j=1}^{q} \frac{du_j}{1 - u_j} \right) \frac{dt_{p+n}}{t_{p+n}}$$

where D is defined as

$$0 < t_1 < \cdots < t_{p+n} < 1 \text{ and } 0 < u_1 < \cdots < u_q < t_{p+n}.$$

Fix t_1 and t_{p+n} and integrate with respect to the remaining variables we obtain the double integral representation of $G_n(p,q)$.

Proposition C.2.1. *For a pair of integers $p, q \geq 0$ and a positive integer $n \geq 2$, we have*

$$(C.2.4) \qquad \begin{aligned} G_n(p,q) = \frac{1}{p!q!(n-2)!} \int_{0 < t_1 < t_2 < 1} & \left(\log \frac{1}{1-t_1} \right)^p \left(\log \frac{1}{1-t_2} \right)^q \\ & \times \left(\log \frac{t_2}{t_1} \right)^{n-2} \frac{dt_1}{(1-t_1)} \frac{dt_2}{t_2}. \end{aligned}$$

In view of both the double integral expressions of a sum and $G_n(p, q)$, we see immediately that

Proposition C.2.2. *For a pair of integers $m, r \geq 0$, we have*

$$\sum_{|\alpha|=m+r+1} \zeta(\alpha_0, \alpha_1, \ldots, \alpha_r + 1) = \sum_{p+q=r} (-1)^p G_{m+2}(p, q).$$

On the other hand, $G_n(p, q)$ has a decomposition as usual multiple zeta values of depth $p + 1, p + 2, \ldots, p + q + 1$:

$$(C.2.5) \quad G_{m+2}(p, q) = \sum_{g=p+1}^{p+q+1} \binom{g-1}{p} \sum_{|\alpha|=p+q+1} \zeta(\alpha_1, \alpha_2, \ldots, \alpha_g + m + 1).$$

Let

$$S_g = \sum_{|\alpha|=r+1} \zeta(\alpha_1, \alpha_2, \ldots, \alpha_g + m + 1), \quad g = 1, 2, \ldots, r + 1.$$

Then the decompositions of $G_{m+2}(p, r - p)$ $(p = 0, 1, \cdots, r)$ are as follow:

$$
\begin{array}{rlllll}
G_{m+2}(0, r) = & S_{r+1}+ & S_r & +\cdots+ & S_3+ & S_2+ & S_1, \\
G_{m+2}(1, r-1) = & rS_{r+1}+ & (r-1)S_r & +\cdots+ & 2S_3+ & S_2, \\
G_{m+2}(2, r-2) = & \binom{r}{2}S_{r+1}+ & \binom{r-1}{2}S_r & +\cdots+ & S_3, \\
& \vdots & & & \\
G_{m+2}(r-1, 1) = & \binom{r}{r-1}S_{r+1}+ & S_r, & & \\
G_{m+2}(r, 0) = & S_{r+1}. & & &
\end{array}
$$

It follows that

$$\sum_{p+q=r} (-1)^p G_{m+2}(p, q) = S_1 = \zeta(m + r + 2).$$

And hence the sum formula is proved once again.

Remark C.2.3. There is another elementary way to prove the identity

$$(C.2.6) \quad \sum_{|\alpha|=m+r+1} \zeta(\alpha_0, \alpha_1, \ldots, \alpha_r + 1) = \sum_{p=0}^{r} (-1)^p G_{m+2}(p, r - p)$$

via the well-known identity:

$$(C.2.7) \quad \sum_{|\beta|=m} a_0^{\beta_0} a_1^{\beta_1} \cdots a_r^{\beta_r} = \sum_{j=0}^{r} \frac{a_j^{m+r}}{\prod_{p\neq j}(a_j - a_p)}$$

for $r+1$ different complex variables $a_0, a_1, \ldots, a_r$. Just substitute $a_0, a_1, \ldots, a_r$ by

$$\frac{1}{k_0}, \frac{1}{k_0 + k_1}, \ldots, \frac{1}{k_0 + k_1 + \cdots + k_r}$$

into (C.2.7). Then multiply both sides by

$$\frac{1}{k_0(k_0 + k_1) \cdots (k_0 + k_1 + \cdots + k_{r-1})(k_0 + k_1 + \cdots + k_r)^2}$$

and sum over all positive integers $k_0, k_1, \ldots, k_r$, we get the identity (C.2.6). Along with the decomposition of $G_{m+2}(p, r - p)$ as

$$\sum_{g=p+1}^{r+1} \binom{g-1}{p} S_g$$

we get

$$\sum_{|\alpha|=m+r+1} \zeta(\alpha_0, \alpha_1, \ldots, \alpha_r + 1) = \sum_{p=0}^{r}(-1)^p \sum_{g=p+1}^{r+1} \binom{g-1}{p} S_g.$$

Exchange the order of summations, the sum is equal to

$$\sum_{g=1}^{r+1} S_g \sum_{p=0}^{g-1} \binom{g-1}{p}(-1)^p.$$

Note that the inner sum is zero unless $g = 1$. So that the sum is just S_1 which is $\zeta(m + r + 2)$.

C.3   Another Application of Euler Sums with Two Branches

For a pair of positive integers p, q and integer $k \geq 0$, we consider the double integral given by

$$\frac{1}{p!q!k!} \int_0^1 \int_0^1 \left(\log \frac{1}{1-t}\right)^p \left(\log \frac{1}{1-u}\right)^q \left(\log \frac{u}{t}\right)^k \frac{dt\,du}{t\,u}.$$

Substitute the factor $\left(\log \frac{u}{t}\right)^k$ by

$$\sum_{j=0}^{k} \binom{k}{j}(-1)^j \left(\log \frac{1}{t}\right)^{k-j} \left(\log \frac{1}{u}\right)^j,$$

we see that the value of the integral is

$$\sum_{j=0}^{k}(-1)^{j}\zeta(\{1\}^{p-1},j+2)\zeta(\{1\}^{q-1},k-j+2).$$

On the other hand, we decompose the region of integration into two simplices:

$$D_1 : 0 < t < u < 1 \text{ and } D_2 : 0 < u < t < 1.$$

The integrations over D_1 and D_2 that yield the values $G_{k+3}(p-1,q)$ and $(-1)^{k}G_{k+3}(q-1,p)$.

Theorem C.3.1. *For a pair of positive integers p, q and integer $k \geq 0$, we have*

$$G_{k+3}(p-1,q) + (-1)^{k}G_{k+3}(q-1,p)$$

$$= \sum_{j=0}^{k}(-1)^{j}\zeta(\{1\}^{p-1},j+2)\zeta(\{1\}^{q-1},k-j+2).$$

Here we mention an application of the above theorem.

In order to evaluate $\zeta(\{1\}^{2k},2n)$ in terms of multiple zeta values of lower depth, we first decompose $G_{2n}(p,q)$ with $p+q = 2k$ into linear combinations of $\zeta(\{1\}^{2k},2n)$ and S_j $(j = 1,2,\ldots,2k)$ as before. Therefore the alternating sum

$$\sum_{p=0}^{2k-1}(-1)^{p+1}G_{2n}(p,2k-p)$$

is equal to

$$\zeta(\{1\}^{2k},2n) - \zeta(2n+2k).$$

However, by the previous theorem, we have for $0 \leq p \leq 2k-1$ that

$$G_{2n}(p,2k-p) - G_{2n}(2k-1-p,p+1)$$

$$= \sum_{\alpha=2}^{2n-1}(-1)^{\alpha+1}\zeta(\{1\}^{p},2n+1-\alpha)\zeta(\{1\}^{2k-p-1},\alpha).$$

The same alternating sum is also equal to

$$\sum_{p=0}^{k-1}\sum_{\alpha=2}^{2n-1}(-1)^{\alpha+p+1}\zeta(\{1\}^{p},2n+1-\alpha)\zeta(\{1\}^{2k-p-1},\alpha).$$

Consequently, we obtain that

$$\zeta(\{1\}^{2k}, 2n)$$

$$= \zeta(2n + 2k) + \sum_{p=0}^{k-1} \sum_{\alpha=2}^{2n-1} (-1)^{\alpha+p+1} \zeta(\{1\}^p, 2n + 1 - \alpha) \zeta(\{1\}^{2k-p-1}, \alpha).$$

In the similar way, we obtain that

$$\zeta(\{1\}^{2k+1}, 2n + 1) = -\zeta(2n + 2k + 2)$$

$$+ \frac{1}{2} \sum_{p=0}^{2k} \sum_{\alpha=2}^{2n} (-1)^{\alpha+p} \zeta(\{1\}^p, 2n + 2 - \alpha) \zeta(\{1\}^{2k-p}, \alpha).$$

Remark C.3.2. We rewrite $G_{k+3}(p - 1, q)$ as

$$\frac{1}{(p-1)!q!(k+1)!} \int_{0<t<u<1} \left(\log \frac{1}{1-t} \right)^{p-1} \left(\log \frac{1}{1-u} \right)^q$$

$$\times \left(\log \frac{u}{t} \right)^{k+1} \frac{dt}{1-t} \frac{du}{u}$$

which is equal to

$$\frac{1}{(p-1)!q!(k+1)!} \int_{0<t<u<1} \left(\log \frac{1-t}{1-u} \right)^{k+1} \left(\log \frac{1}{t} \right)^q$$

$$\times \left(\log \frac{1}{u} \right)^{p-1} \frac{dt}{1-t} \frac{du}{u}$$

by a change of variables. Replace the factor $\left(\log \frac{1}{t} \right)^q$ by

$$\sum_{n=0}^{q} \binom{q}{b} \left(\log \frac{u}{t} \right)^{q-b} \left(\log \frac{1}{t} \right)^b$$

so that the value of integration is

$$\sum_{b=0}^{q} \sum_{|\alpha|=k+q+2-b} \zeta(\alpha_0, \ldots, \alpha_k, \alpha_{k+1} + p + b) \binom{p-1+b}{p-1}$$

or

$$\sum_{|\alpha|=k+p+q+1} \zeta(\alpha_0, \ldots, \alpha_k, \alpha_{k+1} + 1) \binom{\alpha_{k+1}}{p}.$$

Indeed, the previous theorem is equivalent to

$$\sum_{|\alpha|=k+p+q+1} \zeta(\alpha_0,\ldots,\alpha_k,\alpha_{k+1}+1)\left\{\binom{\alpha_{k+1}}{p}+(-1)^k\binom{\alpha_{k+1}}{q}\right\}$$
$$=\sum_{j=0}^{k}(-1)^j\zeta(\{1\}^{k-j},q+1)\zeta(\{1\}^j,p+1).$$

A theorem due to T. Arakawa and M. Kaneko.

C.4 The Restricted Sum Formula

The restricted sum

(C.4.1)
$$\sum_{|\alpha|=q+r+1} \zeta(\{1\}^p,\alpha_0,\alpha_1,\ldots,\alpha_q+1)$$

has the double integral representation
(C.4.2)
$$\frac{1}{p!q!r!}\int_{0<t_1<t_2<1}\left(\log\frac{1}{1-t_1}\right)^p\left(\log\frac{1-t_1}{1-t_2}\right)^q\left(\log\frac{t_2}{t_1}\right)^r\frac{dt_1}{1-t_1}\frac{dt_2}{t_2}.$$

Under the change of variables:

$$x_1=\log\frac{1}{1-t_1},\quad x_2=\log\frac{t_2}{t_1}.$$

The above integral is transformed into

$$\frac{1}{p!q!r!}\int_{D_2} x_1^p x_2^r\left(\log\frac{1}{e^{x_1}+e^{x_2}-e^{x_1+x_2}}\right)^q dx_1 dx_2.$$

Under another change of variables:

$$x_1=\log\frac{1-u_1}{1-u_2},\quad x_2=\log\frac{1}{u_2},$$

the integral is transformed as

$$\frac{1}{p!q!r!}\int_{0<u_1<u_2<1}\left(\log\frac{1-u_1}{1-u_2}\right)^p\left(\log\frac{u_2}{u_1}\right)^q\left(\log\frac{1}{u_2}\right)^r\frac{du_1}{1-u_1}\frac{du_2}{u_2}.$$

In terms of multiple zeta values, it is equal to

(C.4.3)
$$\sum_{|c|=p+q+1} \zeta(c_0,c_1,\ldots,c_p+r+1).$$

Remark C.4.1. The restricted sum can be expressed as

$$\sum_{b=0}^{q} \binom{p+b}{p} (-1)^b G_{r+2}(p+b, q-b).$$

Substitute the decomposition of $G_{r+2}(p+b, q-b)$ into the above sum, we obtain the restricted sum formula. This gives another proof of the restricted sum formula. Here is an elementary proof to the assertion. Substitute the decomposition of $G_{r+2}(p+b, q-b)$ as

$$\sum_{g=p+b+1}^{p+q+1} \binom{g-1}{p+b} S_g, \quad S_g = \sum_{|\alpha|=p+q+1} \zeta(\alpha_1, \ldots, \alpha_g + r + 1)$$

into the identity

$$\sum_{|\alpha|=q+r+1} \zeta(\{1\}^p, \alpha_0, \ldots, \alpha_q + 1) = \sum_{b=0}^{q} \binom{p+b}{p} (-1)^b G_{r+2}(p+b, q-b)$$

we get

$$\sum_{|\alpha|=q+r+1} \zeta(\{1\}^p, \alpha_0, \ldots, \alpha_q + 1) = \sum_{b=0}^{q} \binom{p+b}{p} (-1)^b \sum_{g=p+b+1}^{p+q+1} \binom{g-1}{p+b} S_g.$$

Exchange the order of summation so that the sum is rewritten as

$$\sum_{g=p+1}^{p+q+1} \binom{g-1}{p} S_g \sum_{b=0}^{g-p-1} \binom{g-p-1}{b} (-1)^b.$$

Again, the inner sum is zero unless $g - p - 1 = 0$. Therefore, the total sum is equal to S_{p+1} or

$$\sum_{|c|=p+q+1} \zeta(c_0, c_1, \ldots, c_p + r + 1).$$

Remark C.4.2. The identity produced by the author in [12] is

$$\sum_{|\alpha|=q+r+2} \zeta(\{1\}^p, \alpha_0, \ldots, \alpha_q + 1)$$

$$= G_{r+2}(p, q) - \sum_{j=1}^{q} \binom{p+j}{p} \sum_{|\alpha|=q-j+r+1} \zeta(\{1\}^{p+j}, \alpha_0, \ldots, \alpha_{q-j} + 1).$$

As

$$G_{r+2}(p,q) = \sum_{j=0}^{q} \binom{p+j}{p} S_{p+j+1},$$

the restricted sum formula is proved by an induction on q.

C.5 A Generating Function for Sums of Multiple Zeta Values

Now we are going to investigate a generating function concerning the sum formula.

Proposition C.5.1.

$$\sum_{r=0}^{\infty} \sum_{m=0}^{\infty} \sum_{|\alpha|=m+r+1} \zeta(\alpha_0,\ldots,\alpha_r+1)x^r y^m$$

$$= \sum_{k=1}^{\infty} \frac{1}{(k-x)(k-y)}$$

$$= \sum_{r=0}^{\infty} \sum_{m=0}^{\infty} \zeta(m+r+2)x^r y^m.$$

Proof. The left hand side is equal to

$$\sum_{r=0}^{\infty} \sum_{m=0}^{\infty} \frac{x^r y^m}{r!m!} \int_{0<t_1<t_2<1} \left(\log\frac{1-t_1}{1-t_2}\right)^r \left(\log\frac{t_2}{t_1}\right)^m \frac{dt_1 dt_2}{(1-t_1)t_2}$$

$$= \int_{0<t_1<t_2<1} \left(\frac{1-t_1}{1-t_2}\right)^x \left(\frac{t_2}{t_1}\right)^y \frac{dt_1 dt_2}{(1-t_1)t_2}.$$

Under two consecutive change of variables, the above integral is transformed into

$$\int_{0<u_1<u_2<1} \frac{du_1}{1-u_1} \left(\frac{u_2}{u_1}\right)^x \frac{du_2}{u_2} \left(\frac{1}{u_2}\right)^y$$

which is equal to

$$\sum_{k=1}^{\infty} \frac{1}{(k-x)(k-y)}.$$

$$\square$$

C.6 The Vector Version of the Restricted Sum Formula

Let

$$\mathbf{p} = (p_1, p_2, \ldots, p_n), \quad \mathbf{q} = (q_1, q_2, \ldots, q_n), \quad \mathbf{r} = (r_1, r_2, \ldots, r_n)$$

be n-tuple of nonnegative integers. Define

$$\overrightarrow{\alpha_j} = (\alpha_{j0}, \alpha_{j1}, \ldots, \alpha_{jq_j} + 1) \quad j = 1, 2, \ldots, n$$

with

$$|\overrightarrow{\alpha_j}| = \alpha_{j0} + \alpha_{j1} + \cdots + \alpha_{jq_j}.$$

Also let

$$\mathbf{k} = (\{1\}^{p_1}, r_1 + 2, \{1\}^{p_2}, r_2 + 2, \ldots, \{1\}^{p_n}, r_n + 2)$$

and

$$\mathbf{k'} = (\{1\}^{r_n}, p_n + 2, \{1\}^{r_{n-1}}, p_{n-1} + 2, \ldots, \{1\}^{r_1}, p_1 + 2)$$

the dual of $\mathbf{k}$.

Theorem C.6.1. *For any integer $m \geq 0$,*

$$\sum_{|q|=m} \sum_{|\overrightarrow{\alpha_j}|=q_j+r_j+1} \zeta(\{1\}^{p_1}, \overrightarrow{\alpha_1}, \{1\}^{p_2}, \overrightarrow{\alpha_2}, \ldots, \{1\}^{p_n}, \overrightarrow{\alpha_n})$$

$$= \sum_{|c|=m} \zeta(\mathbf{k} + \mathbf{c}) = \sum_{|d|=m} \zeta(\mathbf{k'} + \mathbf{d}).$$

Outline of the proof:

<u>STEP 1</u>. Expressed the sum of multiple zeta values as an integral over a simplex $E^{2n} : 0 < t_1 < t_2 < \cdots < t_{2n-1} < t_{2n} < 1$,

$$S = \sum_{|q|=m} \int_{E^{2n}} \prod_{j=1}^{n} \frac{1}{p_j! q_j! r_j!} \left(\log \frac{1 - t_{2j-2}}{1 - t_{2j-1}} \right)^{p_j} \left(\log \frac{1 - t_{2j-1}}{1 - t_{2j}} \right)^{q_j}$$

$$\times \left(\log \frac{t_{2j}}{t_{2j-1}} \right)^{r_j} \frac{dt_{2j-1} dt_{2j}}{(1 - t_{2j-1}) t_{2j}}.$$

By convention: $t_0 = 0$, $t_{2n+1} = 1$.

<u>STEP 2</u>. The first change of variables:

$$x_{2j-1} = \log \frac{1 - t_{2j-2}}{1 - t_{2j-1}}, \quad x_{2j} = \log \frac{t_{2j}}{t_{2j-1}}, \quad j = 1, 2, \ldots, n,$$

$$\log \frac{1 - t_{2j-1}}{1 - t_{2j}} = \log \frac{f_{j-1}(\mathbf{x}_{2j-2})}{f_j(\mathbf{x}_{2j})}$$

with

$$\begin{cases} f_0(\mathbf{x}_0) = 1, \quad f_1(x_1, x_2) = e^{x_1} + e^{x_2} - e^{x_1 + x_2} \\ f_{k+1}(\mathbf{x}_{2k+2}) = e^{x_1 + x_3 + \cdots + x_{2k+1}}(1 - e^{x_{2k+2}}) + e^{x_{2k+2}} f_k(\mathbf{x}_{2k}). \end{cases}$$

In particular, we have $f_n(x_1, x_2, \ldots, x_{2n}) = f_n(x_{2n}, x_{2n-1}, \ldots, x_2, x_1)$.

<u>STEP 3</u>. The second change of variables:

$$x_{2j-1} = \log \frac{1 - u_{2j-1}}{1 - u_{2j}}, \quad x_{2j} = \log \frac{u_{2j+1}}{u_{2j}}, \quad j = 1, 2, \ldots, n,$$

so that

$$\log \frac{1}{f_n(\mathbf{x}_{2n})} = \log \frac{u_2}{u_1} + \log \frac{u_4}{u_3} + \cdots + \log \frac{u_{2n}}{u_{2n-1}}$$

and

$$S = \sum_{|\mathbf{q}|=m} \int_{E^{2n}} \prod_{j=1}^{n} \frac{1}{p_j! q_j! r_j!} \left(\log \frac{1 - u_{2j-1}}{1 - u_{2j}} \right)^{p_j} \left(\log \frac{u_{2j}}{u_{2j-1}} \right)^{q_j}$$

$$\times \left(\log \frac{u_{2j+1}}{u_{2j}} \right)^{r_j} \frac{du_{2j-1} du_{2j}}{(1 - u_{2j-1}) u_{2j}}$$

which is equal to

$$\sum_{|\mathbf{c}|=m} \zeta(\mathbf{k} + \mathbf{c}).$$

C.7 Another Generating Function

Consider the generating function

$$\sum_{1 \le k_1 < k_2 < \cdots < k_n} \prod_{j=1}^{n} \frac{1}{(k_j - x_j)(k_j - z)}.$$

It is equal to

$$\sum_{\substack{m_i=0 \\ 1\leq i\leq 2n}}^{\infty} \sum_{1\leq k_1<\cdots<k_n} \left(\prod_{j=1}^{n} k_j^{-(m_{2j-1}+m_{2j}+2)} x_j^{m_{2j-1}}\right) z^{m_2+m_4+\cdots+m_{2n}}$$

or

$$\sum_{\substack{m_i=0 \\ 1\leq i\leq 2n}}^{\infty} \zeta(m_1+m_2+2, m_3+m_4+2, \ldots, m_{2n-1}+m_{2n}+2)$$

$$\times x_1^{m_1} x_2^{m_3} \cdots x_n^{m_{2n-1}} z^{m_2+m_4+\cdots+m_{2n}}.$$

The same generating function comes from the integral

$$\int_{E^{2n}} \prod_{j=1}^{n} \left(\frac{u_{2j}}{u_{2j-1}}\right)^z \left(\frac{u_{2j+1}}{u_{2j}}\right)^{x_j} \frac{du_{2j-1}du_{2j}}{(1-u_{2j-1})u_{2j}},$$

which is equal to

$$\int_{E^{2n}} \prod_{j=1}^{n} \left(\frac{1-t_{2j-1}}{1-t_{2j}}\right)^z \left(\frac{t_{2j}}{t_{2j-1}}\right)^{x_j} \frac{dt_{2j-1}dt_{2j}}{(1-t_{2j-1})t_{2j}}.$$

In terms of multiple zeta values, it is

$$\sum_{\substack{m_i=0 \\ 1\leq i\leq 2n}}^{\infty} \sum_{\substack{|\overrightarrow{\alpha_j}|=m_{2j-1}+m_{2j}+1 \\ 1\leq j\leq n}} \zeta(\overrightarrow{\alpha_1}, \overrightarrow{\alpha_2}, \ldots, \overrightarrow{\alpha_n}) x_1^{m_1} x_2^{m_3} \cdots x_n^{m_{2n-1}} z^{m_2+m_4+\cdots+m_{2n}}$$

with

$$\overrightarrow{\alpha_j} = (\alpha_{j0}, \alpha_{j1}, \ldots, \alpha_{jm_{2j}}+1).$$

Bibliography

[1] T. Arakawa and M. Kaneko, *Multiple zeta values, poly-Bernoulli numbers, and related zeta functions*, Nagoya Math. J. **153** (1999), 189–209.

[2] B. C. Berndt, *Ramanujan's Notebooks, Part I and II*, Springer-Verlag, New York, 1985, 1989.

[3] D. Borwein, J. M. Borwein and R. Girgensohn, *Explicit evaluation of Euler sums*, Proc. Edinburgh. Math. Soc. (2) **38** (1995), no. 2, 277–294.

[4] J. M. Borwein, D. M. Bradley, D. J. Broadhurst and P. Lisoněk, *Special values of multiple polylogarithms*, Trans. Amer. Math. Soc. **353** (2001), no. 3, 907–941.

[5] J. M. Borwein and D. M. Bradly, *Thirty-two Goldbach variations*, Int. J. Number Theory **2** (2006), no. 1, 65–103.

[6] D. Bowman and D. M. Bradley, *Multiple polylogarithms: a brief survey*, Contemp. Math. **291** (2001), 71–92.

[7] J. P. Buhler and R. E. Crandall, *On the evaluation of Euler sums*, Experiment. Math. **3** (1994), no. 4, 275–285.

[8] S.-T. Chang and Minking Eie, *An arithmetic property of Fourier coefficients of singular modular forms on exceptional domain*, Trans. Amer. Math. Soc. **353** (2000), 539-556.

[9] V. G. Drinfel'd, *On quasitriangular quasi-Hopf algebras and on a group that is closely connected with* $\mathrm{Gal}(\overline{\mathbb{Q}}/\mathbb{Q})$, Leningrad Math. J. **2** (1991), no. 4, 829–860.

[10] L. Euler, *Opera Omnia, Ser, Vol. XV* (Teubner, Berlin 1917), 217–267.

[11] M. Eie, *Topics in Number Theory*, Monographs in Number Theory 2, World Scientific Publishing Co. Pte. Ltd, 2009.

[12] M. Eie, W.-C. Liaw and Y. L. Ong, *A restricted sum formula among multiple zeta values*, J. Number Theory **129** (2009), no. 4, 908–921.

[13] _____, *Shuffle relations of two sums of multiple zeta values*, submitted, 2009.

[14] M. Eie, Y. L. Ong, and C.-S. Wei, *Explicit evaluations of quadruple Euler sums*, Acta Arith. **144** (2010), 213–230.

[15] M. Eie and C.-S. Wei, *A short proof for the sum formula and its generalization*, Arch. Math. **91** (2008), no. 4, 330–338.

[16] _____, *A vector version of the restricted sum formula among multiple zeta values*, submitted, 2009.

[17] _____, *Double weighted sum formulas of multiple zeta values*, submitted, 2010.

[18] _____, *Applications of a shuffle product formula*, submitted, 2010.

[19] R.L. Graham, D.E. Knuth and O.Patashnik, *Concrete Mathematics*, Addison-Wesley Publishing Company, (1991).

[20] A. Granville, *A decomposition of Riemann's zeta-function*, Analytic number theory (Kyoto, 1996), 95–101.

[21] P. Flajolet and B. Salvy, *Euler sums and contour integral representations*, Experiment. Math. **7** (1998), no. 1, 15–35.

[22] H. Tsumura, *Combinatorial relations for Euler-Zagier sums*, Acta Arith. **111** (2004), no. 1, 27–42.

[23] L. Guo and B. Xie, *Weighted sum formula for multiple zeta values*, J. Number Theory **129** (2009), no. 11, 2747–2765.

[24] M. E. Hoffman, *Multiple harmonic series*, Pacific J. Math. **152** (1992), no. 2, 275–290.

[25] M. E. Hoffman and C. Moen, *Sums of triple harmonic series*, J. Number Theory **60** (1996), no. 2, 329–331.

[26] C. Markett, *Triple sums and the Riemann zeta function*, J. Number Theory **48** (1994), no. 2, 113–132.

[27] Y. L. Ong, M. Eie, and W.-C. Liaw, *Explict evaluation of triple Euler sums*, Int. J. Number Theory **4** (2008), no. 3, 437–451.

[28] Y. Ohno, *A generalization of the duality and sum formulas on the multiple zeta values*, J. Number Theory **74** (1999), no. 1, 39–43.

[29] Y. Ohno and W. Zudilin, *Zeta stars*, Commun. Number Theory and Phys. **2** (2008) no. 2, 325-343.

[30] J, Pitman, *Probability*, Springer-Verlag, (1993).

[31] D. Zagier, *Values of zeta functions and their applications*, First European Congress of Mathematics, Vol. II (Paris, 1992), 493–512, Progr. Math., **120**, Birkhäuser, Basel, 1994.